THE WILEY BICENTENNIAL–KNOWLEDGE FOR GENERATIONS

Each generation has its unique needs and aspirations. When Charles Wiley first opened his small printing shop in lower Manhattan in 1807, it was a generation of boundless potential searching for an identity. And we were there, helping to define a new American literary tradition. Over half a century later, in the midst of the Second Industrial Revolution, it was a generation focused on building the future. Once again, we were there, supplying the critical scientific, technical, and engineering knowledge that helped frame the world. Throughout the 20th Century, and into the new millennium, nations began to reach out beyond their own borders and a new international community was born. Wiley was there, expanding its operations around the world to enable a global exchange of ideas, opinions, and know-how.

For 200 years, Wiley has been an integral part of each generation's journey, enabling the flow of information and understanding necessary to meet their needs and fulfill their aspirations. Today, bold new technologies are changing the way we live and learn. Wiley will be there, providing you the must-have knowledge you need to imagine new worlds, new possibilities, and new opportunities.

Generations come and go, but you can always count on Wiley to provide you the knowledge you need, when and where you need it!

WILLIAM J. PESCE
PRESIDENT AND CHIEF EXECUTIVE OFFICER

PETER BOOTH WILEY
CHAIRMAN OF THE BOARD

Introduction to Emergency Management

Michael K. Lindell, Carla S. Prater,
and Ronald W. Perry

Credits

Publisher
Anne Smith

Development Editor
Laura Town

Marketing Manager
Jennifer Slomack

Editorial Assistant
Tiara Kelly

Production Manager
Kelly Tavares

Production Assistant
Courtney Leshko

Creative Director
Harry Nolan

Cover Designer
Hope Miller

Cover Photo
©Sam Morris/Las Vegas Sun

This book was set in Times New Roman, printed and bound by R. R. Donnelley. The cover was printed by Phoenix Color.

Library of Congress Cataloging-in-Publication Data
Lindell, Michael K.
Introduction to emergency management/Michael K. Lindell, Ronald W. Perry, and Carla Prater.
p. cm.
Includes bibliographical references and index.
ISBN-13: 978-0-471-77260-6 (pbk.)
ISBN-10: 0-471-77260-7 (pbk.)
1. Emergency management. 2. Emergency management—United States. I. Perry, Ronald W. II. Prater, Carla. III. Title.
HV551.2.L56 2007
363.34'8—dc22 2006023048

ISBN-13 978-0-471-77260-6
ISBN-10 0-471-77260-7

Printed in the United States of America

10 9 8 7 6 5 4 3 2 1

PREFACE

College classrooms bring together learners from many backgrounds with a variety of aspirations. Although the students are in the same course, they are not necessarily on the same path. This diversity, coupled with the reality that these learners often have jobs, families, and other commitments, requires a flexibility that our nation's higher education system is addressing. Distance learning, shorter course terms, new disciplines, evening courses, and certification programs are some of the approaches that colleges employ to reach as many students as possible and help them clarify and achieve their goals.

Wiley Pathways books, a new line of texts from John Wiley & Sons, Inc., are designed to help you address this diversity and the need for flexibility. These books focus on the fundamentals, identify core competencies and skills, and promote independent learning. The focus on the fundamentals helps students grasp the subject, bringing them all to the same basic understanding. These books use clear, everyday language, presented in an uncluttered format, making the reading experience more pleasurable. The core competencies and skills help students succeed in the classroom and beyond, whether in another course or in a professional setting. A variety of built-in learning resources promote independent learning and help instructors and students gauge students' understanding of the content. These resources enable students to think critically about their new knowledge, and apply their skills in any situation.

Our goal with *Wiley Pathways* books—with its brief, inviting format, clear language, and core competencies and skills focus—is to celebrate the many students in your courses, respect their needs, and help you guide them on their way.

CASE Learning System

To meet the needs of working college students, *Introduction to Emergency Management* uses a four-step process: The CASE Learning System. Based on Bloom's Taxonomy of Learning, CASE presents key emergency management topics in easy-to-follow chapters. The text then prompts analysis, synthesis, and evaluation with a variety of learning aids and assessment tools. Students move efficiently from reviewing what they have learned, to acquiring new information and skills, to applying their new knowledge and skills to real-life scenarios. Each phase of the CASE system is signaled in-text by an icon:

- ▲ Content
- ▲ Analysis
- ▲ Synthesis
- ▲ Evaluation

Using the CASE Learning System, students not only achieve academic mastery of emergency management *topics*, but they master real-world emergency management *skills*. The CASE Learning System also helps students become independent learners, giving them a distinct advantage whether they are starting out or seek to advance in their careers.

Organization, Depth and Breadth of the Text

Introduction to Emergency Management offers the following features:

- ▲ **Modular format.** Research on college students shows that they access information from textbooks in a non-linear way. Instructors also often wish to reorder textbook content to suit the needs of a particular class. Therefore, although *Introduction to Emergency Management* proceeds logically from the basics to increasingly more challenging material, chapters are further organized into sections (4 to 6 per chapter) that are self-contained for maximum teaching and learning flexibility.
- ▲ **Numeric system of headings.** *Introduction to Emergency Management* uses a numeric system for headings (for example, 2.3.4 identifies the fourth sub-section of section 3 of chapter 2). With this system, students and teachers can quickly and easily pinpoint topics in the table of contents and the text, keeping class time and study sessions focused.
- ▲ **Core content.** It is critical for a text on emergency management to address all phases of emergency management—the social and environmental processes that generate hazards, hazard/vulnerability analysis, hazard mitigation, emergency response and disaster recovery, emergency response, and disaster recovery. In addition, it should address a number of issues of professional interest—the history of emergency management, organizing an emergency management agency, other countries' approaches, and future trends, for example.

Chapter 1, Introduction to Emergency Management, provides an overview that describes the basic types of hazards threatening the

United States and provides definitions for some basic terms such as hazards, emergencies, and disasters. The chapter also provides a brief history of emergency management in the federal government and a general description of the current emergency management system—including the basic functions performed by local emergency managers. The chapter concludes with a discussion of the *all-hazards* approach and its implications for local emergency management.

Chapter 2, Emergency Management Stakeholders, will introduce the many actors in emergency management and examine some of the problems inherent in dealing with the complex emergency management policy process. The first section will address four basic issues. First, how is a "stakeholder" defined, especially in the context of emergency management? Second, who are the stakeholders emergency managers should be concerned about? Third, at what level in the system and by which different stakeholders are different types of emergency management decisions made? Fourth, how can emergency managers involve these stakeholders in the emergency management process? Last, what types and amounts of power do different stakeholder groups have and how do they influence the emergency management policy process.

Chapter 3, Building an Effective Emergency Management Organization, describes the activities needed to build effective emergency management organizations, beginning with the fundamentals of running a local emergency management agency. The most important concept in this chapter is the development of a local emergency management committee (LEMC) that establishes horizontal linkages among a local jurisdiction's government agencies, non-governmental organizations, and private sector organizations relevant to emergency management. In addition, an LEMC can provide vertical linkages downward to households and businesses, and upward to state and federal agencies.

Chapter 4, Risk Perception and Communication, explains how people perceive the risks of environmental hazards and the actions they can take to protect themselves from those hazards. Addressing such perceptions is the most common way for emergency managers to change the behavior of those at risk from long-term threats or imminent impacts of disasters. This chapter describes the Protective Action Decision Model, which summarizes findings from studies of household response to disasters, and concludes with recommendations for risk communication during the continuing hazard phase, escalating crises, and emergency response.

Chapter 5, Principal Hazards in the United States, describes the principal environmental hazards that are of greatest concern to emergency managers in communities throughout the United States. Each of these hazards will be described in terms of the physical processes that generate them, the geographical areas that are most commonly at risk, the types of impacts and typical magnitude of hazard events, and hazard-specific issues of emergency response.

Chapter 6, Hazard, Vulnerability and Risk Analysis, describes how pre-impact conditions act together with event-specific conditions to produce a disaster's physical and social impacts. These disaster impacts can be reduced by emergency management interventions. In addition, this chapter discusses how emergency managers can assess the pre-impact conditions that produce disaster vulnerability within their communities. The chapter concludes with a discussion of vulnerability dynamics and methods for disseminating hazard/vulnerability data.

Chapter 7, Hazard Mitigation, explains what mitigation is, and how it fits in with the other classic phases of emergency management. Next, the chapter will describe the most widely used mitigation strategies and the ways they are applied to the most common types of environmental hazards. The following section will describe the legal basis for hazard mitigation as it stands in the United States today. Problems in the adoption and implementation of mitigation policies will be described and some methods of addressing them will be offered. Finally, the chapter will conclude with a discussion of the relationship between hazard mitigation and sustainable development.

Chapter 8, Myths and Realities of Disaster Response, challenges the vision of what happens during and because of disaster impacts shape the way that one thinks about emergency response actions. Community vulnerability judged from accurate observations of events is the basis for defining emergency response strategies and functions that are manifest as emergency preparedness. The quality of the preparedness is a function of many factors, but it is particularly important that observations of disaster consequences and reactions be based upon accurate information. This chapter examines myths that have arisen and persist about behavior in disasters, reviews documented demands imposed by disasters and closes with discussions of organizational emergency response functions and household emergency response.

Chapter 9, Preparedness for Emergency Response, examines the readiness of an organization or jurisdiction to constructively react to threats from the environment in a way that minimizes the negative consequences of impact for the health and safety of individuals and the integrity and functioning of physical structures and systems. The achievement of emergency preparedness takes place through a process of planning, training and exercising accompanied by the acquisition of equipment and apparatus to support emergency action (Gillespie and Colignon, 1993). The objective of this chapter is to examine the emergency planning process, review the content of an emergency plan and describe the organizational structures—the incident management system and emergency operations center—through which emergencies are managed.

Chapter 10, Organizational Emergency Response, examines the four principal functions the community emergency response organization must perform during disaster response. These are emergency assessment, expedient hazard mitigation, population protection, and incident management. Community organizations, especially government agencies, perform these functions to assist households and businesses in situations where they need help or where all households and businesses lack the resources. In addition, this chapter describes the ways in which households respond to disasters. In particular, this section emphasizes the relationship between households and the community emergency response organization in connection with the population protection function—especially warning and the implementation of protective actions such as evacuation and sheltering in-place.

Chapter 11, Disaster Recovery, defines disaster recovery in terms of the activities that take place during this phase of the emergency management cycle and explains how disaster recovery is related to emergency preparedness, emergency response, and hazard mitigation. Disaster victims pass through four stages of housing recovery—emergency shelter, temporary shelter, temporary housing, and permanent housing, but the rate at which this process occurs depends upon the vulnerability of housing in the community and the speed with which reconstruction takes place. Business recovery is another important aspect of community recovery; the amount of loss that businesses experience depends upon indirect losses due to business interruption as well as direct damage. Finally, local government has an important role in disaster recovery partly because it links households and businesses with higher levels of government and also because it can experience losses to its buildings and infrastructure. However, it is important to recognize that most disasters do not receive Presidential Disaster Declarations or even state disaster declarations, so local government must usually take responsibility for guiding the community's recovery from disaster.

Chapter 12, Evaluations, discusses evaluation in emergency management, beginning with performance appraisals for individual members of the local emergency management agency (LEMA). Next, the chapter addresses the procedures for periodic evaluation of the local emergency management agency and local emergency planning committee (LEPC). The discussion then turns to procedures for evaluating drills, exercises, and incidents. The chapter concludes with a discussion of procedures for evaluating organizational training and community risk communication programs.

Chapter 13, International Emergency Management, discusses research on emergency management suffers from the same problems as much public policy research, in that most of it has been done in the English-speaking countries, and much of the remaining work has been done in former colonies of these nations (Heady, 1996). Students of emergency management are thus exposed to a great deal of information on what is being done in the English-speaking world, yet often are unaware of the different approaches to emergency management used in other regions. A few scholars have examined the applicability of emergency management principles developed in rich countries to other areas, and have concluded that the principles of an all hazards, integrated and comprehensive approach covering all phases of emergency management and integrating relevant agencies, together with a focus on building community resilience at the local level, are viable and useful in a wide variety of settings (Martin, Capra, van der Heide, Stoneham, and Lucas 2001). However, resources, both human and technical, are frequently lacking for the development of adequate programs (Vaste and Joseph 2003).

Chapter 14, Professional Accountability, discusses how it has only been in the past two decades that government departments changed names from emergency services to emergency management agencies; much scrutiny and deliberation were required before government personnel officers permitted employee titles to change from emergency planner to emergency manager. There remains less than full consensus regarding whether emergency management is appropriately labeled a profession, with many academics still preferring the concept of occupation. These issues notwithstanding, this chapter examines the concept of a profession of emergency management and the processes through which it is moving toward further establishing itself as recognized profession.

Chapter 15, Future Directions in Emergency Management, discusses future directions in emergency management. These can be classified as challenges and opportunities at the global, national, and professional levels. Many of the trends identified by Drabek (1991a) and Anderson and Mattingly (1991) continue to dominate emergency management, including increasing exposure to environmental hazards, increased capabilities offered by advanced emergency management information technology, increasing recognition of the need for pre-impact action (hazard mitigation, emergency preparedness, and recovery preparedness) in the face of inertia or outright resistance, and increased professionalization of emergency management. Nonetheless, there are some new issues, including the potential for changes in the nature of environmental hazards and the increased salience of terrorism as a threat to communities throughout the United States.

Pre-Reading Learning Aids

Each chapter of *Introduction to Emergency Management* features the following learning and study aids to activate students' prior knowledge of the topics and orient them to the material.

▲ **Pre-test.** This pre-reading assessment tool in multiple-choice format not only introduces chapter material, but it also helps students anticipate the chapter's learning outcomes. By focusing students' attention on what they do not know, the self-test provides students with a benchmark against which they can measure their own progress. The pre-test is available online at www.wiley.com/college/Lindell.

▲ **What You'll Learn in This Chapter and After Studying This Chapter.** These bulleted lists tell students what they will be learning in the chapter and why it is significant for their careers. They also explain why the chapter is important and how it relates to other chapters in the text. "What You'll Learn..." lists focus on the *subject matter* that will be taught (e.g. what emergency response is). "After Studying This Chapter..." lists emphasize *capabilities and skills* students will learn (e.g. how to respond to a disaster).

▲ **Goals and Outcomes.** These lists identify specific student capabilities that will result from reading the chapter. They set students up to synthesize and evaluate the chapter material, and relate it to the real world.

▲ **Figures and Tables.** Line art and photos have been carefully chosen to be truly instructional rather than filler. Tables distill and present information in a way that is easy to identify, access, and understand, enhancing the focus of the text on essential ideas.

Within-text Learning Aids

The following learning aids are designed to encourage analysis and synthesis of the material, and to support the learning process and ensure success during the evaluation phase:

- **Introduction.** This section orients the student by introducing the chapter and explaining its practical value and relevance to the book as a whole. Short summaries of chapter sections preview the topics to follow.
- **"For Example" Boxes.** Found within each section, these boxes tie section content to real-world organizations, scenarios, and applications.
- **Self-Check.** Related to the "What You'll Learn" bullets and found at the end of each section, this battery of short answer questions emphasizes student understanding of concepts and mastery of section content. Though the questions may either be discussed in class or studied by students outside of class, students should not go on before they can answer all questions correctly. Each *Self-Check* question set includes a link to a section of the pre-test for further review and practice.
- **Summary.** Each chapter concludes with a summary paragraph that reviews the major concepts in the chapter and links back to the "What you'll learn" list.
- **Key Terms and Glossary.** To help students develop a professional vocabulary, key terms are bolded in the introduction, summary and when they first appear in the chapter. A complete list of key terms with brief definitions appears at the end of each chapter and again in a glossary at the end of the book. Knowledge of key terms is assessed by all assessment tools (see below).

Evaluation and Assessment Tools

The evaluation phase of the CASE Learning System consists of a variety of within-chapter and end-of-chapter assessment tools that test how well students have learned the material. These tools also encourage students to extend their learning into different scenarios and higher levels of understanding and thinking. The following assessment tools appear in every chapter of *Introduction to Emergency Management:*

- **Summary Questions** help students summarize the chapter's main points by asking a series of multiple choice and true/false questions that emphasize student understanding of concepts and mastery of chapter content. Students should be able to answer all of the Summary Questions correctly before moving on.

- ▲ **Review Questions** in short answer format review the major points in each chapter, prompting analysis while reinforcing and confirming student understanding of concepts, and encouraging mastery of chapter content. They are somewhat more difficult than the *Self-Check* and *Summary Questions,* and students should be able to answer most of them correctly before moving on.
- ▲ **Applying This Chapter Questions** drive home key ideas by asking students to synthesize and apply chapter concepts to new, real-life situations and scenarios.
- ▲ **You Try It Questions** are designed to extend students' thinking, and so are ideal for discussion or writing assignments. Using an open-ended format and sometimes based on Web sources, they encourage students to draw conclusions using chapter material applied to real-world situations, which fosters both mastery and independent learning.
- ▲ **Post-test** should be taken after students have completed the chapter. It includes all of the questions in the pre-test, so that students can see how their learning has progressed and improved.

Instructor and Student Package

Introduction to Emergency Management is available with the following teaching and learning supplements. All supplements are available online at the text's Book Companion Website, located at *www.wiley.com/college/lindell.*

- ▲ **Instructor's Resource Guide.** Provides the following aids and supplements for teaching:
  - ▲ *Diagnostic Evaluation of Grammar, Mechanics, and Spelling.* A useful tool that instructors may administer to the class at the beginning of the course to determine each student's basic writing skills. The Evaluation is accompanied by an Answer Key and a Marking Key. Instructors are encouraged to use the Marking key when grading students' Evaluations, and to duplicate and distribute it to students with their graded evaluations.
  - ▲ *Sample syllabus.* A convenient template that instructors may use for creating their own course syllabi.

  - ▲ *Teaching suggestions.* For each chapter, these include a chapter summary, learning objectives, definitions of key terms, lecture notes, answers to select text question sets, and at least 3 suggestions for classroom activities, such as ideas for speakers to invite, videos to show, and other projects.

- ▲ **Test Bank.** One test per chapter, as well as a mid-term and a final. Each includes true/false, multiple choice, and open-ended questions. Answers and page references are provided for the true/false and multiple choice questions, and page references for the open-ended questions. Available in Microsoft Word and computerized formats.
- ▲ **PowerPoints.** Key information is summarized in 10 to 15 PowerPoints per chapter. Instructors may use these in class or choose to share them with students for class presentations or to provide additional study support.

ACKNOWLEDGMENTS

Taken together, the content, pedagogy, and assessment elements of *Introduction to Emergency Management* offer the career-oriented student the most important aspects of the emergency management field as well as ways to develop the skills and capabilities that current and future employers seek in the individuals they hire and promote. Instructors will appreciate its practical focus, conciseness, and real-world emphasis. We would like to thank all of the reviewers for their feedback and suggestions during the text's development. Their advice on how to shape *Introduction to Emergency Management* into a solid learning tool that meets both their needs and those of their busy students is deeply appreciated.

ABOUT THE AUTHORS

Michael K. Lindell is the former Director of the Hazard Reduction & Recovery Center (HRRC) at Texas A&M University and has 30 years of experience in the field of emergency management, conducting research on community adjustment to floods, hurricanes, earthquakes, volcanic eruptions, and releases of radiological and toxic materials. He worked for many years as an emergency preparedness contractor to the U.S. Nuclear Regulatory Commission and has provided technical assistance on radiological emergency preparedness for the International Atomic Energy Agency, the Department of Energy, and nuclear utilities. In addition, he has trained as a Hazardous Materials Specialist at the Michigan Hazardous Materials Training Center and worked on hazardous materials emergency preparedness with State Emergency Response Commissions, Local Emergency Planning Committees, and chemical companies. In the past few years, Lindell directed HRRC staff performing hurricane hazard analysis and evacuation planning for the entire Texas Gulf coast. He has made over 120 presentations before scientific societies and short courses for emergency planners, as well as being an invited participant in workshops on risk communication and emergency management in this country and abroad. Lindell has also written extensively on emergency management and is the author of over 120 technical reports and journal articles, as well as 5 books.

Carla S. Prater is currently a Lecturer in the Department of Landscape Architecture and Urban Planning, and the Associate Director at the Hazard Reduction & Recovery Center (HRRC). She was educated at private schools in Brazil, and received her Bachelor of Arts in Foreign Languages from Pepperdine University in 1975. She then returned to Brazil where she spent most of the next 12 years as a missionary. In 1991 she joined the Hazard Reduction & Recovery Center as a graduate student, and received a Master of Science degree in Urban and Regional Planning from Texas A&M University in 1993, followed by a PhD in Comparative Political Science, also from Texas A&M, in 1999. She has worked on several National Science Foundation projects covering various disaster phases, has published articles in disaster journals, and presented papers at conferences around the world. She has done consulting on emergency management and hazard mitigation for international agencies and foreign governments, and has taught courses on Comparative Politics, Risk Analysis, and Organizational & Community Planning & Response for Disasters in several countries. Her research interests include hazard mitigation policy and emergency management institutions viewed from a cross-national perspective. Current projects include a study of the response to the Indian Ocean tsunami of December 2004 and a comparison of preparations for tsunamis and hurricanes in the United States.

Ronald W. Perry joined Arizona State University in 1983 as Professor of Public Affairs. He has studied natural and technological hazards and terrorism since 1971. His principal interests are incident management systems, citizen warning behavior, public education and community preparedness. He has published more than a dozen books and many journal articles. Perry currently serves on the Steering Committees of the Phoenix Urban Areas Strategic Initiative and the Phoenix Metropolitan Medical Response System. He also serves on the Arizona Council for Earthquake Safety and on the Fire Chief's Advisory Committees for the Arizona Cities of Gilbert, Mesa, Phoenix and Tempe. He holds the Award for Excellence in Emergency Management from the Arizona Emergency Services Association and the Pearce Memorial Award for Contributions to Hazardous Incident Response from the Phoenix Fire Department. He also holds both the Award for Outstanding Environmental Achievement by a Team from the U.S. Environmental Protection Agency and a Certificate of Recognition from Vice President Gore's National Partnership for Reinventing Government.

BRIEF CONTENTS

CONTENTS

1

INTRODUCTION TO EMERGENCY MANAGEMENT

The Role of the Emergency Manager

Starting Point

Go to www.wiley.com/college/lindell to assess your knowledge of the basic role of the emergency manager.
Determine where you need to concentrate your effort.

What You'll Learn in This Chapter

- ▲ The differences between hazards, emergencies, and disasters
- ▲ The role of emergency management
- ▲ The importance of hazard mitigation

After Studying This Chapter, You'll Be Able To

- ▲ Identify what hazards, emergencies, and disasters have a potential impact on a community
- ▲ Determine what steps to take before a disaster strikes
- ▲ Describe how to respond to a disaster

Goals and Outcomes

- ▲ Compare and contrast hazards, emergencies, and disasters
- ▲ Design a disaster preparedness program
- ▲ Design a plan for disaster response

INTRODUCTION

Hazards, emergencies, and disasters threaten the United States daily. Whether it's a natural disaster, technological accident, or terrorist act, emergency managers must prepare for every type of event. As an emergency manager, it is your job to put systems in place to prevent and reduce losses. To do this, you must study and then respond to events.

This chapter introduces you to emergency management and the role of the emergency manager in dealing with a variety of emergency threats. You will learn to compare and contrast hazards, emergencies, and disasters. With this knowledge, you will assess how to prepare for and respond to threats.

1.1 Defining Hazards, Emergencies, and Disasters

Hazard, *emergency*, and *disaster* seem like different words that mean similar things; however, differences do exist. In emergency management, it is important to distinguish the meaning of these three terms.

A **hazard** is a source of danger or an extreme event that has the potential to affect people, property, and the natural environment in a given location. We are exposed to a variety of risks in our natural environment and through technology. These risks include both health and safety dangers, and they vary by location. For example tsunami (seismic sea wave) hazards are nonexistent in Ames, Iowa, because it is far from the run-up zones at the ocean shore. However, tsunami hazards are significant on the Pacific coast. Other hazards include accidents involving nuclear and chemical technologies that threaten our health and safety. Hazardous materials released from nuclear power plants or chemical facilities could cause many casualties and significant damage. To protect public health and safety, we must adjust to both natural and technological processes. To reduce the potential for casualties and damage from hazards, we must change the physical processes that generate hazardous events or change our behavior by living in less dangerous locations, building hazard-resistant structures, or improving our ability to respond and recover from extreme events. This is the classic definition of hazard adjustment (Lindell and Perry, 2003b).

The term **emergency** is used in two slightly different ways. First, we use the term to describe minor events that cause a few casualties and a limited amount of property damage. Common emergencies include car crashes, house fires, and heart attacks. Fire departments, police departments, and emergency medics are the first responders to these events. These events affect few people, so only a few community agencies need to respond. In addition, these events are well understood, so communities have standard operating procedures for responding to them (Quarantelli, 1987). However, it is important to understand that each emergency situation can present unique elements; experts caution that there is no such thing as a "routine" house fire. Believing that each new fire will be like previous ones increases firefighter deaths and injuries (Brunacini, 2002).

Second, the term emergency can refer to an imminent event. For example, a hurricane that is 48 hours from landfall creates an emergency situation because there is little time to respond. The urgency of the situation requires prompt and effective action. Unlike with the previous use of the term emergency, the event has not occurred, but the consequences are likely to be major, so many community agencies need to mount a coordinated response.

The term **disaster** is reserved for events that produce more losses than a community can handle. A community struck by disaster can cope only with help from other communities, state government, or the federal government. Disasters cause many casualties, much property damage, or significant environmental damage. There are many different types and causes of disasters. As we will discuss, strategies to deal with disasters vary with the causes of the disasters.

1.1.1 Natural Hazards

The 1995 earthquake in Kobe, Japan, killed more than 6000 people and injured 30,000. The 1994 earthquake in Northridge, California, resulted in $33 billion in damages. The losses are even greater over time. Between 1989 and 1999, the United States lost $1 billion each week due to natural disasters. The victims absorb most of the costs as only about 17% of losses are insured. These events, however, pale in comparison to potential future losses. Major earthquakes in Los Angeles and in the Midwest's New Madrid Seismic Zone are bound to occur at some point. These earthquakes could cause thousands of deaths, tens of thousands of injuries, and tens of billions of dollars in economic losses.

We see **natural disasters** nearly every day on the news. Large-scale natural disasters such as earthquakes, floods, hurricanes, volcanic eruptions, and wildland fires occur all over the globe. When we add severe storms, mudslides, lightning strikes, and tornadoes, we can see that natural disasters are very common.

Is Mother Nature out to get us? This idea might make an exciting movie, but it isn't true. Disaster movies are recurrent box-office successes but are often filled with many scientific errors. The natural environment is, of course, not getting some sort of revenge. Natural systems are behaving just as they always have. It is difficult to identify meaningful changes in event frequency for the short time period in which scientific records are available on geological, meteorological, and hydrological processes. However, more people are being affected by natural disasters, and losses are becoming progressively greater. More people now choose to live in hazardous places, building houses on picturesque cliffs, on mountain slopes, in floodplains, near beautiful volcanoes, and along seismic faults. And because of this, more people are exposed to disasters.

Another problem is that people sometimes choose hazardous building designs and inadequate structural materials that fail under extreme stress.

Figure 1-1

Earthquakes are a natural hazard.

One example of this is people failing to install window shutters in hurricane-prone areas. Another example is the use of nonreinforced, brick construction in seismically active areas (see Figure 1-1). These two patterns are among the many that add to rising disaster losses in the United States.

According to Mileti (1999), other factors in the rising disaster losses in the United States include an increase in human population and an increase in the value of property in hazard prone areas.

1.1.2 Technological Disasters

The pattern among **technological disasters** is somewhat different from the pattern among natural disasters. Certainly more people are affected simply because there are more people living close to known technological hazards, and we often choose to live in structures that cannot resist hazard impact. However, the types of threats are also changing. The potential for human loss often increases with the growth and change of existing technologies and the introduction of new ones. Risks are also rising due to the increasing quantity and variety of hazardous materials used. Threats also have risen from the use of energy technologies such as nuclear power plants and liquefied natural gas facilities, posing risks for both employees and those who live nearby. We also sometimes discover that what we thought was safe is actually hazardous. The use of asbestos is an example of this. As technologies grow, diversify, and become integrated into our lives, the variety of risks also grows. Fortunately, advancing technology often produces an improved capability to detect and control the release of hazardous materials.

1.1.3 Terrorist Disasters

Terrorist disasters are recently recognized additions to the types of threats we must confront. Unlike natural disasters and technological accidents, terrorist attacks involve deliberate human causality. Terrorists use some of the same materials as are involved in technological disasters. However, Unlike terrorist activities such as political assassinations and kidnappings, terrorist disasters are intended to cause many casualties and inflict major damage (see Figure 1-2). Emergency managers respond to terrorist attacks using the same basic approach that is used in other disasters. They must rapidly detect and assess the situation, mobilize relevant organizations and facilities, take action to limit casualties and damage, and coordinate the organizations responding to the incident.

Figure 1-2

The terrorist attacks on September 11, 2001, were the worst attacks ever carried out on American soil.

FOR EXAMPLE

England's "9/11"

On July 7, 2005, terrorists unleashed three bombs on the London underground and one on double-decker bus during rush hour. Fifty-six people were killed in the attacks, and 700 were injured. The terrorists were also killed in the explosion. The media dubbed this attack "England's 9/11," as it was the deadliest attack on British soil since the Pan Am bombing in 1988, which killed 270 people. It was also the deadliest bombing in London since World War II.

In addition, emergency managers must work with law enforcement agencies that assess the terrorists' capabilities. For example, the 1995 Aum Shinrikyo attack did not produce high casualties. The terrorists involved used the nerve agent sarin. Cult members placed bags of liquid sarin on Tokyo subway cars and cut the containers. Although sarin is extremely lethal, the attack resulted in only 12 deaths and 1046 patients admitted to hospitals (Reader, 2000). If the terrorists had been more effective in turning the liquid into a vapor, the death and injury rates could have been much higher. Japanese emergency managers had difficulty determining what chemical agent had been used and, therefore, how to treat patients. This incident should be an important lesson for emergency managers, and it highlights the difficulty in rescuing chemically contaminated patients and transporting them to hospitals where they can be treated.

SELF-CHECK

- Define **hazard, disaster, and emergency.**
- Define **technological disaster** and **natural disaster.**
- Provide an example of a technological disaster and a natural disaster.
- Provide an example of a hazard, a disaster, and an emergency.

1.2 The Role of the Emergency Manager

Emergency management is "applying science, technology, planning and management to deal with extreme events that can injure or kill large numbers of people, do extensive damage to property, and disrupt community life" (Drabek, 1991a, p. xvii).

The emergency manager's role is to prevent or reduce losses that occur due to hazards, disasters, and emergencies. This is a big responsibility, and it requires continuing education.

Emergency managers save lives and property. Many people recognize that they are exposed to natural disasters, technological accidents, and deliberate attacks such as terrorist disasters. These threats require preventive measures, but only a few people realize how many different types of threats there are. Losses from disasters—in the United States and the rest of the world—have been growing over the years and are likely to continue to grow (Berke, 1995; Mileti, 1999; Noji, 1997). Losses are measured in a variety of ways, including in terms of the number of deaths and injuries and in terms of the property damage.

Given the increasing toll from disasters, we must decide whether the risks are acceptable. Moreover, given the limited amount of time and resources, we must decide which risks to address (Lowrance, 1976). When we agree on what risks should be addressed, we can focus our efforts and money. We can use these resources to eliminate the danger or change the way people relate to the source of danger. For example, building dams or channeling streams can eliminate the risk of floods. Alternatively, we can relocate people and dwellings outside the floodplain. Or, we can devise a warning and evacuation system that moves people (but, of course, not their property) when disasters threaten. Emergency management is about identifying risks, assessing weaknesses, and devising strategies for reducing risks. Emergency management has traditionally been seen as the sole responsibility of government. This is changing, with households and businesses playing a more active role. Emergency management is now best conceived as relying on alliances among all levels of government and the private sector.

Four factors have led to the increased importance of emergency management:

1. Public awareness of hazards, emergencies, and disasters has increased as the cost of disasters has increased dramatically in recent years.
2. Businesses understand that disasters can disrupt their operations and even cause bankrupcy.
3. Rapid population growth in the most hazardous geographical areas of the country has created increased exposure to disaster impacts.
4. Emergency managers have undergone more and more specialized training, leading to the development of emergency management as a profession.

The United States still does not have a completely integrated emergency management system. However, we do have a collection of organizations that perform roles in planning for, responding to, and recovering from disasters. There have been intense efforts to improve the system since the 9/11 attacks. We are making progress, but much work remains to be done. Emergency managers agree on how to assess and respond to hazards. There is also increasing agreement on the goals and structures for an integrated emergency management system. Nonetheless, there are 50 states, 3000 counties, and thousands more cities and towns. It will be a challenging task to devise emergency response organizations that meet the needs of all and to allocate resources effectively.

1.2.1 A Brief History of Emergency Management

To understand our present, we must understand the past. The trend has been for legislation to focus on recovery from, not prevention of, disasters. The following list highlights the major events that have influenced emergency management in the United States:

- ▲ **1803:** The Fire Disaster Relief Act made funds available to Portsmouth, New Hampshire
- ▲ **1928:** The Lower Mississippi Flood Control Act was passed.
- ▲ **1930s:** The Reconstruction Finance Company (RFC) was created in 1933. The RFC gave loans for public buildings damaged by earthquakes, and its creation marked the beginning of federal involvement in disaster management. The Army Corps of Engineers was created in 1936.
- ▲ **1950s:** The Federal Civil Defense Administration was created in response to the Soviet Union's testing of an atomic bomb in the summer of 1949. Both the Federal Civil Defense Act and Disaster Relief Act of 1950 were passed. Both laws left disaster relief to the states but also spelled out federal responsibilities.
- ▲ **1970s:** The Federal Emergency Management Agency (FEMA) was created in 1978 in response to widespread recognition that emergency management was too fragmented across many federal agencies.
- ▲ **1990s:** FEMA's response to Hurricanes Hugo and Andrew was criticized as inadequate. President Clinton appointed James Lee Witt to be the Director of FEMA in 1993. This marks the only time a professional emergency manager has held the post.
- ▲ **2002:** President Bush creates a Department of Homeland Security. FEMA is merged into the Department of Homeland Security.

1.2.2 Local Emergency Management

Who really performs emergency management? The history in the preceding list focuses on federal efforts. However, in keeping with FEMA's practice of attempting to manage events locally whenever possible, emergency management is a *local* job expected to influence events with *local* consequences. Of course, this places a major burden on state and local government agencies. In major disasters,

FOR EXAMPLE

Homeland Security

FEMA's duties now fall to an undersecretary for Emergency Preparedness and Response. Among other duties, this undersecretary is responsible for recovery from terrorist attacks and major disasters.

such as Hurricane Katrina, federal resources are also needed. There is a time lag, however. The current National Response Plan states that local jurisdictions must be able to operate without external help for 72 hours after hazard impact. When help does arrive, it works best if there is a strong local structure in place (Perry, 1985).

Figure 1-3 shows a local emergency management system. You can see the tasks and the tools available. Although this chart does not show everything, it does show the critical elements. The processes take place at every level of government.

Figure 1-3

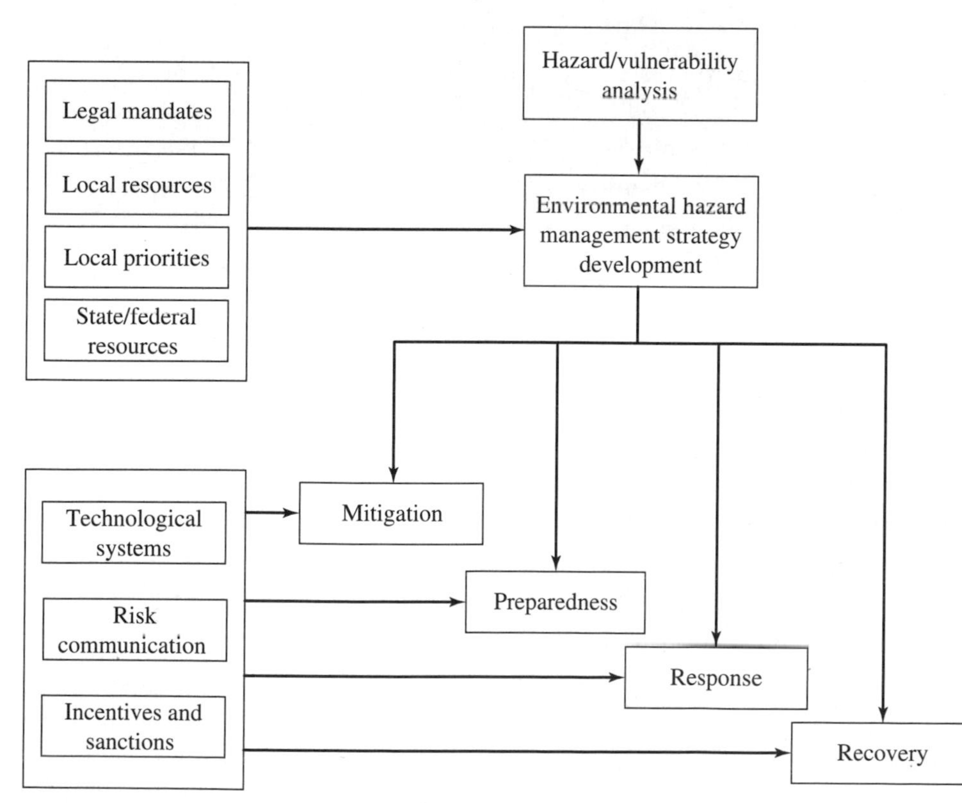

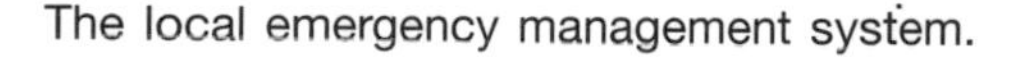
The local emergency management system.

SELF-CHECK

- Name two methods of measuring losses.
- Name three factors that led to the increased importance of emergency management.
- Name the agency formed in response to widespread recognition that emergency management was too fragmented across federal agencies.
- Describe the role of an emergency manager.

1.3 Studying and Responding to Hazards

As the top box in Figure 1-3 indicates, emergency management begins with a careful study of local hazards. This study is referred to as a hazard vulnerability analysis. This process helps emergency managers decide which hazards require active management. There are three steps in performing a hazard vulnerability analysis (Greenway, 1998; Ketchum and Whittaker, 1982).

Step 1: Identify hazards: Each community has its own set of hazards. For example, a community could be in a hurricane-prone area or contain a manufacturing plant that uses a large quantity of toxic chemicals.

Step 2: Estimate the probability: How likely is it that the hazard will occur?

Step 3: Project the consequences: What are the consequences for each geographic area? each population segment? each sector of the local economy?

As the left side of Figure 1-3 indicates, multiple considerations influence hazard management decisions. These include the following:

- ▲ Legal mandates, defined by federal, state, and local laws and regulations
- ▲ Local resources, such as a large tax base, an active local emergency planning committee (LEPC), and cooperative industry
- ▲ Local priorities, defined by the emphasis local officials give to emergency management
- ▲ State and federal resources, such as guidance manuals, technical training courses, and financial grants

Once a decision has been made to actively manage one or more hazards, the community needs to develop its hazard management strategy from four basic elements—hazard mitigation, emergency preparedness, emergency response, and disaster recovery. Mitigation and preparedness activities take place before disasters strike. Response and recovery activities take place after the disaster has occurred.

1.3.1 Hazard Mitigation

Hazard mitigation addresses the causes of a disaster, reducing the likelihood it will occur or limiting its impact. When we involve households, businesses, and government agencies, we can work on hazard mitigation.

The focus is to stop disasters before they happen. Changing either the natural event or human behavior or both reduces the impact of a natural event such as a flood, hurricane, or earthquake. In the case of floods, for example, changing the natural event system by using dams or levees that confine floodwater reduces the loss of life and property. Changing the human use system, such as not allowing construction on a floodplain, can also reduce losses.

The amount of control over natural event systems is often limited. This is not true with technological hazards. Dangerous chemicals can be produced, stored, and transported in ways that safely contain them. However, containment failure can cause hazardous materials to be released to the air, surface, or ground water. The choice of whether to reduce technological hazards by controlling the hazard agent or by controlling the human use depends on political and economic decisions. These decisions are made based on the costs and benefits of using these two types of control. Specific questions include Who controls the hazards? What degree of control is maintained? and What incentives are there for the maintenance of control?

1.3.2 Disaster Preparedness

Disaster preparedness protects lives and property and facilitates rapid recovery. Preparedness consists of plans, procedures, and resources that must be developed in advance. These are designed not only to support a timely and effective emergency response to the threat of imminent impact, but also to guide the process of disaster recovery. A disaster preparedness program needs to answer four questions:

1. **Which agencies will participate in preparedness?** Managers must know what the needs will be and which agencies can respond to these needs.
2. **What emergency response and disaster recovery actions are feasible for each community?** Managers must study the plans the community has adopted. For example, if an evacuation is needed, are the evacuation routes well planned?
3. **How will the response and recovery organizations function and what resources do they need?** An emergency operations plan and a recovery operations plan should be written. These define the role of each agency. While developing the plans and procedures, emergency managers also need to identify the needed resources. Such resources include facilities, trained personnel, equipment, materials and supplies, and information.
4. **How will disaster preparedness be established and maintained?** The plans should define the methods and schedule for plan maintenance,

training, drills, and exercises. First responders in fire, police, and emergency medical services should be trained. Training is also often needed for hospital, nursing home, and school employees.

1.3.3 Emergency Response

Emergency response begins when the event occurs. In some cases, hazard-monitoring systems alert authorities of an imminent disaster. Warnings, such as weather forecasts, can provide time to activate the emergency response organization before impact. In other cases, such as earthquakes, preimpact prediction is not available. However, a rapid assessment of the impact area can quickly direct resources to the most damaged areas.

Emergency response has three goals:

1. Protect the population.
2. Limit damage from the primary impact.
3. Minimize damage from secondary impacts.

Consequently, emergency response activities include

- ▲ Securing the impact area.
- ▲ Evacuating threatened areas.
- ▲ Conducting search and rescue for the injured.
- ▲ Providing emergency medical care.
- ▲ Sheltering evacuees and other victims.

Secondary impacts are "disasters caused by the disaster" and include such events as hazardous materials releases initiated by earthquakes (Lindell and Perry, 1997). Secondary impacts are different from the primary impacts. Thus, a secondary impact is different from the repeated impact of the hazard agent, as occurs in connection with aftershocks from earthquakes and repeated volcanic eruptions (Perry and Lindell, 1990).

Operations mounted to counter secondary threats include:

- ▲ Fighting urban fires after earthquakes.
- ▲ Identifying contaminated water supplies following flooding.
- ▲ Identifying contaminated wildlife or fish in connection with a toxic chemical spill.
- ▲ Preparing for flooding following a glacier melt during a volcanic eruption.

During the response stage, emergency managers must constantly assess damage. They must also coordinate the arrival of equipment and supplies, so they can be sent to those areas with the greatest need. Emergency managers work with professionals and volunteers. These activities are managed through an emergency operations center (EOC). Local emergency responders dominate the response period, which is characterized by uncertainty and urgency. Minutes of delay can cause the loss of life and property. Speed is essential, but actions that are impulsive and lead to mistakes must be avoided. Finally, emergency response actions need to anticipate the recovery phase. For example, emergency managers must perform damage assessments to support their requests for presidential disaster declarations. The emergency response phase ends when the situation is stable, and the threat to life and property has returned to its normal level.

1.3.4 Disaster Recovery

Recovery begins as the disaster is ending and continues until the community is back to normal. In some cases, the recovery period may be a long time. The immediate goal is to restore the infrastructure of the community. The basic infrastructure consists of systems for delivering water/wastewater, electric power, fuel, telecommunications, and transportation. The ultimate goal is to return the community's quality of life to the same level it was before the disaster. Recovery measures are both short-term and long-term. Short-term measures are relief and rehabilitation. Long-term measures include reconstruction.

Relief and rehabilitation activities usually include

- ▲ Clearing debris for access to the impact area.
- ▲ Renewing economic activities.
- ▲ Restoring government services.
- ▲ Providing housing, clothing, and food for victims.

Reconstruction activities tend to be dominated by:

- ▲ Rebuilding major structures, including buildings, roads, bridges, and dams.
- ▲ Revitalizing the area's economic system.

Leaders may use reconstruction to make changes they wanted before the disaster. After the eruption of Mt. Usu in Japan, local leaders convinced the government to improve towns to attract tourists (Perry and Hirose, 1982). Leaders may also improve the community beyond its pre-disaster state by revitalizing dilapidated residential and commercial areas.

Most of the resources used during recovery come from outside the community. Some resources come from private organizations and state governments. However, the majority of resources in a major disaster come from the federal government.

FOR EXAMPLE

Mitigation and Guam Memorial Hospital

A typhoon devastated Guam Memorial Hospital in 1997. The hospital is critically important because it is one of the few places that can offer oxygen and dialysis on the island. Through a grant from FEMA, Guam officials took steps to reduce typhoon damage to the hospital. Corridors were enclosed and the barriers around the oxygen storage unit were strengthened. In 2002, when another typhoon hit Guam, the hospital sustained the damage and still provided oxygen and dialysis to patients (FEMA, March 5, 2003). See link at www.fema.gov/news/newsrelease.fema?id=2223

1.3.5 The Mix of Strategy Elements

Historically, more of the resources for emergency management have been allocated to response and recovery than to mitigation and preparedness. Recent events, such as Hurricane Katrina, have called attention to the flaw in this policy. Thus, a community's hazard management strategy should combine some degree of mitigation, preparedness, response, and recovery. The relative emphasis on each of these elements depends on three principal implementation mechanisms:

1. **Available technological systems:** Building dams, installing warning systems, and expanding highways for quick evacuation are all examples of creating technological systems.
2. **Risk communication:** Risk communication highlights the consequences of certain behaviors. Risk communication explains the personal risks associated with living in a hazard-prone area. Changes that can be made to reduce vulnerability to hazards are also communicated. For example, explaining that beach homes built on stilts are less likely to be destroyed by storm surges might lead some people to elevate their homes.
3. **Sanctions and incentives:** Sanctions punish actions that increase hazard vulnerability. Incentives reward actions that reduce hazard vulnerability. Sanctions are usually achieved through regulations (e.g., zoning and subdivision regulations) that will reduce the impact of a hazard. A violation of these regulations can be punished by fines or imprisonment, or both.

SELF-CHECK

- Define **hazard mitigation.**
- Compare and contrast response and recovery.
- Name the three steps for performing a hazard vulnerability analysis.
- List three considerations that influence hazard management decisions.
- Describe the primary purpose of hazard mitigation.
- Name the goals of **emergency response.**

SUMMARY

Today, we can't just hope for the best. Unfortunately, we must be prepared for the worst. This chapter identifies the threats that we deal with: disasters, emergencies, and hazards. It describes how the history of emergency management has changed (and evolved) just as the threats of today have changed with advances in technology, increases in terrorist activities, and development of environmental hazards. This chapter also showed how you must prepare for threats through use of mitigation strategies, for preparedness through disaster response and recovery plans, and for response and recovery processes that focus on prevention, population protection, and minimization of damage. In the event of a disaster, emergency management also deals in relief, rehabilitation, and reconstruction—activities important to communities that have suffered from a disaster. Finally, this chapter emphasized the need to *combine* mitigation, preparedness, response, and recovery methods by implementing technological systems, risk communication, and sanctions and incentives.

KEY TERMS

Disaster — An event that produces greater losses than a community can handle, including casualties, property damage, and significant environmental damage.

Emergency — A minor event that can cause a few casualties and a limited amount of property damage *or* an imminent event that requires prompt and effective action.

Emergency Response — A hazard management strategy that has the goal of protecting the population, limiting damage from the impact of an event, and minimizing damage from secondary impacts. Response begins when a disaster event occurs.

Hazard A source of danger. Hazards have the potential to affect people's health and safety, their property, and the natural environment.

Hazard Mitigation A hazard management strategy that takes place before disasters strike that addresses the causes of a disaster, reducing the likelihood it will occur or limiting its impact.

Natural disaster An event that occurs in nature that results in casualties, property damage, and environmental damage. Natural disasters include earthquakes, floods, hurricanes, volcanic eruptions, and wildland fires.

Recovery A hazard management strategy that has the goal of restoring the normal functioning of a community. Recovery begins as a disaster is ending and continues until the community is back to normal.

Secondary impacts Disasters caused by a disaster, including events such as hazardous materials releases caused by earthquakes.

Technological disasters Events that result from the accidental failures of technologies, such as the release of hazardous materials from facilities where they are normally contained.

Terrorist disaster A deliberate attack that is intended to achieve political objectives by inflicting damage and casualties. Also referred to as terrorism.

ASSESS YOUR UNDERSTANDING

Go to www.wiley.com/college/lindell to evaluate your knowledge of the basic role of the emergency manager.
Measure your learning by comparing pre-test and post-test results.

Summary Questions

1. An increase in the size of the human population contributes to the rising number of losses as a result of disaster in the United States. True or False?
2. Sanctions punish actions that increase hazard vulnerability. Incentives reward actions that reduce hazard vulnerability. True or False?
3. All of the resources used during recovery come from outside the community. True or False?
4. Which of the following are methods used to counter secondary threats?
   (a) Fighting urban fires after earthquakes
   (b) Identifying contaminated water supplies following flooding
   (c) Preparing for flooding following a glacier melt during a volcanic eruption
   (d) All of the above
5. Which of the following organizations merged with the Department of Homeland Security in 2002?
   (a) local emergency planning committees (LEPCs)
   (b) state emergency response commissions (SERCs)
   (c) Federal Emergency Management Agency (FEMA)
   (d) the Defense department
6. Hazard mitigation involves working with
   (a) home owners.
   (b) government agencies and business owners.
   (c) schools and businesses.
   (d) businesses, households, and government agencies.

Review Questions

1. What is emergency management?
2. What is the difference between a hazard, an emergency, and a disaster?
3. Why is hazard mitigation important?

Applying This Chapter

1. You are an emergency manager for New York City. How do you prepare for a possible terrorist attack? What would you ask residents to do?
2. You are hired as an emergency manager in an industrial town. The town wants you to determine what hazards it is vulnerable to. How do you do this?
3. You are an emergency manager for a small town in Florida that will be hit with a hurricane in two days. What steps do you take to prepare?

YOU TRY IT

The Terrorists Try to Strike Again

Two weeks after July 7, 2005, on July 21, terrorists in England tried to strike again. A second series of four explosions took place. This time, however, only the detonators of the bombs exploded, and all four bombs did not fully detonate. What does the incident illustrate about the nature of terrorism?

Local Hazard Mitigation

You are hired as an emergency manager of a small town. The town has continuous flooding in one area. What steps can you take to reduce the impact of the flooding?

Financial Impact of Terrorism

As discussed, four factors have affected emergency management. One of the factors is that businesses understand that disasters can affect them negatively and even lead them into bankruptcy. Give some specific examples of how terrorist attacks affect businesses negatively.

2

EMERGENCY MANAGEMENT STAKEHOLDERS

Influencing the Decision-Making Process

Starting Point

Go to www.wiley.com/college/lindell to assess your knowledge of the basics of the emergency management decision-making process.
Determine where you need to concentrate your effort.

What You'll Learn in This Chapter

▲ The types of stakeholders: social, economic, and governmental
▲ How stakeholders are involved in emergency management
▲ The different power bases of stakeholders
▲ How stakeholders can influence policy
▲ The elements of an emergency management policy process
▲ How policies are adopted and implemented

After Studying This Chapter, You'll Be Able To

▲ Examine social, economic, and governmental groups and explain how they affect the emergency management process
▲ Examine how to use focusing events and windows of opportunity as ways to promote mitigation
▲ Examine the effects of business interruption on a community
▲ Examine ways to work with stakeholders in the emergency management process
▲ Differentiate between different types of power
▲ Analyze and diagram an emergency management policy process

Goals and Outcomes

▲ Compare and contrast stakeholders from social, economic, and governmental groups
▲ Support the appropriate stakeholders in emergency response planning
▲ Network with appropriate stakeholders to obtain local, regional, and national resources for emergency response planning
▲ Evaluate how to involve communities in emergency management
▲ Plan and develop an emergency management policy process
▲ Implement and evaluate an emergency management policy process

INTRODUCTION

Disasters affect all of us, and we all have a stake in how well communities prepare. We are all stakeholders. **Stakeholders** are people who have, or think they have, something to lose or gain. An emergency management stakeholder is affected by the decisions made (or not made) by emergency managers and policy makers. Community stakeholders can be divided into three categories:

1. Social groups
2. Economic groups
3. Governmental groups

This chapter discusses each of these groups and how each is involved in emergency management.

As emergency managers, you need to know how to involve each of these groups of stakeholders in emergency planning processes. This chapter explains how to do this. This chapter also explains how to develop communication and negotiation skills to help you network with stakeholders in order to obtain resources: local, regional, and federal. Much of your skill in dealing with stakeholders will depend on power, how it plays in your relationships with stakeholders and how it influences policies.

Finally, this chapter walks you through the important elements of an emergency management policy: how to formulate, adopt, implement, and evaluate one. You will see how each of the topics discussed in this chapter influences the decisions you will make.

2.1 Social Groups

The basic **social group** unit is the household. Households:

- Try to prevent accidents.
- Prepare for natural disasters.
- Evacuate.
- Suffer economic losses.

Households can take actions, called **hazard adjustments**, which can reduce their vulnerability to disasters by:

- Living in less hazard-prone locations.
- Renting or buying residences that are more resistant to wind, water, and ground-shaking.

▲ Taking precautions (such as boarding up their houses) to lessen the impact of the disaster.
▲ Purchasing hazard insurance.

As a group, households control a substantial amount of the social assets (buildings and their contents) at risk from disasters. However, not all households take the same precautions. Homeowners differ in their perceptions of risk and their perceptions of the effectiveness of appropriate hazard adjustments. Homeowners have more to lose than renters, because they own buildings as well as building contents. Homeowners also vary in their ability to take precautions. Not everyone can afford to buy needed supplies. Also, not all homeowners are educated about hazards. Other stakeholders, such as government, have little direct influence. However, government agencies can provide information and financial incentives for taking precautions. Unfortunately, information and incentives don't always convince households to take precautions.

2.1.1 Community Emergency Response Teams

Homeowners can organize as groups to develop an emergency management policy in their neighborhoods. In some communities, **community emergency response teams (CERTs)** are beginning to fill this role. CERTs may also be known as neighborhood emergency response teams, or other similar names, but they share a common origin and many other characteristics (Simpson, 2001). CERTs train emergency response volunteers at the neighborhood level and organize them in groups capable of providing basic services such as:

▲ Performing triage.
▲ Administering first aid.
▲ Organizing urban search and rescue.
▲ Suppressing fire.
▲ Estimating damage and casualties.

Local emergency service agencies train and support these groups.

2.1.2 Private Sector Groups

In addition to households, we have larger private sector groups. Private sector groups include:

▲ Religious organizations
▲ Nongovernmental organizations (NGOs)
▲ Nonprofit organizations (NPOs)

FOR EXAMPLE

Aid from Private Groups

Churches are often used as mass care facilities during evacuations. Churches also help provide recovery funding. Emergency managers should partner with churches in the early stages of the response and recovery process. The Salvation Army is also an important player in response and recovery activities. The United Way serves to channel local funds to those needing help during the recovery period. The American Red Cross has an official role in this country as the provider of emergency shelter.

- Community based organizations (CBOs)
- Businesses

All of these groups are different sizes and have different budgets. The functions they perform vary, as does their level of interest in emergency management activities. Nonetheless, all are potential partners in formulating emergency management practices and policies. Private sector groups can be important resources. Some play key roles in specific phases of emergency management.

ROLE OF ENVIRONMENTAL ORGANIZATIONS

Environmental groups, such as the Sierra Club and WorldWatch Institute, have limited involvement with emergency management. Yet there is an overlap between protecting the environment and working on hazard mitigation. This presents an opportunity to work with environmental groups. Both groups want to encourage sound land use practices. Both groups can work on issues like the prevention of floods through watershed management.

SELF-CHECK

- Define **stakeholder.**
- Name three ways that households can reduce their vulnerability to disasters.
- Name three private sector groups.
- Define **CERT.**

2.2 Economic Groups

Economic groups, or businesses, organize the flow of goods and services. The economy is affected anytime there is an interruption to business. Businesses range from small mom-and-pop businesses to large corporations, employing tens of thousands of people. Businesses have different needs and resources. Small businesses are the most at risk. However, small businesses are close to the community and are more likely to respond to appeals for assistance. Large corporations have vast personnel and money. However, local store managers may not have authority to decide how or if to use the resources for local emergencies.

Business owners control their resources like homeowners do and should take the same precautions. However, it can be difficult to convince owners to take precautions. Instead, businesses tend to focus on response and recovery. Some businesses, however, are active supporters of emergency management. These businesses include:

- ▲ Insurance companies
- ▲ Real estate developers
- ▲ Bankers
- ▲ Home improvement retailers

Business interruption is the loss of revenue due to a disruption. Disasters cause businesses interruptions. For example, a flooded store loses money every day it is closed. Once businesses realize the costs of a disaster, they prepare. The key is for businesses to understand their relationship to suppliers, customers, and employees. Businesses must also understand their dependence on information technology (IT) systems (Lindell & Prater, 2003). If any of these relationships is disrupted, businesses can suffer financial losses.

Public utility companies are critical business stakeholders and their services include:

- ▲ Electricity
- ▲ Water
- ▲ Sewage treatment and disposal
- ▲ Solid waste management
- ▲ Telecommunications (telephone, television, internet, etc.)

Such businesses are active in emergency management because they are responsible for restoring service quickly. All other stakeholders depend on utilities so business, household, governmental, and health care interruptions are minimized.

FOR EXAMPLE

Project Impact

Project Impact was started by FEMA. It is a model involving businesses aimed at reducing hazards and preparing for disasters. It met with great success in cities like Tulsa and Seattle. The suspension of federal funding has slowed the spread of Project Impact. However, its success makes it a valuable method for managers to develop a relationship with their local business communities.

MEDIA'S ROLE IN DISASTERS

The media are important to the success of emergency management programs. The media cover all phases of emergency management. The media warn the public of coming natural disasters and educate the public about hazards. The media both consume and create the news. They consume "hard news" by describing disasters and create "soft news" by reporting about emergency preparation measures. This "soft news" builds support for emergency management. Emergency managers should know their local news media and create relationships with reporters and producers.

SELF-CHECK

- Define business interruption.
- Name three utility stakeholders.
- Identify the role the media play in disasters.
- Name some businesses that are active supporters of emergency management.

2.3 Governmental Groups

There are various types of **governmental groups.** The foundation of the governmental structure is the town or the city followed by the county. The third level is the state. Cities and counties have varying levels of power from one state to another because states differ in the powers they grant. The majority of emergency management policies are set at the state level.

In addition to the different levels of government, there are different agencies within each level. These agencies vary widely in:

- ▲ Size
- ▲ Organizational complexity
- ▲ Human resources
- ▲ Financial resources
- ▲ Technical resources

At each level of government, agencies differ in their functions. Municipal fire and police departments are the first responders to most emergencies. In many jurisdictions, emergency management is attached to one of these departments. However, in larger communities, emergency management might be an independent agency. In some communities, there is a separate emergency medical services agency. Working together, fire departments and hospitals can also provide this function.

2.3.1 Regional Stakeholders

Regional and state-level stakeholder agencies include:

- ▲ City and county councils
- ▲ Flood control districts
- ▲ State-level coastal zone agencies
- ▲ Geological services agencies
- ▲ Soil conservation agencies

The most important stakeholders are the state emergency management agencies. These agencies vary in:

- ▲ Levels of expertise
- ▲ Staffing
- ▲ Budgets
- ▲ Other organizational resources

These agencies provide direction for local emergency managers. Together with state legislatures, these agencies provide the legal framework within which managers work. These agencies also link local governments with FEMA regional offices.

ROLE OF ACADEMIC INSTITUTIONS

Academic institutions are also stakeholders. They provide the science for policy making. There are several research centers around the country. Some of these centers focus on one type of hazard. An example of this is the Multidisciplinary Center for Earthquake Engineering. Others study all hazards. An example of this is the Hazard Reduction and Recovery Center at Texas A&M University. In addition to these academic institutions, there are emergency management consultants.

2.3.2 National Stakeholders

FEMA was the lead national agency for emergency management until the **Homeland Security Act (HSA)** was signed in November 2002. The HSA caused a restructuring of emergency management to begin. FEMA's role remains to be determined.

FEMA is not the only national stakeholder. Other agencies with responsibilities for hazards and disasters include:

- ▲ U.S. Geological Survey (USGS)
- ▲ Army Corps of Engineers
- ▲ National Weather Service (NWS)
- ▲ Environmental Protection Agency (EPA)

Some federal agencies, such as the Department of Transportation, provide training materials for emergency responders. Other agencies, such as the National Science Foundation, support basic research.

Emergency managers must know the different types of stakeholders and those stakeholders' local, state, or federal roles. Their roles can be understood by examining the levels at which decisions are made. Families make decisions about the level of preparedness for each household, and emergency managers can support families' good practices by educating the public. Managers can also enhance local government support for organizations like CERTs.

Local governments determine what resources to devote to emergency management. Outside agencies, such as state agencies and FEMA, influence policies. However, emergency management remains a local issue. Cities control their first responders. Emergency responders compete for resources that are also needed by schools and for road repairs and maintenance.

Local governments control land use. Local governments develop land use planning and zoning programs. In addition, local governments establish building code requirements for hazard resistance. This is especially true for wind and earthquake hazards. In some jurisdictions, emergency management operates jointly with transportation and police departments to integrate many functions. This is the case in Harris County, home to Houston, Texas.

State Government

State governments have a number of important functions. State governments pass legislation that affects the decisions that local governments can make. For example, some states require local governments to engage in land use planning whereas other states do not (Burby, 1998). Moreover, state support for local emergency managers varies in terms of technical resources and funding.

Federal Government

In the case of a major disaster, local governments request aid from the state. If a state believes the response and recovery will require more resources than are available, it requests a presidential disaster declaration for access to federal assistance. Most, but not all, requests for presidential disaster declarations are approved. If a request is denied it is because FEMA may disagree that local and state resources have been exceeded. Between the passage of the Stafford Act in 1988 and 1998, only about one-fourth of the requests were denied (Sylves, 1998). The federal government tries to use an objective set of criteria for issuing declarations. However, the process still includes many subjective decisions. Also, there are political considerations that affect the process. Very few presidents are willing to deny resources to a state during a disaster.

ST. LOUIS, POLITICS, AND HAZARD MITIGATION

St. Louis Missouri lies in the New Madrid Fault Zone. Most of the buildings are nonreinforced masonry structures, and these buildings will be severely damaged if there is an earthquake. In 1976, the Department of Housing and Urban Development (HUD) decided to give construction loans for only those buildings that met strict earthquake-resistant building codes (Drabek, Mushkatel & Kilijanek, 1983). Concerned about the effect on new construction, local developers, contractors, and officials challenged the policy. HUD officials saw this as a threat to their policy. They felt their policy was justified by possible threats to residents' safety. Experts attacked the scientific basis for HUD's policy. They argued that including St. Louis in Zone II was in error. They also argued that the projected damage from a repeat of the 1811–1812 earthquakes was exaggerated. The city lobbied the local HUD office to exempt St. Louis from the building requirements. The city also asked its congressional delegation, the Home Builder's Association and public interest groups, to support this request. By 1981, the strict building requirements were used for all structures except multi-family housing rehabilitation projects. The impact of this was minimal because it was enforced by HUD's regional and local offices. It was not enforced by the city or county of St. Louis. As a result, most engineers and developers were uncertain about which building standards to use.

SELF-CHECK

- **Name three regional stakeholders.**
- **Name three national stakeholders.**
- **Describe the function of state and federal governments during a disaster.**
- **Define social groups, economic groups, and governmental groups.**

2.4 Involving Stakeholders in Emergency Management

To develop an effective emergency management system, the local emergency manager must involve all relevant stakeholders in the process.

A networking with stakeholders checklist includes the following actions:

- ▲ Encourage relationships among stakeholders to improve the flow of information, services, and supplies.
- ▲ Consult with all relevant agencies when making mitigation, response, and recovery plans.
- ▲ Coordinate the stakeholders as emergency operations plans are made.
- ▲ Coordinate the stakeholders as recovery operations plans are made.
- ▲ Coordinate stakeholders during the emergency exercises.
- ▲ Stage exercises frequently.
- ▲ Ask the state emergency management agency for assistance in evaluating exercises.

City planners have many ways of involving the public in policy development. Emergency managers can use these methods as well. Citizen committees, which consider policy changes, can be recruited to contribute to the local emergency plans. Local emergency planning committees (LEPCs) already exist, and are valuable forums for input.

Zoning changes require public hearings. These hearings serve as forums for public participation. The partnerships developed in Project Impact provide a model for using the expertise and resources of local business in improving emergency plans. Such partnerships are a useful way to involve the private sector.

Stakeholders have different needs. Local officials need quick, positive answers to their requests. Federal officials need to enforce national policies and remain within budget. These different needs cause conflict. The manager must ensure all requests for disaster relief are well-documented. The requests must also be consistent with federal reimbursement policies. Establishing local demographic and economic information expedites preparation of disaster relief requests.

Working with groups interested in related issues is a good strategy. Emergency managers can work with groups to ensure that policies that are adopted have several purposes. This way, managers will have a base of support. For example, environmental groups are interested in preserving wetlands or riverine corridors for their scenic value and other reasons. These same lands can perform valuable functions by absorbing floods or by keeping housing developments out of a floodplain. To be successful, emergency managers must constantly find ways to work with stakeholders.

2.4.1 Getting the Community Involved

There are four simple ways to get the community involved in hazard prevention:

1. **Discuss your work with your friends and neighbors.** Discuss potential threats and the emergency management plans. Get informal reactions. Creating a buzz about emergency management is an inexpensive and valuable way to get community support.
2. **Set up a hazard hotline.** Advertise the hotline. This is an effective way to receive information. You can also use the hotline to warn and inform the public of hazards. The FBI, for example, has a hotline for tips on criminals and terrorist activities.
3. **Speak at schools, neighborhood organizations, and community organizations.** Inform people in the community that you don't personally know about your work. Discuss potential threats and the emergency management plans.
4. **Form citizen committees.** Advise community members on the emergency management plans and gather volunteers to carrying out the plans. For example, volunteers can fill sandbags, direct traffic, and serve on search-and-rescue teams.

FOR EXAMPLE

America's Most Wanted

In 1981, John Walsh was a hotel developer when his son, Adam Walsh, was kidnapped and killed. The violent death of his six-year-old son led John Walsh to become an advocate for crime victims. In 1987, *America's Most Wanted,* a reality television show about crime was born. The show features stories about the crimes and a hotline that encourages people to call in to provide tips. This very successful technique of getting the public involved in fighting crime has led to the arrest of over 300 suspects.

SELF-CHECK

- Name three ways to get stakeholders involved in emergency management.
- Name two ways to the get the community involved in hazard prevention.
- Identify the abbreviation **LEPC.**
- Describe the benefits of emergency managers working with groups.

2.5 Types of Power

We discussed different types of stakeholders. Stakeholders also vary in the types of resources and power they bring to the emergency management process. Organizational theorists (French and Raven, 1959; Raven, 1965) describe six bases of power:

- ▲ Reward
- ▲ Coercive
- ▲ Legitimate
- ▲ Expert
- ▲ Referent
- ▲ Information

2.5.1 Reward and Coercive Power

Reward and coercive power is frequently referred to as the "carrot and the stick" approach. For example, if children clean their rooms for money, the parents have exercised reward power (the carrot). If children don't clean their rooms and are punished, the parents have exercised coercive power (the stick). Reward and coercive power require a socially dependent relationship. In our example, the children's behavior depends on the parents' continual monitoring. The parents must frequently inspect the rooms' cleanliness to determine whether to reward or punish their children. Coercive power can produce deception to avoid the punishment and hostility as a result of the punishment.

2.5.2 Legitimate, Expert, and Referent Power

Power holders have to follow up to see if their carrot and stick approach is working. Legitimate, expert, and referent power bases are more attractive because they involve little follow-up. **Legitimate power** arises from one person's relationship to another and can come from a formal position. For example, any official elected by a fair voting process has legitimate power. **Expert power** is based on someone's extensive knowledge of cause and effect relationships in a specific subject area. Physicians have expert power because they can diagnose illnesses from specific symptoms, and they know how to treat those illnesses. **Referent power** is based on one's desire to be like the power holder. For example, many individuals want to look like a glamorous celebrity. That celebrity, therefore, has referent power. Millions of magazines are sold with the headline "Beauty Secrets from the Stars".

2.5.3 Information Power

Information power involves true, new, and relevant facts or arguments about a situation. Information power is exercised by either introducing or withholding information (Mechanic, 1962). Information power is, in many respects, the most effective basis of power because it is socially independent. That is, once the new information is understood and accepted, its source becomes inconsequential. As a result, one does not need to monitor the target's desired behavior. However, information power does require the information to be checked for accuracy, which can be time-consuming.

2.5.4 Direction of Power

Households exert their power up the chain to the federal government through voting, lawsuits, and boycotts. Likewise, the federal government exerts power to households by passing laws. With multiple bases of power, power operates in an upward or downward direction.

Figure 2-1 (adapted from Lindell et al., 1997) shows the relationships among stakeholders. The figure shows relationships between federal, state, and local governments. The solid arrows indicate the downward direction in which most power is exerted in the relationship. The HSA, for example, was imposed on the

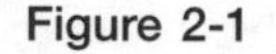

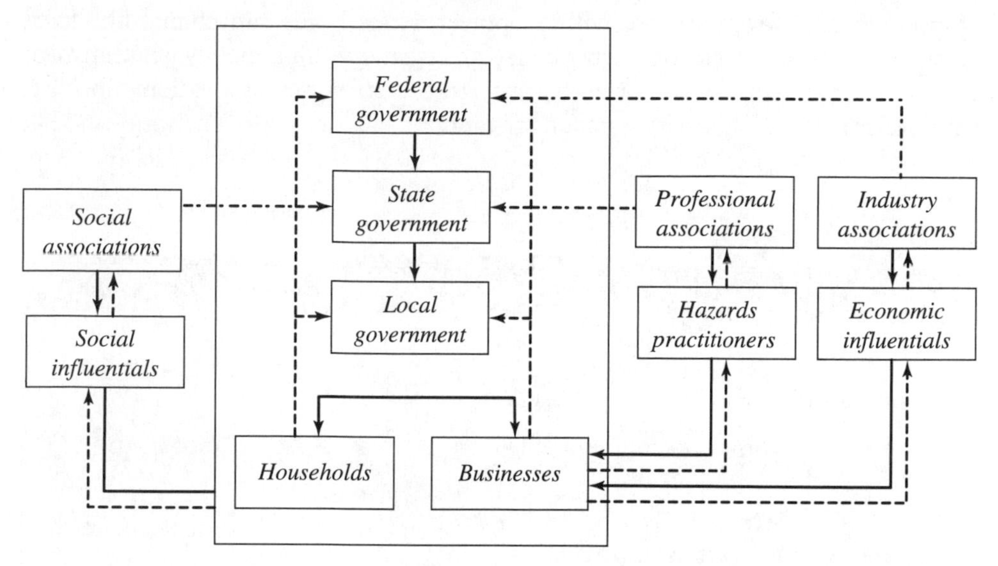

Power relationships among emergency management stakeholders.

state governments by the federal government. No federal funding was given for this legislation. This is an example of an unfunded mandate and of power being asserted in a downward direction. Information and influence flow from the bottom up as well as from the top down. Information and influence also flow between groups of stakeholders.

A relationship might change over time. It could change from coercive power to information power. Stakeholders at the top must gain the support of local government officials to accomplish anything at the lower levels. In the local

FOR EXAMPLE

The Federal Government and Reward Power

In 1968, congress created the National Flood Insurance Program (NFIP) in response to the rising burden of taxpayers for flood victims. Twenty thousand communities participate in NFIP by adopting floodplain-management ordinances. These ordinances reduce future flood damage. In exchange, the NFIP makes federally backed flood insurance available to homeowners, renters, and business owners in these communities.

government-households relationship, the local government has more power. Households, however, are not without power. Households can change the local government through elections, boycotts, or lawsuits. Other policy relationships have similar dynamics. Importantly, this is a complex set of relationships that the emergency manager needs to understand.

SELF-CHECK

- Define **reward and coercive power** and provide an example that relates to emergency management.
- Define **legitimate power** and provide an example that relates to emergency management.
- Define **expert power** and provide an example that relates to emergency management.
- Define **referent power** and provide an example that relates to emergency management.
- Define **information power** and provide an example that relates to emergency management.

2.6 The Emergency Management Policy Process

The basic steps of the policy process (adapted from Anderson, 1994) are presented in Table 2-1. This table presents five stages through which policies move. Of course, the actual policy process is not as clear. However, it is still useful to consider the various stages, recognizing that they may occur at the same time. Also, the process may occur several times for one policy as feedback leads to adjustments.

2.6.1 Agenda Setting

Getting people to prepare for disasters is difficult. They tend to pay attention only when a disaster happens. Unfortunately, this is too late to do anything but react. The time to think about disasters is before they happen so planning can occur. The emergency manager's first task is to put hazards on the political agenda.

There are three types of political agendas: the systemic, the governmental, and the institutional. The systemic agenda includes hot topics that concern voters. The

Table 2-1: The Standard Policy Process Stages

Policy terminology	*Stage 1: Agenda setting*	*Stage 2: Policy formulation*	*Stage 3: Policy adoption*	*Stage 4: Policy implemention*	*Stage 5: Policy evaluation*
Definition of policy stage	Establishing which problems will be considered by public officials	Develop-ing ideas for solving problem	Developing support and author-ization for a specific proposal	Applying the policy through the government	Determin-ing whether the policy was effective and why
Emergency management question	How can I get officials to consider action?	What should I propose?	How can I convince officials to accept my solution?	How can the adopted policy be applied?	Did the policy work? Why? How can it be improved?

governmental agenda includes issues on which the government is working. The institutional agenda includes issues on which institutions are working. Agendas are unstable. The public shifts attention from one issue to another as events occur. In response, the government and institutions change their agendas to respond to voters' concerns. They also attempt to shape voters' concerns (Baumgartner and Jones, 1993).

People avoid discussing disasters in their communities for many reasons. First, many local governments and business leaders believe calling attention to potential disasters may discourage investment or tourism. Second, problems arise that directly compete for attention and resources. Every year, a new class of children enters the public school system, even though a disaster does not strike. We cannot predict with certainty when disasters will strike. This makes it easy for local officials to push emergency management to the back burner. Third, hazard prevention is controversial, because many developers believe land use and building construction restrictions will reduce their profits.

There are various ways to shape the policy agenda. First, use current events. A natural or technological disaster is a **focusing event** that draws public attention to the need for local disaster planning and hazard mitigation (Birkland, 1997; Lavell, 1994). For example, after the devastating tsunami of 2004, federal officials improved their early warning system. This **window of opportunity** is not open for long (Prater and Lindell, 2000). The challenge for local emergency managers is to use this policy window while it is open. It is unknown how long such a policy window will stay open or what will close it. Windows close because of any of the following situations (Kingdon, 1984):

▲ The problem is solved.
▲ Persistent failure to take action.
▲ Another event occurs that shifts the public's attention.
▲ Key stakeholders or advocates for that policy leave, or are pushed out of, their positions in a policy making body.
▲ No possible course of action seems available.

Because of the short amount of time available to effect policy change, individuals must work aggressively to set issues on the agenda and to keep them there. Such individuals always champion important issues and are called policy entrepreneurs. Policy entrepreneurs might be

▲ Elected or appointed officials
▲ Local media personalities
▲ Educators
▲ Business owners
▲ Interested citizens

Whoever they are, however, they need three qualities to be successful. Keep in mind, however, that policy change is possible even if no single individual has all three qualities. A group of individuals can be effective if they collectively have:

▲ Technical expertise in hazards, acquired either through education or experience.
▲ Political expertise, necessary for any successful policy change effort.
▲ Personal commitment, necessary because it can take years to overcome opposition to new policies.

The emergency manager needs data on potential hazards and on which specific populations are most at risk. With this data in hand, the manager can make a case that such an event could "happen here." Second, the manager should have policy ideas that are relevant to the local situation. These ideas are ones that the local legislative body can quickly adopt and enact. Managers should push the agenda, rather than assume that the community will follow their lead when an event occurs.

Introducing new emergency management practices can raise opposition during other phases of the disaster cycle. Consequently, emergency managers must be familiar with the sources of this resistance for emergency management practices to work effectively.

What features of emergency management can arouse opposition? Why? Disaster relief is rarely opposed. It is an example of distributive policy that benefits a deserving population. It is difficult to oppose disaster relief without appearing unsympathetic. Not all emergency management policies are distributive, however. Land use controls and building codes are examples of regulatory policy, which imposes limits and higher costs to some stakeholders. Such policies frequently generate conflict because there are obvious losers. For example, prohibiting construction on barrier islands produces benefits for the entire community when it is protected from hurricane damage. However, these benefits seem hypothetical, and they require money for each household. Few of those who benefit are likely to fight for policy adoption. By contrast, the "losses" are concentrated among a few powerful people who will fight against policy adoption. This leads to conflict that emergency managers must manage.

As in any policy debate, there are voters that will be just as interested in keeping emergency management off the public agenda as emergency management professionals are to put it on (Bacharach and Baratz, 1962). This is especially true when it comes to hazard mitigation. In some cases, there may be ideological opposition to any government control over private land use decisions.

Managers can support hazard mitigation efforts by increasing the number of groups involved in the process. Since hazard mitigation and emergency preparedness are meant to protect lives and property, it is possible to form a strong coalition.

2.6.2 Policy Formulation

Managers should have a set of proposed solutions before they attempt to shape the agenda. If not, policy makers may be overwhelmed. During policy formulation, many options emerge. Different stakeholders will propose different solutions (Kingdon, 1984; Anderson, 1994). This is a critical stage in the process because careful drafting of legislation is crucial to a policy's success. Poorly drafted laws are difficult to implement. Some even make the situation worse. The goal is to minimize court challenges and unintended consequences.

A policy formulation checklist includes the following actions:

- ▲ Identify the hazards.
- ▲ Assess the probability and seriousness of each threat.
- ▲ Design policies with full awareness of the local politics.
- ▲ Define the targets of a policy clearly (e.g. what *types* of households and businesses).

- ▲ Define what activities are to be regulated (e.g. land-use practices and building construction practices).
- ▲ Define which influence mechanisms are to be used (e.g. technological advances, risk information, economic incentives, and legal penalties). The government has many alternatives. One option is to control lot sizes to limit the population at risk in hazard-prone areas. Also, the government can mandate that streets be wide enough for large emergency vehicles such as fire trucks. Alternatively, building codes can restrict construction designs and materials.
- ▲ Create public awareness campaigns.
- ▲ Encourage governments to promote the adoption of hazard-resistant land-use and construction practices. Two incentives are low interest loans and tax credits. Poor jurisdictions might not be able to provide these incentives.
- ▲ Encourage governments to require hazard-resistant land-use and building construction practices for construction permits. This requires on-site inspections.

A combination of risk communication, land-use regulations, building codes, and hazard insurance is an excellent way to address environmental hazards (Burby, 1998). Successful implementation requires the policy to be consistent with the agencies' commitment and **capacity.** Capacity includes budget allocations, staffing levels, and staff members' knowledge and skills.

When developing any public policy, stakeholders must be included. This is especially important for hazard policies, because these policies often require

- ▲ Certain present investment (e.g. increased taxes to develop effective emergency management capabilities).
- ▲ Certain opportunity cost (e.g. a lucrative land development project that cannot be pursued).
- ▲ Uncertain future benefit, which are reduced disaster losses that will only be realized much later and, even then, will be difficult to measure.

Emergency managers should consider involving business leaders. For example, business owners must plan ways to keep their businesses running in case of disaster. Economic incentives, such as offering tax credits, will help involve business leaders.

Emergency managers should address other considerations. Local officials may be threatened when new community groups participate in the decision-making process. Some officials are not used to being held accountable for individual decisions. Also, they may view citizen participation as causing trouble. Some neighborhoods might have lower income or ethnic minority residents who lack knowledge about the political system or actively mistrust it. Emergency managers must anticipate all of these problems. Any perceived unfairness in the policy will cause implementation problems. Even after a policy has been developed, there are many points that can cause a policy to be vetoed.

2.6.3 Policy Adoption

Policy adoption involves getting stakeholders to urge elected officials to pass a policy. Emergency managers should have a strategy for presenting the policy so to avoid procedural issues. Presenting a policy in the correct manner and at the right time lessens the likelihood that the policy adoption process is derailed. It is important to have a policy adopted and on the books, for that is what gives it legal authority.

2.6.4 Policy Implementation

Adoption is not the end of the story. All policies must be implemented to be effective. This stage is tricky. Opponents who have failed to block a policy often undermine it as it is put into practice. All policies are filtered through individuals who interact with the public (e.g., land use planners, building inspectors, and emergency medical technicians) and those individuals' support of the policy and its goals is especially important. There are three questions that affect policy implementation (Mazmanian and Sabatier, 1989):

1. How easy is it to solve the problem?
2. Is there a clear link between the solution and the problem (e.g. building dams to stop floods)?
3. What level of technology and amount of resources are available to solve the problem?

The way policy is implemented depends on the nature of the government. The United States has a federal government. As a result, strong state and local governments can support or thwart federal policy. Conversely, the federal government can strengthen local emergency management by providing information or technical support. The federal government can also undermine local goals by withholding funding. If all stakeholders are included in the early stages, it is more likely that a policy will have a smooth implementation.

The stronger the commitment to a policy's goals, the more likely it is that an agency will devote the necessary resources to implementation. The agency needs enough tools, in the form of incentives and sanctions, to implement a policy. If lawmakers are convinced of the seriousness of the problem, they will provide adequate authority and capacity to the agency. This is especially important if the target population is powerful and resists the policy.

FOR EXAMPLE

Evaluating Project Impact

Evaluating Project Impact was difficult. One requirement for selection as a Project Impact community was a history of commitment to hazard mitigation. However, the goal was to increase hazard mitigation efforts at the local level. This confusion of selection criteria and desired results made it difficult to determine how much the community improved because of the program and how much the community improved because of its commitment to hazard mitigation.

2.6.5 Policy Evaluation

Finally, as in any system, the policy process includes evaluation. A policy should be evaluated periodically and either improved or terminated. The most effective programs include feedback and allow for clear evaluation. Each program has unique, specific criteria. Criteria for evaluating hazard management policies include present and future reduction of losses and reduction of expenses.

SELF-CHECK

- Name three types of political agendas.
- List three situations in which the "window of opportunity" of policy setting might be closed.
- Name three things you must do to formulate a policy.
- Define **focusing event** and **window of opportunity.**

SUMMARY

How do you measure your stake in preparing for a disaster? Would a disaster affect you economically? Would your family be affected? As this chapter illustrates, we are all stakeholders. Your job and the decisions you make as an emergency manager affects stakeholders. In turn, stakeholder and policy makers can affect how you do your job. Everyone has a voice, and that voice is often heard in the power we hold over each other. This chapter discusses stakeholders, the power relationships you develop with them, and the roles they play in emergency management. It also looks at how you can involve stakeholders (including your neighbors) in emergency management planning and processes. To do this, you must practice negotiation skills and understand how policies are influenced. Whether you are trying to obtain resources from local, regional, or federal sources, or trying to influence policy, you need other people to help you.

This chapter also shows you the important elements of an emergency management policy and describes how to formulate, adopt, implement, and evaluate one. As you develop your policies, consider the checklists presented in this chapter. In addition, recognize your own power to affect stakeholders and policies.

KEY TERMS

Business interruption The loss of revenue due to disruption of a business's normal production of goods and services in exchange for money.

Capacity A measurement of an organization's ability to implement policy that includes budget allocations, staffing levels, and staff members' knowledge and skills.

Community Emergency Response Teams (CERTs) Homeowners organized as groups to perform emergency management tasks in their neighborhoods. CERTs may also be known as neighborhood emergency response teams, or other similar names, but they all organize and train neighborhood volunteers to perform basic emergency response tasks, such as search and rescue and first aid.

Economic groups Business stakeholders that organize the flow of goods and services and who are affected anytime there is an interruption to business caused by a disaster.

Expert power Power that is based on someone's expertise on a particular topic.

Focusing event A natural or technological disaster that draws public attention to the need for local disaster planning and hazard mitigation.

Governmental groups Stakeholders who are part of the government's structure. The foundation of the government structure is the town or the city, followed by the county. The third level is the state. Cities and counties have varying levels of power from one state to another because states differ in the powers they grant. Most emergency management policies are set at the federal and state levels.

Hazard adjustments Actions that can reduce vulnerability to disasters. These include actions such as purchasing hazard insurance, living in safer locations, and renting or buying homes that are resistant to disaster.

Homeland Security Act (HSA) An act signed in November 2002 that restructured emergency management by integrating many agencies having emergency- or security-related functions into the Department of Homeland Security.

Information power Power that involves true, new, and relevant facts or arguments. Information power can be exercised by either introducing or withholding information.

Legitimate power Power that arises from one person's relationship to another and can come from a formal position. Any official elected by a fair voting process has legitimate power.

Referent power Power that is based on a person's desire to be like the power holder.

Reward and coercive power Power frequently referred to as the "carrot and the stick" approach. Coercive power can produce deception to avoid punishment. Moreover, punishment typically produces continuing hostility.

Social groups Stakeholders that are primarily defined by households, who control a substantial amount of the assets (buildings and their contents) that are at risk from disasters. Social groups also include neighborhood, service, and environmental organizations.

Stakeholder Someone who has, or thinks they have, something to lose or gain in a situation. An emergency management stakeholder is affected by the decisions made (or not made) by emergency managers and policy makers.

Window of opportunity The time during which local emergency managers are most likely to be able to influence policy. A window of opportunity usually opens immediately after a focusing event has drawn attention to hazard and closes after attention moves on to other public issues.

ASSESS YOUR UNDERSTANDING

Go to www.wiley.com/college/lindell to evaluate your knowledge of the basics of the emergency decision-making process.
Measure your learning by comparing pre-test and post-test results.

Summary Questions

1. A stakeholder is someone who has nothing to lose. True or False?
2. CERTs train emergency response volunteers at the neighborhood level. True or False?
3. Which of the following was started by FEMA and is a model that that involves businesses in reducing hazards and preparing for disasters?
   (a) CERTs
   (b) Project Impact
   (c) Department of Homeland Security
   (d) *America's Most Wanted*
4. Which of the following is power that involves true, new, and relevant facts or arguments?
   (a) referent power
   (b) legitimate power
   (c) coercive power
   (d) information power
5. Which of the following is not an agenda type?
   (a) economic
   (b) governmental
   (c) institutional
   (d) systemic
6. If a state believes the response and recovery from a disaster will require more resources than it has available, it should
   (a) request a presidential disaster declaration.
   (d) request access to state funds.
   (c) request military troops.
   (d) request local volunteers.
7. Zoning changes require public hearings. True or False?
8. Only business people can be policy entrepreneurs. True or False?

Review Questions

1. What is a stakeholder?
2. Who are stakeholders you influence to get involved in emergency management (name three)?
3. What types of power do stakeholders have (name three)?
4. What must you consider when formulating a policy?
5. What factors affect policy implementation?
6. What is important about a focusing event?
7. Successful implementation requires the policy to be consistent with the agencies' commitment and what?
8. What is one way in which expert power differs from information power?
9. What 1968 program did Congress create in response to the rising taxpayer burden for flood victims?

Applying This Chapter

1. You are the emergency manager of a small town in the panhandle of Florida. As your town is vulnerable to hurricanes, you need to create a local hazard mitigation program. Who do you involve and who do you ask to help you create this program?
2. You have just been appointed to be the director of FEMA. You were appointed, in part, because you spent years as a first responder. What type of power base or power bases do you have and why?
3. You are the director of FEMA and the President of the United States has appointed a commission to evaluate the National Flood Insurance Program which has been in place since 1968. What criteria would you ask the commission to use in the evaluation and why?
4. You are the emergency manager for a small town that is vulnerable to tornadoes. You do not feel that your town properly plans for tornadoes. How can you influence policy?
5. You are the emergency manager of an area that is home to a nuclear power plant. You have an emergency plan in case there is a major accident involving the reactor. What steps do you take to get the community more involved with the emergency plan?
6. You are in charge of coordinating response efforts after a terrorist bombing at the local university. First responders must arrive quickly. What other agencies, businesses, and teams do you involve and why?

YOU TRY IT

Organizing a Relief Effort

Remember the terrible tsunami devastation in Asia in 2005? What groups collected money and supplies for the affected populations? If a major disaster hits your state, who would you call upon for help for both short- and long-term recovery efforts?

Securing Support from Businesses

Think about your community. If a major disaster strikes (e.g. a large tornado), how would local businesses be affected? How would you expect businesses to help the local population? Name five businesses and list ways they could help the community in the event of such a disaster.

Mobilizing the Community

You are working with a group of citizens who are concerned about toxic chemicals in your community. The local chemical facilities have a good safety record and so do the truck and rail companies that transport chemicals through the community. Consequently, the local population has become complacent. How would you use different bases of power to mobilize support for toxic chemical emergency preparedness?

3

BUILDING AN EFFECTIVE EMERGENCY MANAGEMENT ORGANIZATION

Planning for Emergencies

Starting Point

Go to www.wiley.com/college/lindell to assess your knowledge of the basics of planning for emergencies.
Determine where you need to concentrate your effort.

What You'll Learn in This Chapter

- ▲ The role of local emergency management agencies (LEMAs)
- ▲ How emergency management organizations can be effective
- ▲ The stages of the emergency planning process
- ▲ How to write an emergency operations plan (EOP)

After Studying This Chapter, You'll Be Able To

- ▲ Define the role of LEMAs, including job descriptions, staffing issues, program plans, budget and funding issues, and individual outcomes
- ▲ Develop effective emergency management organizations
- ▲ Identify a planning process that includes desirable individual and organizational outcomes
- ▲ Identify the components of an EOP

Goals and Outcomes

- ▲ Organize and staff local emergency planning committees (LEPCs)
- ▲ Design an effective emergency management organization
- ▲ Create a planning process
- ▲ Create an EOP

INTRODUCTION

Effective emergency management organizations save lives and prevent losses. To build effective local emergency management agencies (LEMAs), you must understand the role of LEMA, including job descriptions, staffing issues, program plans, budget and funding issues, and individual outcomes. As an emergency manager and a LEMA member, you also need to understand how to develop an effective emergency management organization that supports your position and responsibilities. This involves creating a planning process that produces an operations plan as well as positive individual and organizational outcomes.

This chapter starts with a discussion of LEMAs. Then, you are tasked with creating an emergency management organization. Finally, you will develop a planning process and an effective emergency operations plan (EOP). As important as the overall concepts of this chapter are, it's also important that you learn the basics, such as how to conduct an effective meeting and how to solicit community support and resources. The larger tasks may seem overwhelming if you don't effectively master the smaller tasks.

3.1 The Local Emergency Management Agency (LEMA)

In practice, LEMAs might be known by other names, such as the Office of Civil Defense, emergency management, emergency services, or Homeland Security. LEMAs might be separate departments, part of another department, or an individual working with the chief administrative officer's (CAO) office. The CAO can be either a mayor or city manager. This person has the authority to hire, fire, allocate funds, and evaluate performance. In many communities, one person staffs a LEMA. This is especially true for small cities and cities that have minimal hazard vulnerability. As the leader of a LEMA, you are referred to as the local emergency manager. In larger cities, you will have a large staff. This is especially true for cities, such as New York City, that are exposed to major hazards. You will also form a committee, a local emergency management committee (LEMC), which will help formulate and implement emergency management policies.

As the local emergency manager, you will report to the CAO during emergencies. However, during the normal workweek, you will report to the head of a major agency, such as the local police or fire department. You may be a:

- ▲ Full-time employee
- ▲ Part-time employee
- ▲ Volunteer

Your status depends on the:

- ▲ Size of community
- ▲ Financial resources of the community
- ▲ Community's vulnerability to hazards

Emergency managers vary in their training and experience. Larger communities can afford to pay more so they tend to attract personnel with greater qualifications. Of course, there are many well-qualified personnel in smaller communities.

3.1.1 Job Description and Reporting Structure

As a local emergency manager you must understand:

- ▲ Your duties as defined by a job description (FEMA, 1983).
- ▲ Who you will report to.
- ▲ Who, if anyone, will report to you.
- ▲ Specific qualifications, such as education, training, and experience, that you must have.

If there is not a current job description, you should draft one and discuss it with your boss.

You must also understand the relationships among various agencies within the local government. You will work mostly with the following departments:

- ▲ Police
- ▲ Fire
- ▲ Emergency medical services
- ▲ Public works

All of these departments report to the CAO. The CAO's job is to make sure the departments:

- ▲ Perform their duties within the requirements of the law.
- ▲ Complete their duties within the time allowed.
- ▲ Complete their duties within budget.

Usually, the CAO is not an expert in public safety, emergency medicine, or emergency management. It is difficult for the CAO to provide technical guidance. Because of this, local agencies work with the agencies at the state (and sometimes federal) level. These state and federal agencies provide technical, and sometimes financial, assistance. State and federal agencies lack authority over local agencies. Their relationship is represented as a dotted line in organizational charts (see Figure 3-1). The state agencies report to the governor, just as the local agencies report to their CAO.

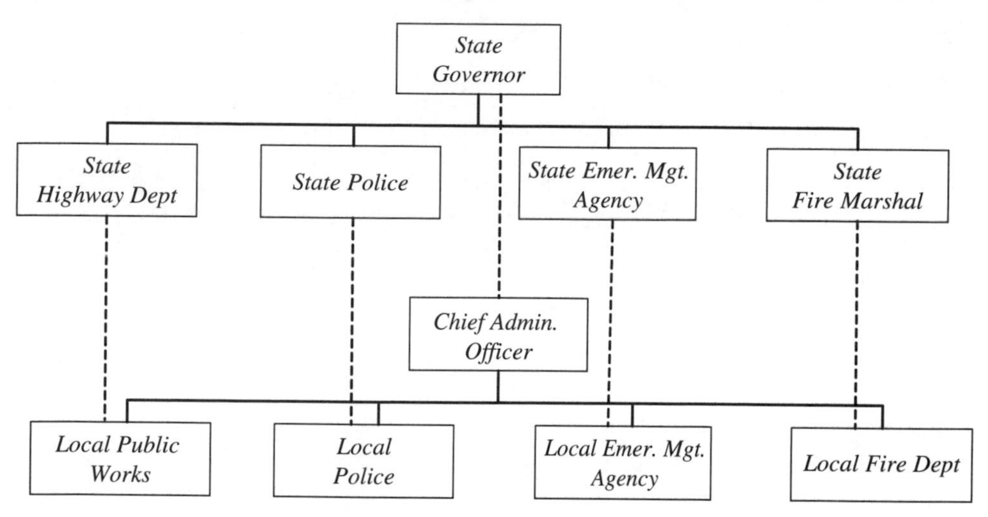

Figure 3-1

Organizational Chart for State and Local Agencies

The relationship between the local and state levels also extends to the federal level. In addition, there are two other relationships that should be noted.

1. You will provide support to nearby communities during emergencies. This agreement is formalized with a memorandum of agreement (MOA).
2. You will have a close relationship with the **local emergency management committees (LEMCs).** A LEMC is a disaster-planning network that increases coordination among local agencies. Some of these LEMCs inform and prepare their communities for accidental releases of toxic chemicals. LEPCs are established by law. However, some emergency managers have established similar committees to address all hazards. Some of these LEMCs assume responsibility for:

   - ▲ Hazard mitigation.
   - ▲ Disaster preparedness.
   - ▲ Emergency response.
   - ▲ Disaster recovery.

3.1.2 LEMA Staffing

Many LEMAs have administrative staff, such as clerks or secretaries. Some LEMAs also have professional staff, such as an emergency management analyst. All staff members need job descriptions that include:

- ▲ A title
- ▲ A supervisor's name

- A list of functions
- A list of duties
- A list of qualifications

Support staff:

- Receive and tracks correspondence.
- Draft plans and procedures.
- Maintain databases.
- Schedule meetings.
- Maintain meeting minutes.

Most communities require that paid staff:

- Receive one job review each year.
- Receive training and performance objectives each year.

In many cases, the budget will not support enough staff to perform all of the activities, so volunteers are recruited. These volunteers can help achieve a LEMA's goals. Some of these volunteers have valuable skills (e.g., computing, radio communications) that you may lack. Although not mandatory, performance reviews for volunteers are important to improve their performance effectiveness.

3.1.3 LEMA Program Plan

You will need to develop a program plan that directs your efforts over the course of a year. FEMA (1983, 1993) has advised emergency managers to set annual goals in each of the major areas for which they are responsible, such as:

- Hazard and vulnerability analysis
- Hazard mitigation
- Emergency preparedness
- Recovery preparedness
- Community hazard education

After setting these goals, you must determine your ability to meet these goals. This **capability assessment** will likely show you that you can easily meet some goals while other goals will be more difficult. You should document this **capability shortfall.** Due to limited money available for emergency management, you will not be able to eliminate the shortfall in a single year. This means you must write a **multiyear development plan** to reduce the capability shortfall. The development plan is typically based on five years. Because of its long planning horizon, your multiyear development plan should identify specific annual milestones to see if you are on target.

3.1.4 LEMA Budget Preparation

Your budget will:

- ▲ Categorize anticipated expenses.
- ▲ Detail the amount of money allotted to each category.
- ▲ Cover the *fiscal year* (a 12-month period), which might not be the same as the *calendar year* (January 1 to December 31).

Typical budget categories include:

- ▲ Staff salaries.
- ▲ Office space.
- ▲ Office equipment (e.g., copiers, computers, fax machines).
- ▲ Telephone service (local and long-distance).
- ▲ Travel.
- ▲ Materials and supplies (e.g., paper, toner).

You should anticipate the need to replace worn out equipment or to purchase new equipment that will increase your capabilities. This expense must be in the budget. The budget should also contain a **contingency fund** that addresses the costs of resources that will be needed in case of an emergency.

Your challenge is to make sure expenses do not exceed the budget. This is not difficult to do for the routine, continuing items because those items are fixed expenses. Repairs to office equipment can be unpredictable, but you can sign a service contract that establishes a fixed fee for maintenance. Long-distance telephone service and travel expenses for training are less predictable; however, these expenses are discretionary, so they can be reduced if needed.

The contingency fund for emergency response is difficult to estimate. The scope of the emergency (or even whether it occurs) is unpredictable. Nonetheless, past records or discussions with emergency managers in nearby communities can provide some insight into the appropriate amount to request. When preparing a budget, you must justify each of the budget items. Records of previous years' expenses are useful guides, but it is important to make adjustments for inflation as well as for changes in the program plan. As new needs arise that were not addressed in the previous budget, you must request additional money. This is documented in a **budget narrative** that accompanies the budget request. The budget and accompanying narrative are submitted in written form and, in many cases, are presented orally to the funding sources as well. Using graphics can explain how each of the budget items contributes to the achievement of the program plan. It is essential that you submit the new year's budget in the format that is being used by your jurisdiction. Your jurisdiction's budget office will provide assistance in this area.

You will probably receive a monthly report on your budget. If your budget was based on accurate information, you should be on target. If you have overspent your budget, you will need to correct the deficit. Many people do this by cutting expenditures for travel and training. This situation can produce training problems if the budget is cut several times in a row.

Senior elected and appointed officials typically require periodic reports of progress on the program plan and budget. Graphics are valuable in demonstrating your progress. You should also explain what percentage of each budget line has been spent to date. You should compare this to what percentage of the year has passed. For example, you would find it easy to explain why you spent 0% of the budget for computer replacement in the first three months (25%) of the year. However, it would be difficult to explain why 40% of the budget for salaries had been spent in that same period. In either case, you must explain why the expenses are different from the projected budget. You will also have to explain how you are going to correct the budget.

3.1.5 LEMA Funding Sources

Your most obvious source of funding is the CAO. There are other funding sources as well. FEMA has a range of programs that provide financial assistance. For example, you can receive matching funds through your state emergency management agency. Each state has slightly different requirements. For example, Texas requires you to have a plan that meets a specific standard of quality and provides competitive awards based on planning, equipping training, and exer-

FOR EXAMPLE

Drought in Texas

In 2000, small north Texas towns faced terrible drought. Residents of small towns like Electra and Throckmorton were not allowed to flush their toilets and were given a $500 fine for watering their lawns. George W. Bush, then Governor of Texas, declared the area a disaster area. The state of Texas gave Throckmorton an $800,000 grant to build a 21-mile pipeline to the neighboring community of Graham, which agreed to provide a backup water supply.

cising activities. Continued financial support is based on meeting performance and financial requirements. You must also achieve the annual objectives. FEMA also supports programs for managing chemical hazards. Applications must list the objectives and how they will be achieved. LEMAs submit applications through their state emergency management agencies. The FEMA regional offices review the applications.

Contact local sources for assistance. Industrial facilities, such as nuclear power plants, can help defray the costs of preparing their facilities for an emergency. Truck and rail carriers can provide training assistance. Commercial businesses can contribute to hazard awareness programs.

SELF-CHECK

- Describe the reporting structure of LEMA.
- Define **LEMC.**
- Define **capability assessment** and **capability shortfall.**
- Explain how you acquire funding for LEMAs.

3.2 Designing Effective Emergency Management Organizations

How effective is your LEMA? How about your LEMC? What makes an emergency management organization effective? Money? Trained staff? Community support? The effectiveness of a local emergency management organization is measured by the quality, timeliness, and cost of the community hazard adjustments that have been adopted and implemented.

An effective LEMA has the following components:

▲ **Organizational outcomes:** Desirable outcomes for effective LEMAs include high quality and quantity, timely, and low cost delivery of products such as hazard vulnerability analyses (HVA), EOPs, and recovery operations plans.

▲ **Individual outcomes:** Desirable outcomes for the individual members of an effective LEMA include high job satisfaction, organizational commitment, individual effort and attendance, and organizational citizenship behaviors.

▲ **Planning process:** An effective planning process includes productive planning activities, team climate development, situational analysis, and strategic choice.

Figure 3-2

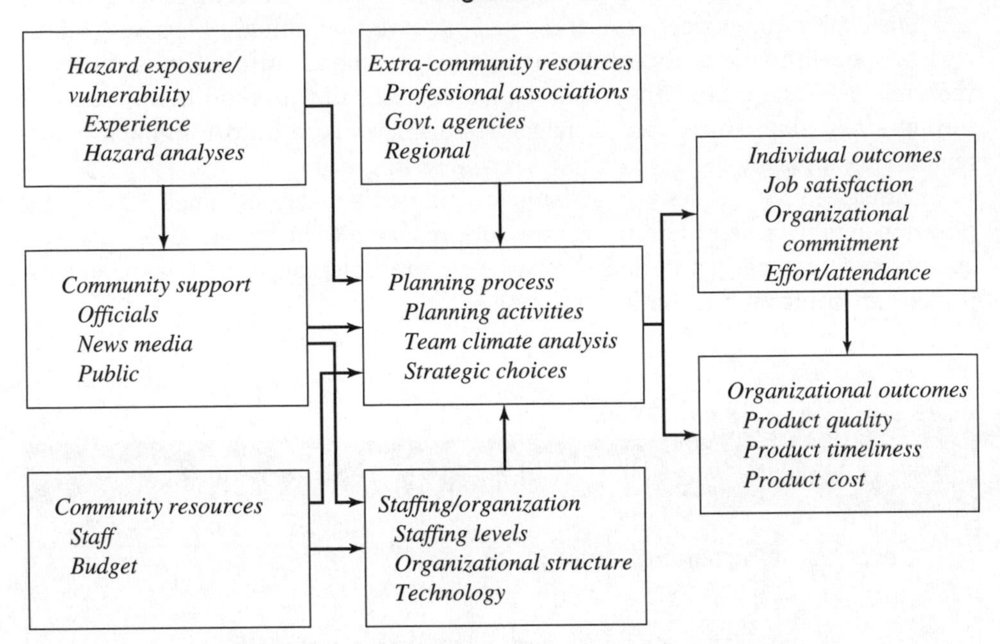

A model of local emergency management effectiveness

As Figure 3-2 indicates, an effective planning process is determined by:

- Hazard exposure and vulnerability, either experience being involved in a hazard or knowledge of the hazard's possible impact.
- The level of community support from officials, news media, and the public.
- Community resources, such as level of staff and budget allocated to emergency response organizations.
- Extra-community resources from governmental and nongovernmental sources.
- The staffing and organization of a LEMA.

The planning process is dynamic because success tends to breed success. High levels of individual and organizational outcomes produce increased levels of vicarious experience with disaster demands (through emergency training, drills, and exercises), community support, better staffing and organization, and increased emergency planning resources.

3.2.1 Hazard Exposure and Vulnerability

Many times, communities that have suffered through disasters resolve to become better prepared in the future. Frequent, recent, and severe impacts can lead to a **disaster subculture**, in which residents adopt routines to prepare for disasters (Wenger, 1978). The community will not be focused on hazard preparation when disasters are infrequent, long past, or minor in terms of losses. Cities that were high in experience adopted 1.5 more preparedness practices than those that were low in experience (Kartez and Lindell, 1990). However, the community's exposure to hazards can also seem real through reading or hearing about other communities' experiences. This experience can be gained through media accounts. Disaster experience is more powerful through firsthand accounts—especially if they come from peers (Lindell, 1994a). For example, a local fire chief will be influenced by other fire chiefs' experiences.

3.2.2 Community Support

Community support from senior elected and appointed officials, the news media, and the public is important because it affects the resources that are allocated to LEMAs and LEMCs. Emergency management is a low priority for the local elected and appointed officials who control budgets and staffing allocations (Labadie, 1984; Sutphen and Bott, 1990). As one police chief said

> My number one priority is getting the uniforms out in response to calls. The public judges me on that performance, not whether I'm planning for an earthquake that may never happen. If left alone, disaster planning would get even less attention from my office. It requires that the executive clearly make this a priority. (Kartez and Lindell, 1990, p. 13)

Two-thirds of the inactive LEMCs blamed community indifference and more than one-third blamed lack of funding for their lack of achievement. Community information requests, media coverage, local support, and the backing of local officials are strongly and significantly correlated with LEMC effectiveness.

3.2.3 Community Resources

Differences among jurisdictions in the effectiveness of their LEMAs and LEMCs can be partly attributed to the difference in the communities' resources. For example, Adams and his colleagues (1994) and Lindell and his colleagues (1996) found that compliance with emergency planning mandates was significantly related to:

- Jurisdiction size
- Median household income
- Percentage of urban population
- Jurisdiction budget
- Police and fire department staffing

3.2.4 Extra-Community Resources

Extra-community resources also contribute to effectiveness of emergency planning. Such resources include:

- Federal agency technical reports.
- State emergency planning agency technical support.
- Industry training programs.
- Computer software.
- Membership in statewide emergency management associations.
- New ideas, plans, procedures, and equipment from private industries and neighboring jurisdictions (Kartez and Lindell, 1987).

3.2.5 Staffing and Organization

A number of studies have shown that how an LEMC is staffed and organized impacts its effectiveness. For example, the characteristics of effective emergency management organizations include (International City Management Association, 1981):

- Defined roles for elected officials.
- Clear internal hierarchy.
- Good interpersonal relationships.
- Commitment to planning as a continuing activity.
- Member and citizen motivation for involvement.
- Coordination among participating agencies.
- Public/private cooperation.

Emergency management network effectiveness is greater in communities with recent disaster experience. If there have been no recent disasters, the emergency management network can still be effective if there is agreement on what the most likely and dangerous hazard is (Caplow, Bahr and Chadwick, 1981). The more effective networks:

- Have members with more experience.
- Have a wider range of local contacts.
- Have written plans and were familiar with them.
- Have personal experience in managing routine natural hazards such as floods.
- Are more familiar with the policies and procedures of emergency-relevant state and federal agencies.

FOR EXAMPLE

The Big Apple's Experience with Terrorism

Even though the events of 9/11 were catastrophic and caused more losses than any other terrorist event on American soil, New York was relatively well prepared. The evacuation of the World Trade Center was surprisingly successful. Tens of thousands of people worked in the towers, yet there were only about 3,000 casualties, which included firefighters and other personnel who did not work in the towers. The success in limiting the damage can be attributed to planning and experience. New York City had extensive plans for disasters. Also, it wasn't the first time the Towers were hit. In 1993, a truck bomb was detonated in the World Trade Center.

Other factors that contribute to the effectiveness of an emergency management network include (Lindell and Meier, 1994):

- ▲ The number of members.
- ▲ The number of hours worked by paid staff.
- ▲ The number of agencies represented on the LEMC.
- ▲ Organization into subcommittees.
- ▲ Representation by elected officials and by citizens' groups.

Surprisingly, having representatives from the news media was least important for overall emergency planning effectiveness. Establishing subcommittees helps because this seems to allow members to focus on specific tasks and thus avoid feeling overwhelmed by all the work that needs to be done (Lindell et al., 1996a).

SELF-CHECK

- Describe what makes an effective LEMA.
- Define **disaster subculture.**
- Name three extra-community resources.
- Describe how a LEMA's staffing affects its success.

3.3 The Planning Process

The emergency planning process consists of five principal functions.

1. Planning Activities
2. Providing a Positive Work Climate
3. Analyzing the Situation
4. Acquiring Resources
5. Choosing a Strategy

3.3.1 Step 1: Planning Activities

Superior planning practices affect the adoption of good emergency preparedness practices more than disaster experience does (Kartez and Lindell, 1990). Cities with a better planning process adopt 2.5 more preparedness practices. Interestingly, as Table 3-1 indicates, planning activities such as interdepartmental training, reviews with senior officials, and establishment of interdepartmental task forces were especially important. By contrast, more routine activities such as procedure updates, plan updates, and reviews of mutual aid agreements had little effect.

Characteristics of meetings are important influences on organizational effectiveness. To run an effective meeting:

▲ Schedule meetings on a regular basis. If possible schedule the meetings on the same day of the week and the same time of day.

Table 3-1: Planning Activities and Their Importance in the Adoption of Good Emergency Preparedness Practices

Largest difference	*Smallest difference*
Interdepartmental training	Procedure updates
Reviews with senior officials	Plan updates
Interdepartmental task force	Review mutual aid agreements with neighboring cities
Community disaster assistance council	
After action critiques	
Training exercises	
Vulnerability analyses	
Meetings with TV/radio managers	

- ▲ Circulate an agenda before the meeting.
- ▲ Keep written minutes.
- ▲ Set goals and review progress at the meetings.
- ▲ Schedule meeting times convenient for all staff (full-time employees, part-time employees, and volunteers).

3.3.2 Step 2: Providing a Positive Work Climate

Planning effectiveness is highest in LEMCs that have positive organizational climates. To provide an effective climate:

- ▲ Be clear about what tasks you think need to be done and how to perform them.
- ▲ Know and recognize each team member's strengths and weaknesses.
- ▲ Be supportive of the needs of the team members.
- ▲ Seek agreement from others about which tasks they think need to be done.
- ▲ Give people enough time to complete their tasks.
- ▲ Give members enough independence to perform tasks that make a meaningful contribution.
- ▲ Share information and coordinate individual efforts.
- ▲ Reward good job performance with recognition by the group.
- ▲ Foster team pride.

3.3.3 Step 3: Analyzing the Situation

In planning for an emergency, you must analyze potential hazard impacts on the community. There are five factors you must examine.

1. **Hazard exposure:** Identify the hazards threatening the community. Determine the locations that would be affected by the impact. Also, determine how intense the impact needs to be to damage the area.
2. **Physical vulnerability:** Examine the community's buildings and determine if they can withstand the predicted impact from the hazard. Examine the community's infrastructure and determine how it will be affected by the hazard.
3. **Social vulnerability:** Examine the community's population to determine how different segments are exposed to hazards. Some people may live in homes that will be destroyed by tornadoes. For example, mobile homes are extremely vulnerable to tornadoes. Also, look at the amount and type of resources available to different population segments. These resources

will not only help in preparation for a hazard, they will also help during recovery.

4. **Alternative hazard adjustments:** Identify ways to reduce losses and speed recovery from a disaster. For example, installing window shutters is a hazard adjustment for hurricanes. An example of a hazard adjustment for emergency response is making sure local hospitals have backup generators. Examine the effectiveness of hazard adjustments in protecting persons and property. Assess the resources needed in terms of money, specialized knowledge and equipment, time and effort, and social cooperation.
5. **Capability assessment:** Discuss hazard adjustments with the community. Talk to household owners, business leaders, government agency officials, and nongovernmental organizations (NGOs). Determine if they have the capacity and commitment needed to adopt the available hazard adjustments. For example, few people in Miami took Hurricane Andrew seriously until it was close to landfall. Many households did not have the commitment needed to adopt any hazard adjustments. Many people did not even board up their homes.

3.3.4 Step 4: Acquiring Resources

You need to acquire many resources to effectively handle hazards. You need to obtain staff, equipment, and information from a variety of sources. One important resource, a microcomputer, is available to almost all emergency managers. However, the high speed/high storage capacity computers needed for analyzing hazards may not be available. You can find these computers at the land-use planning department (Lindell, Sanderson and Hwang, 2002). You can obtain data from Web sites maintained by federal agencies such as FEMA and the National Weather Service (NWS). You can also use state Web sites (Hwang, Sanderson and Lindell, 2002). These organizations also provide software, manuals, and training courses to help you assess community vulnerability.

3.3.5 Step 5: Choosing a Strategy

Community-wide disasters differ from routine emergencies that can be handled by a single agency (Dynes, Quarantelli, and Kreps, 1972). To prepare for a disaster, there are six effective strategies to involve others. You can use multiple strategies, and the extent to which you use each one depends on the size of the community, available funding, and your own personal characteristics (Drabek, 1987, 1990; Mulford, Klonglan, and Kopachevsky, 1973).

1. **Resource building strategy:** If you choose this approach, you will spend your time acquiring resources. Resources include staff, technical expertise, and equipment. To be effective, you must actively increase resources of

all local agencies. To do this, seek agreement with other agencies on the mission of the LEMA.

2. **Emergency resource strategy:** This approach emphasizes working with all emergency agencies. For example, if you choose this approach, you will work closely with fire and police departments.
3. **Elite representation strategy:** This approach is about building relationships. If you choose this approach, you will ask LEMA members to interact with influential members of other emergency-relevant organizations. Personal and professional contacts are important. It is also important to work together and have routine meetings, drills, and exercises.
4. **Constituency strategy:** This is another approach centered on building relationships. However, this relationship is between the LEMA and one other organization. Both organizations must benefit from the cooperation. Organizations are more likely to work with you if there are good reasons for them to do so. Relationships are often based on an awareness of potential disaster demands. Relationships also occur because both organizations recognize the needs for avoiding gaps in services or duplicating efforts.
5. **Cooptation strategy:** This approach is about tapping into the knowledge of other people. This approach emphasizes asking key personnel, especially those from other organizations, to become part of LEMA's formal structure as directors or advisors.
6. **Audience strategy:** This is a public relations approach. If you choose this approach, you will spend your time educating the public about emergency preparedness.

3.3.6 Individual Outcomes

For LEMAs to be effective, you need dedicated individuals. Studies show that people are committed to organizations that benefit them by meeting certain needs:

- Personal needs, such as receiving a salary and benefits.
- Social needs, such as having friends at work.
- Purposive needs, such as the feeling of doing something positive for the community and of having an identity.

Volunteers are even more likely than paid workers to enjoy their work (Pearce, 1983). According to Porter, Steers, Mowday, and Boulian (1974), volunteers often have a

- strong belief in, and acceptance of, an organization's goals and values;
- willingness to exert considerable effort on behalf of the organization;
- strong desire to maintain organizational membership.

Figure 3-3

Drills are an important preparedness activity.

Regardless of whether someone is a paid staff member, or a volunteer, they are likely to have some type of commitment to the organization. There are two types of commitment: affective and continuance (Meyer and Allen, 1984). Affective commitment is an emotional bond to the organization. This type of commitment leads to high employee performance. A person's affective commitment is influenced by:

- ▲ Organizational leadership
- ▲ Their perceptions of their own competence
- ▲ Role clarity
- ▲ Identification with organization's goals
- ▲ Opportunity for reward

By contrast, continuance commitment is a commitment made to preserve tangible benefits. For example, staff members may not want to leave the organization because they don't want to lose health insurance, a pension, or some other benefit. Continuance commitment motivates employees to remain in the job but does not motivate them to do a good job. Organizational commitment is important in understanding LEMA and LEMC effectiveness. If people are not committed to the organization, then they will leave and expertise will be lost (Mathieu and Zajac, 1990).

3.3.7 Organizational Outcomes

The success of an organization is based primarily on two factors. As discussed, an organization can only be as successful as its people are. Second, to be a success, an organization must have a well thought-out planning process. Emergency management organizations are often judged based on:

- ▲ Effectiveness of plans
- ▲ Timeliness
- ▲ Cost of plans
- ▲ Hazard vulnerability analyses
- ▲ Public information briefings
- ▲ Education efforts, such as brochures and Web sites

Performance of LEMCs varies from activity to activity. LEMCs are generally effective at:

- ▲ Collecting and filing hazard data.
- ▲ Taking inventory of local emergency response resources.
- ▲ Acquiring emergency communications equipment.
- ▲ Developing training for local emergency responders (see Figure 3.3).

FOR EXAMPLE

Ineffective Response to Hurricane Andrew

Three days after Hurricane Andrew slammed into South Florida in 1992, Dade County's emergency operations director made a plea on national television. "Where in the hell is the cavalry on this one?" asked an exasperated Kate Hale. After one of the worst hurricanes, a Category 5, in America's history, hurricane victims went days without food, water, or financial assistance. Consistent with established procedures, the federal government waited for local governments to document their resource needs and forward a request for a presidential disaster declaration through the state governor. However, the devastation was so widespread that local government was unable to rapidly assess the disaster impacts. Was local government at fault for failing to protect its emergency management organization? Were the state and federal governments at fault for failing to recognize local governments' inability to respond? Would you want the federal government to come in when it decided rather than when requested?

By contrast, LEMCs are ineffective at

▲ Developing protective action guides.
▲ Analyzing air infiltration rates for local structures.
▲ Analyzing evacuation times for vulnerable areas.
▲ Promoting community toxic chemical hazard awareness.

SELF-CHECK

- Name the five major functions of the emergency planning process.
- Describe how to run an effective meeting.
- Name six strategies you can use to deal with a disaster.
- Compare individual outcomes to organizational outcomes. How do they differ? How are they the same?

3.4 Developing an Emergency Operations Plan

The previous section describes the factors that influence emergency planning effectiveness and later chapters provide recommendations for the content of EOPs. However, there is an important intermediate step that needs to be addressed—the *process* of plan development. The development of an EOP is a multistage process that encompasses eight steps.

3.4.1 Step 1: Establish a Preliminary Planning Schedule

Table 3-2 is an example of how to identify the principal tasks and the expected amount of time required to perform them. You will provide accurate time estimates. However, the LEMC members need to confirm that the deadline is feasible.

3.4.2 Step 2: Publish a Planning Directive

You don't control all the needed resources, so you must coordinate the efforts of other agencies. Since you do not have direct authority over them, you will need to gain their cooperation by having the CAO delegate authority to you. Ask the CAO to sign a planning directive stating expectations for the planning process. The planning directive contains three sections. The first section states the purpose of the planning process, the legal authority under which it is being conducted, and the specific objectives. The second section describes the planning process, the LEMC organization, the other participating organizations, and your authority as the CAO's representative. The third section addresses the process for plan approval and the anticipated deadline for publication of the final plan. Even though the CAO signs the directive, you should draft it to ensure all the necessary elements are present.

Table 3-2: Sample Preliminary Planning Schedule

Time (months)	*0*	*2*	*4*	*6*	*8*	*10*	*12*
Organize the LEMC	[—]						
Conduct the capability assessment		[——]					
Assign responsibility for plan components		[—]					
Finalize the planning schedule			[—]				
Write plan components				[————]			
Evaluate/revise the draft plan					[——]		
Obtain community review						[—]	
Revise/publish the final plan						[———]	

3.4.3 Step 3: Organize the LEMC

You build a committee by asking others to serve. To do this, ask for volunteers from each organization that is able to respond to hazards. Also, ask for volunteers from each organization that is especially vulnerable to hazards. This is important because public safety agencies, such as the police and fire departments, will always participate, but other local organizations may only participate if the CAO tells them they have to (Kartez and Lindell, 1990). These organizations are listed in the planning directive. A typical list of such organizations is in Table 3-3. Members of the committee will work a few hours a month for the LEMC while working their normal jobs. You and the members will select officers such as the:

- ▲ Chair
- ▲ Vice-Chair

Table 3-3: Organizations Typically Participating in the LEMC
Fire Department
Local Utilities (Gas, Electric, Telephone)
Police Department
Red Cross
Emergency Medical Services
Hospitals
Public Works Department
Nursing Homes
Land-Use Planning Department
Schools
Building Inspection Department
News Media
Chief Administrative Officer's Office
Environmental Groups
Public Health Department
Local Industry
Local Elected Officials
Labor Unions

- ▲ Secretary
- ▲ Information Coordinator
- ▲ Subcommittee chairs

As with other organizations, the Chair:

- ▲ Presides over meetings.
- ▲ Represents the LEMC to government officials.
- ▲ Represents the LEMC to private sector organizations.
- ▲ Represents the LEMC to the news media and the public.
- ▲ Represents the LEMC to state and federal agencies.

The Vice-Chair performs the Chair's duties when the Chair is absent. However, the Vice-Chair's primary role is to manage the internal affairs of the LEMC. The Secretary schedules meetings and keeps written meeting minutes. The Information Coordinator is the contact for information about hazards, the planning process, and planning products. The Information Coordinator can also be responsible for monitoring the LEMC's budget.

LEMCs are more effective when members are assigned to specific tasks rather than having everyone contribute to all tasks. This is why LEMCs should have subcommittees. Each committee must determine the appropriate division of labor for its own situation. Listed below are the typical subcommittees and their duties.

The Hazard Vulnerability Analysis committee:

- ▲ Identifies the hazards to which the community is exposed.
- ▲ Analyzes the vulnerability of residential, commercial, and industrial structures to hazards.
- ▲ Analyzes the vulnerability of the infrastructure (fuel, electric, water, sewer, telecommunications, and transportation) to hazards.
- ▲ Identifies any secondary hazards that could be caused by an initial disaster impact.
- ▲ Identifies the locations of facilities, such as schools, hospitals, nursing homes, and jails, whose populations are vulnerable because of the limited mobility of their resident populations.
- ▲ Identifies the locations of other facilities with vulnerable nonresident populations.

The Planning, Training, and Exercising committee:

- ▲ Writes the emergency operations plan.
- ▲ Develops a training program to improve emergency responders' capabilities.

- ▲ Develops training materials for disaster-related tasks that are not performed during normal operations or routine emergencies.
- ▲ Develop training that provides an overview of disaster response.
- ▲ Develops training that improves skills required for tasks that are infrequently performed, difficult, and critical to the success of the emergency response organization.
- ▲ Develops the necessary training materials or obtains them from other sources.
- ▲ Tests the plan through drills and exercises.
- ▲ Recruits representatives from the primary emergency response and public health agencies.

The Recovery and Mitigation committee:

- ▲ Develops a preimpact recovery plan that will facilitate a rapid recovery.
- ▲ Identifies mitigation projects that will reduce the community's vulnerability to hazards.
- ▲ Identifies projects to be completed before a disaster.
- ▲ Identifies projects that will be implemented during recovery.
- ▲ Recruits help from representatives from public works, community development, land-use planning, and building construction agencies.

The Public Education and Outreach committee:

- ▲ Communicates with the news media and the public.
- ▲ Explains how the activities of the Planning, Training and Exercising committee will provide an effective response to disasters.
- ▲ Explains how the activities of the Recovery and Mitigation committee will provide an effective recovery plan.
- ▲ Writes nontechnical summaries that can be understood by household owners and business members.
- ▲ Develops slides or other graphic presentations to support talks to community groups.
- ▲ Develops brochures to be distributed to the public.

The Executive Committee (Chair, Vice-Chair, Secretary, and subcommittee chairs):

- ▲ Ensures the LEMC sets specific, achievable objectives each year.
- ▲ Ensures the LEMC accomplishes those objectives through an efficient expenditure of resources.

▲ Obtains the resources to support the LEMC's activities.

▲ Conducts a planning orientation so the members of the LEMC will develop a common understanding of the process.

There are two big hurdles for emergency planning (Daines, 1991). First, planning agencies do not have experience with emergency response. Second, emergency response agencies do not have planning experience. These agencies usually work with minor emergencies. For minor emergencies, they have standard operating procedures and do not need detailed plans. In addition, they are not aware of the planning resources available from state and federal agencies. You need to introduce LEMC members to the main points in the state's emergency, recovery, and mitigation plans. You also need to introduce LEMC members to the main points of the Federal Response Plan, as well FEMA response, recovery, and mitigation programs and planning guidance.

3.4.4 Step 4: Assess Disaster Demands and Capabilities

LEMC members must identify the tasks that need to be performed in a community-wide emergency. To do this, they must study the hazard vulnerability analysis (HVA). In addition to knowing the response and recovery tasks that need to be performed, the LEMC members must also address how the public will respond in the face of a disaster. For example, will everyone try to evacuate using the same highways? Will there be shortage of canned food and bottled water at the grocery stores? These are the types of questions the LEMC must consider and plan for.

In addition, LEMC members may expect outside agencies to perform tasks for which they lack the resources. For example, local officials might assume that adjacent jurisdictions will manage traffic coming from your jurisdiction, but this might not be the case. You need to assist the LEMC and ensure the plans are based upon realistic assumptions about what needs to be done and who will be able to do it (Dynes et al., 1972).

3.4.5 Step 5: Write Plans

Every LEMA needs an emergency operations plan, a recovery operations plan, and a hazard mitigation plan. However, not all the work falls to you, as you have committees to help you.

When you begin to write these plans, you should:

▲ Ensure the committees have talented people drafting each section of the plans.

▲ Ensure each plan has the following sections: basic plan, annexes, and hazard-specific appendices.

- ▲ Ask representatives from each organization to draft their own sections. For example, the police department should draft the section on law enforcement.
- ▲ Provide guidance regarding the structure and content of the plans.
- ▲ Provide resources for committee members to use.
- ▲ Set performance goals and deadlines.
- ▲ Draft the basic plan.
- ▲ Discuss goals annually with the CAO.

3.4.6 Step 6: Evaluate and Revise Draft Plans

Make sure that all drafts are reviewed by committees within the LEMC to identify potential problems. Potential conflicts include gaps between agency tasks and their capabilities. There could also be conflicts between the provisions of one plan and another. For example, the EOP might set priorities for infrastructure restoration that conflict with those in the recovery operations plan.

3.4.7 Step 7: Obtain Community Review

After the draft plans have been reviewed within the LEMC, release the plans for review throughout the community. Work with the Public Education and Outreach committee. Make copies available at libraries and other public places throughout the community. Notify the public that the draft plans are available for review. The Public Education and Outreach committee should meet with neighborhood groups (e.g., community councils, parent-teacher associations) and service organizations (e.g., Rotary, Kiwanis, Chamber of Commerce) to discuss the plans. Give the public an adequate amount of time to review the plans.

FOR EXAMPLE

Testing the EOP

In 2003, Pennsylvania practiced their response to a terrorist attack, by acting out the following scenario. Three terrorists armed with a radioactive bomb took control of a Port Authority bus during the morning commute and held the passengers hostage. Emergency responders had hazardous materials to contain, hostages to rescue, and the terrorists to take into custody. This drill revealed weaknesses in the capabilities of the emergency response crews (police personnel, firefighters, paramedics, and bomb squad personnel) without endangering anyone.

In addition, you ensure at least one public meeting is held for residents to provide feedback. Such comments should be transcribed and retained in the LEMC's archives.

3.4.8 Step 8: Publish Plans in Final Form

Give all input from the community to the appropriate committees. The committees must address any problems in the final versions of the plans. Each final plan should include a document that categorizes the comments received and explains how they were incorporated. If the comments are not addressed, explain why. Forward copies of the final plans and accompanying documents to all government agencies and other participating organizations that have roles in the plans. Deposit additional copies of the final plans and documents with the draft plans. Make these documents accessible to households and businesses throughout the jurisdiction.

SELF-CHECK

- **Name the eight steps of developing an EOP.**
- **Describe the three sections of the planning directive.**
- **Name three LEMC subcommittees.**
- **List four things you must do to write the plans.**

SUMMARY

Organization, leadership, and excellent communication skills are key to your success as an emergency manager. As you can see, LEMAs are organized groups, and they rely on organizational structure, clearly assigned tasks, effective management, and thorough planning to succeed. Just as LEMAs should be organized and managed well, so should the tasks of developing emergency organizations, planning processes, and creating emergency operating plans. This chapter outlined the major steps of how to do these tasks. All of these tasks are based on outcomes and all of them rely on your ability to organize resources, find funding, and communicate well. Develop these skills by following the steps outlined in this chapter.

KEY TERMS

Budget narrative A document that accompanies the budget and includes a request for additional money. The narrative is submitted in written format and can include graphics to explain budget items.

Capability assessment An evaluation of the degree to which your jurisdiction's resources are sufficient to meet the disaster demands identified in the hazard vulnerability analysis.

Capability shortfall The difference between the level of resources a jurisdiction currently has and the level it will need to meet the disaster demands identified in the hazard vulnerability analysis.

Contingency fund A sum of money in the budget that addresses the costs of resources that will be needed in case of an emergency.

Disaster subculture Behavioral patterns among groups of residents who adopt routines to prepare for disasters. These groups have usually experienced disasters and have resolved to better prepare for them in the future.

Local Emergency Management Committee (LEMC) A disaster-planning network that increases coordination among local agencies.

Multiyear development plan A plan that documents the specific steps for reducing the capability shortfall. The development plan is typically based on five years and should identify specific annual milestones and specific, measurable achievements to keep emergency managers on target.

ASSESS YOUR UNDERSTANDING

Go to www.wiley.com/college/lindell to evaluate your knowledge of the basics of planning for emergencies.
Measure your learning by comparing pre-test and post-test results.

Summary Questions

1. Which of the following is not a factor in determining your status as an LEMA emergency manager?
   - **(a)** Size of community
   - **(b)** Financial resources of the community
   - **(c)** Community's vulnerability to hazards
   - **(d)** Your age
2. Most communities do not require paid staff to receive job reviews. True or False?
3. What must you do to determine your ability to meet the goals of the LEMA?
   - **(a)** Create a new LEMA
   - **(b)** Conduct a capability assessment and document the capability shortfall
   - **(c)** Conduct goal-planning meetings
   - **(d)** Determine the risks involved in meeting the goals
4. What documents the requests for new budget needs?
   - **(a)** Budget request form
   - **(b)** Budget review
   - **(c)** Budget narrative
   - **(d)** Budget allowance form
5. Communities focus on hazard preparation when disasters are infrequent. True or False?
6. Staffing affects the effectiveness of LEMCs. True or False?
7. To run an effective meeting, you should *not* do which of the following?
   - **(a)** Schedule meetings on a regular basis
   - **(b)** Circulate an agenda before the meeting
   - **(c)** Schedule meetings on different days and at different times to provide variety
   - **(d)** Keep written minutes
8. Which of the following is not a factor in determining a strategy for dealing with disasters?

(a) The potential disasters
(b) The size of the community
(c) The availability of funding
(d) Personal characteristics

9. LEMCs are more effective when members are assigned to specific tasks rather than having everyone contribute to all tasks. True or False?

Review Questions

1. What does the organizational chart for state and local agencies look like? Draw a diagram of the chart.
2. How is the effectiveness of an emergency management organization measured?
3. What are the five steps in the planning process?
4. Name three LEMC subcommittees and list their responsibilities.
5. What are three ways you can ensure to provide an effective response to a disaster?

Applying This Chapter

1. You head up the LEMA in a small community, and you are the only person on the staff. What do you include in your budget?
2. You are responsible for emergency planning in Miami, an area that is vulnerable to hurricanes. You devise a plan that will limit damage from a hurricane. How can you aquire the money to fund the plan? Name three sources.
3. You have recently taken over an LEMA that has eight individuals on staff. Morale among the staff is low because staff members don't feel they have accomplished anything. You feel you can improve morale by showing staff members tasks that you believe they can complete on a schedule. What specific steps would you take to give the staff members a sense that they can accomplish something meaningful?
4. A CAO has asked the agencies in your community to provide a report that outlines the steps they are taking to provide an emergency plan. What would you provide to the CAO to show what your LEMC is going to implement when writing your plan?

YOU TRY IT

Seeking Help in Writing the Emergency Response Plan

You are working with the LEMC to write hazard-specific appendices for different terrorist scenarios. The first scenario is one that is similar to what happened on 9/11, when terrorists took over airplanes and flew them into skyscrapers. What agencies/personnel would you ask to be part of the plan and why?

Obtaining Community Review

Your LEMC has written an EOP for your jurisdiction. Write three paragraphs on how you would obtain community review and feedback on the plan.

Educating the Public

You are in charge of emergency planning in an area that hasn't had any disasters in recent memory. The Department of Homeland Security has informed you that chatter was picked up and your city is a potential terrorist target. The terrorists have threatened to release a "dirty bomb" (an explosive device for dispersing radioactive materials) in your area. How do you garner community support for disaster preparation procedures?

4

RISK PERCEPTION AND COMMUNICATION

Saving Lives

Starting Point

Go to www.wiley.com/college/lindell to assess your knowledge of the basics of risk perception and communication.
Determine where you need to concentrate your effort.

What You'll Learn in This Chapter

▲ Responses to warnings
▲ Risk communication during the hazard phase
▲ Risk communication during crisis and emergency response

After Studying This Chapter, You'll Be Able To

▲ Analyze how people respond to warnings
▲ Involve the media and the public in risk communication
▲ Experiment using local and national information channels

Goals and Outcomes

▲ Design a risk communication plan
▲ Create and implement a risk communication plan
▲ Perform a protective action assessment

INTRODUCTION

As we have seen in many hurricanes, floods, and other disasters, people will not protect themselves if they don't believe their lives are at risk. Changing the way people perceive danger is an important way to save lives. To change the way people think, you must have specific plans for communicating the risks they face.

Risk is the possibility that people or property could be hurt. Risk is defined as the likelihood that an event will occur at a given location within a given time period and will inflict casualties and damage. This risk must be effectively communicated to the people who are likely to be affected. You must share information about hazards and hazard adjustments. Sharing is important because you must find out from different population segments how they think about hazards. Regardless of whether a hazard is natural, technological, or terrorist, the same basic principles of risk communication apply.

In this chapter you will examine how people respond to warnings and includes an outline and discussion of the eight stages of information processing. It also shows how you influence perceptions by building credibility with those you need to influence. This chapter also discusses risk communication during the continuing hazard phase and during a crisis. This chapter shows you how to save lives by communicating. The best communication involves clarity, trust, and timing.

4.1 Household Response to Warnings

A **warning** is a risk communication about an imminent event and is intended to produce an appropriate disaster response. Examples of disaster responses include evacuating and sheltering in-place (Drabek, 1986; Mileti, Drabek and Haas, 1975). There are eight stages of a person's information processing during a warning. However, before these stages begin, people must receive, heed, and comprehend information about the risks. Let's take the case of an approaching tornado and examine what needs to happen before people seek shelter.

1. **People must *receive* information.** Warnings transmitted through television and radio are only effective if people receive them. Consequently, these warning mechanisms are much less effective between 11:00 pm and 6:00 am when most people are asleep. Of course, most televisions and radios are completely ineffective when power is lost.
2. **People must *heed* (pay attention to) available information.** Many people in tornado-prone areas know spring is the peak season for tornado activity. During those months, they should check weather forecasts more frequently. They should look for environmental cues, such as

cloud formations. However, others may not pay attention to their environment. People who engage in tasks requiring intense concentration are less likely to notice gathering storm clouds and might not notice warnings.

3. **People must *comprehend* the information.** Environmental cues must be correctly processed; that is, people must know a funnel cloud is a sign of a tornado. Warnings and communication efforts must be understood as well. Warnings given in English will not help Spanish speakers. A tornado siren will not mean anything to someone who doesn't understand what the signal means. Only a few people will understand highly specialized technical terms such as millirem and pyroclastic flow.

4.1.1 Step 1: Risk Identification

Decisions about how to respond to a hazard begin with risk identification. As noted earlier, this process begins with the detection of environmental cues. However, the most important sources of risk identification are warning messages from authorities, the media, and peers. The first step you must take is to disseminate your message widely. Try to attract the attention of those at risk and inform them of the potential for disaster that threatens their health, safety, and property.

Those at risk must answer the basic question of risk identification, "Is there a real threat that I need to pay attention to?" Those who do not believe the threat is real are likely to continue their normal activities.

4.1.2 Step 2: Risk Assessment

Risk assessment involves evaluating the personal consequences if the disaster occurs (Otway, 1973; Perry, 1979a). The primary question at this stage is "Do I need to take protective action?" A positive response to this question results in **protection motivation.** People's personal risk assessment—their risk perception—is critical in understanding their disaster response (Mileti and Sorensen, 1987). If people think they are in danger, then they are more likely to protect themselves.

Peoples' risk assessments include the perceived probability, magnitude, and immediacy of the disaster impact. Perceived probability of impact affects people's judgments of the likelihood that they will be affected, whereas perceptions of event magnitude increase their perceptions of the severity of personal consequences, including death, injury, and property damage. As perceived probability and magnitude increase, so do a person's likelihood of taking protective action. The perceived immediacy of disaster impact affects people in a different way.

Instead of affecting a person's *likelihood* of acting, perceived immediacy increases a person's *urgency* to act.

4.1.3 Step 3: Protective Action Search

The primary question in protective action search is "How can I protect myself?" Residents' first attempts to answer this question often involve a search for what can be done by *someone else* to protect them against the hazard. However, when disaster impact is imminent, household owners must rely mostly on their own resources to achieve protection. In many instances, an individual's knowledge of the hazard suggests what type of protection to seek. People are likely to recall actions they have taken on previous occasions if they have had experience with that hazard. Alternatively, they might consider actions they took in similar hazards. For example, they might recognize that the impact of a volcanic mudflow is similar to that of a flood and, thus, they might take the protective responses that they took for a flood during a mudflow.

Information is also received from outside sources. For example, people might observe neighbors packing their cars in preparation for a hurricane evacuation. People also are likely to consider actions they have read or heard about. Such vicarious experiences are frequently transmitted by the news media and relayed by peers. Finally, people are also aware of appropriate protective actions when warnings include guidance about what to do. However, do not assume warning recipients will follow the recommendation even if the warning mentions only one protective action. People will always recognize that continuing their normal activities is an option; however, they might invoke other alternatives by remembering or observing the actions of others.

4.1.4 Step 4: Protective Action Assessment

At this point, the primary question is "What is the best method of protection?" The answer to this question is an **adaptive plan.** Those at risk generally have at least two options: taking protective action or continuing normal activities. Sometimes, those at risk must choose between two alternatives, but they don't really like either of them. During a hurricane, for example, evacuation protects people, but abandons property to storm damage (Perry, Lindell, and Greene, 1981; Lindell and Perry, 1990). On the other hand, emergency measures to protect property (e.g., sandbagging) require the property owner to remain in a hazardous location. When there is even a moderate amount of forewarning, households can engage in a combination of actions. For example, if a flood is forecast to arrive within a few hours, people could perform emergency flood proofing by placing sandbags around the building. They could also elevate the building's contents to higher floors. Finally, they could evacuate family members before floodwater reaches a dangerous level.

People are unlikely to consider protective action unless the action is considered to be effective. Thus, *efficacy*, which is measured by the degree of reduction in vulnerability to the hazard, refers to success in protecting both persons and property (Cross, 1980; Kunreuther et al., 1978). In some cases, such as sandbagging during floods, property protection is the goal. In other cases, people protect buildings because this also protects the people inside those buildings. People also consider the *safety* of the recommended action. For example, some people are reluctant to evacuate because they are concerned about the traffic accident risks involved.

Protective actions are also assessed in terms of perceived *time requirements*. Evacuation is time consuming. By contrast, time requirements for in-place protection are small. Occupants must shut off sources of outside air and the HVAC system (Lindell and Perry, 1992). A major problem in large-scale evacuations such as those for hurricanes is people's underestimation of the time needed to reach their destinations. Residents have accurate expectations about the time required to pack their bags and other tasks, but they underestimate the amount of travel time needed to clear the risk area. People take the typical routes out of the city and assume it will take the usual amount of time. People fail to account for immense traffic, which can turn a two-hour trip into a twenty-hour trip.

The perceived implementation barriers inhibiting residents from taking protective action include:

- **Lack of knowledge and skill.** In the case of evacuation, this may include a lack of knowledge of a safe place to go and a safe route to travel.
- **Lack of access to a personal vehicle.** Many evacuations require traveling long distances to reach safety, so those who don't have their own vehicles must rely on other means. Some evacuees who lack their own vehicles are able to find rides with friends, relatives, neighbors, or coworkers, but others must rely on buses organized by their local governments.
- **Lack of personal mobility due to physical handicaps.** A small but significant percentage of the population requires assistance because they (and, frequently, other members of their households) are unable to evacuate themselves (see Figure 4-1).
- **Separation of family members.** Some family members may be away from home when an evacuation occurs and the other family members do not want to leave until they return. Until family members establish communication contact and agree upon a place to meet, evacuation is unlikely to occur (Killian, 1952; Drabek and Boggs, 1968; Drabek and Key, 1976; Haas, Cochrane and Eddy, 1977).

Figure 4-1

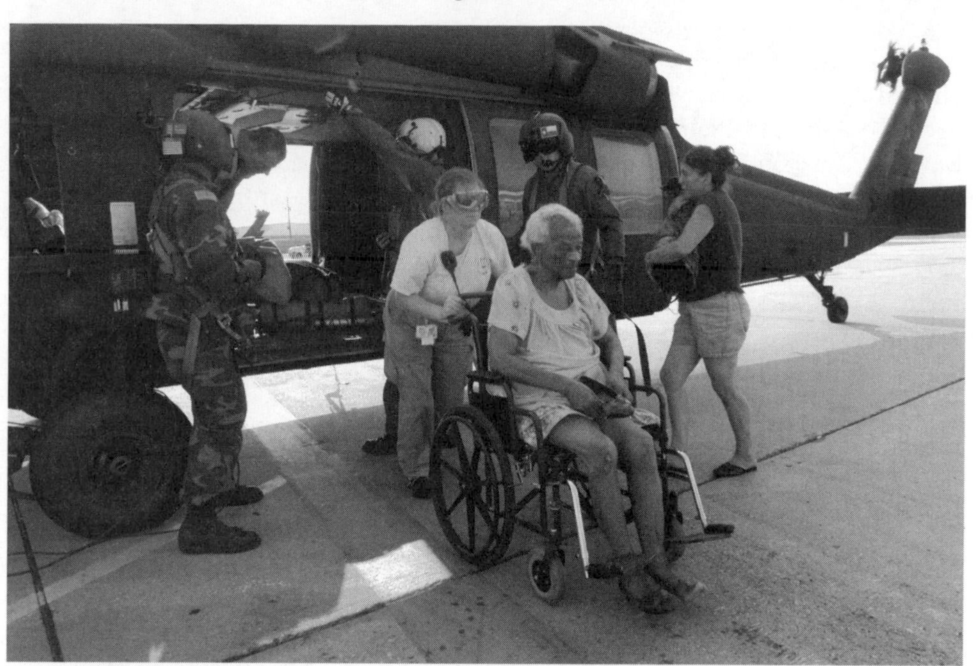

Those who have a lack of physical mobility need assistance evacuating.

- **Perceived cost of actions to protect personal safety.** Such costs include out-of-pocket expenses, opportunity costs (e.g., lost pay), and effort. The high cost can lead people to delay taking protective action until they are certain it is necessary.

When no one option seems better than other options or continuing normal activities, it is difficult for people to decide what to do. For example, evacuation is a superior protective action than seeking shelter during a hurricane, but it also costs more in terms of money and time. For people on the fringes of an evacuation area, the risk of staying may be offset by the cost of evacuating. This can cause people to wait for further information about the hurricane to see if the risk has changed enough to push the balance more clearly one way or the other.

The result of protective action assessment is an adaptive plan. People's adaptive plans vary widely, with some plans being only vague goals and others being extremely detailed. At minimum, a specific evacuation plan includes a destination, a route of travel, and a means of transportation. More detailed plans include:

- A procedure for reuniting families if members are separated.
- Advance contact to confirm the destination is available.

▲ Alternative routes.
▲ Alternative methods of transportation.

Those who do not have a detailed plan are more likely to suffer negative consequences. A classic example is an interview with the recipient of an evacuation warning that contained no information on safe evacuation routes or safe destinations: "We couldn't decide where to go. So we grabbed our children and were just starting to move outside ... if it had just been ourselves, we might have taken out. But we didn't want to risk it with the children" (Hamilton, Taylor, and Rice, 1955, p. 120).

4.1.5 Step 5: Protective Action Implementation

Protective action implementation occurs when those at risk know they have to take action (see Figure 4-2). A primary question at this stage is "Does protective action need to be taken now?" The answer is crucial because people sometimes postpone taking action even when faced with danger. For example, some recipients of hurricane warnings often endanger their safety by waiting until the last minute to evacuate. Unfortunately, they fail to recognize that bad weather and a high

Figure 4-2

Sandbagging is a protective action.

traffic volume reduces the speed of evacuating vehicles. These conditions may lead to an incomplete evacuation before the arrival of storm conditions (Baker, 1979, 1991, 1993; Dow and Cutter, 2002; Prater, Wenger, Grady, 2000). The problem of procrastination is worse for long-term hazard adjustment without specific timetables. For example, an earthquake prediction might indicate a 75% chance of a severe earthquake within the next 20 years. This type of prediction often fails to motivate immediate protective action because people can rationalize that it is quite reasonable to worry about the problem later.

4.1.6 Step 6: Information Needs Assessment

People who are taking protective action need information. Before taking action, they must decide if they have enough information. Any confusing messages or expressed doubts from officials will cause people to seek more information. If the answer to the questions at any of the previous stages cannot be answered with a definite *yes*

Table 4-1: Warning Stages and Actions

Steps	*Activity*	*Question*	*Outcome*
1	Risk identification	Is there a real threat that requires my attention?	Threat belief
2	Risk assessment	Do I need to take protective action?	Protection motivation
3	Protective action search	What can I do to achieve protection?	Decision set (alternative actions)
4	Protective action assessment	What is the best method of protection?	Adaptive plan
5	Protective action implementation	Do I need to take protective action now?	Threat response
6	Information needs assessment	What information do I need to answer my questions?	Identified information need
7	Communication action assessment	Where and how can I obtain this information?	Information search plan
8	Communication action implementation	Do I need the information now?	Decision information

Adapted from Lindell and Perry (2004).

or *no,* people will ask, "What information do I need to answer my question?" Through this process, people identify an **information need.** Take, for example, the case of someone who does not know the answer to the question "What is the best method of protection?" They can search for additional information about alternative protective actions to make it clearer which option is best. People frequently seek additional information because the consequences of a decision error are very serious (e.g., failing to evacuate in time can result in death or injury) and they rarely have all the information they need to make a confident decision.

4.1.7 Step 7: Communication Action Assessment

The next question is "Where and how can I obtain this information?" Addressing this question leads to an **information search plan.** Uncertainty about risk identification and risk assessment can stimulate questions directed to officials and, more likely, the news media (Lindell and Perry, 1992). People often rely on the news media to confirm information they received about the hazard from other sources. However, people often consult their peers about what to do and how to do it. It is difficult for people to reach authorities because they are usually busy handling other calls. People are often forced to rely on the media and their peers even when they would prefer to contact authorities.

4.1.8 Step 8: Communication Action Implementation

The last question is "Do I need the information now?" If the answer is *yes,* then people will seek the information. People will go to great lengths, contacting

FOR EXAMPLE

Risk Assessment and Hurricanes

After every hurricane, researchers find that some people failed to evacuate because they didn't believe the threat was likely to affect them. Part of this is due to the difficulty in predicting hurricanes. Sometimes forecasters predict that a strong hurricane will make landfall in one place but it turns and strikes somewhere else. This can lower people's perceived probability of a hurricane strike. Other times people expect the hurricane to strike but they don't expect it to affect them. For example, it is common to hear people say, "I survived the last storm, I can survive this one." In this case, the perceived severity of the event will tend to be low. Although many authorities are concerned about a "cry wolf" effect, these types of experiences do not seem to decrease people's intentions to evacuate in future hurricanes (Dow and Cutter, 1998). The most likely explanation is that people understand that hurricane behavior is inherently uncertain, so forecast errors are inevitable.

many people over a short period of time (Drabek and Stephenson, 1971). However, people will not work as hard to get information unless the threat is imminent. For example, many residents close to Mt. St. Helens checked radio news bulletins several times a day after the initial ash and steam eruptions led authorities to believe there was a high probability of a larger eruption. Without specific threat information, people tend to *passively monitor* the situation. For example, residents might only check the morning paper or the evening news for information about the hazard rather than checking many times a day. This passive monitoring continues until the threat escalates and people need to resume active monitoring.

Communication action implementation can have one of three outcomes. First, people confirm the threat and proceed to take protective action. Second, if the information source is unavailable, people try to find different sources. Third, if the new information contradicts previous information, then people try to resolve the conflict. Often this involves considering the relative credibility of the information sources.

SELF-CHECK

- Define **adaptive plan** and **information search plan. List the reasons for needing each.**
- Define **protection motivation.**
- Define **risk and risk assessment.**
- Define **warning.**
- **Name the three possible outcomes of a communication action implementation.**

4.2 Risk Communication during the Continuing Hazard Phase

The continuing hazard phase is marked by a low probability that a catastrophic incident will threaten public safety, property, and the environment. During this phase, you should engage in hazard mitigation, emergency preparedness, and recovery preparedness actions. In addition, you should also pursue an active program of risk communication.

There are five basic risk communication functions to address in the continuing hazard phase. Table 4-2 identifies these as strategic analysis, operational analysis, resource mobilization, program development, and program implementation.

Table 4-2: Tasks for the Continuing Hazard Phase
Strategic analysis
Conduct a community hazard vulnerability analysis
Analyze the community context
Identify the community's prevailing perceptions of the hazards and hazard adjustments
Set appropriate goals for the risk communication program
Operational analysis
Identify and assess feasible hazard adjustments for the community and its households/businesses
Identify ways to provide incentives, sanctions, and technological innovations
Identify the available risk communication sources in the community
Identify the available risk communication channels in the community
Identify specific audience segments
Resource mobilization
Obtain the support of senior appointed and elected officials
Enlist the participation of other government agencies
Enlist the participation of nongovernmental (nonprofit) and private sector organizations
Work with the mass media
Work with neighborhood associations and service organizations
Program development for all phases
Staff, train, and exercise a crisis communications team
Establish procedures for maintaining an effective communication flow in an escalating crisis and in emergency response
Develop a comprehensive risk communication program
Plan to make use of informal communication networks
Establish procedures for obtaining feedback from the news media and the public
Program implementation for the continuing hazard phase
Build source credibility by increasing perceptions of expertise and trustworthiness

Use a variety of channels to disseminate hazard information
Describe community or facility hazard adjustments being planned or implemented
Describe feasible household hazard adjustments
Evaluate program effectiveness

Adapted from Lindell and Perry (2004).

The tasks are listed in the table as if they should be performed in sequence. However, you will perform some tasks at the same time. In addition, you will repeat some steps frequently.

4.2.1 Step 1: Conduct a Strategic Analysis

You must understand who is at risk. You must understand the likelihood of different hazards. Knowing what types of hazards are a threat helps identify what actions people should take to protect themselves. Identifying the geographic areas at greatest risk makes it possible to identify the most vulnerable population segments and types of businesses. Knowing the vulnerable population segments and types of businesses provides information about how to communicate the risk. This information also helps you pick which incentives and sanctions will get people to adopt hazard adjustments.

Analyze the Community Context

You should know the following information about your community:

- ▲ Ethnic composition.
- ▲ Availability of communication channels.
- ▲ Perception of authorities.
- ▲ Levels of education.
- ▲ Income distribution.

Build support within your community. If managing hazards is not a community priority, show how easy it can be. Begin with a small hazard management program, demonstrate effectiveness, and build support (Lindell, 1994b). In developing a risk communication program, determine how much money the community can afford to spend. Talk to other agencies and explore how you can work together to develop a comprehensive risk communication program.

Identify the Community's Perception of Hazards and Hazard Adjustments

The hazards that produce the greatest community conflict are those having a potential for inflicting significant harm on bystanders. These hazards include nuclear power plants and chemical facilities. People perceive these risks as greater than those of other technologies and other natural hazards (Lindell and Earle, 1983; Slovic, 1987). Some of the reasons, called "outrage factors" by Hance et al. (1988) are because people believe the risk is

- ▲ Not natural.
- ▲ Not familiar.
- ▲ Not understood by scientists.
- ▲ Difficult to detect.
- ▲ Associated with untrustworthy information sources.
- ▲ Not controllable by those exposed.
- ▲ Characterized by involuntary exposures.
- ▲ Unfair in its distribution of risks and benefits.

Most people believe the risks of technological facilities are greater than those of natural hazards. And yet, the annual fatality rate is the same for both types of hazards (Slovic, 1987). You can address this problem by explaining the concepts of risk analysis. However, you should recognize that it is difficult even for experts to understand small probabilities of occurrence such as one in a million. Some experts believe that if people accept risks having higher fatality rates, like driving, they also should accept risks having a lower fatality rate such as having a nuclear power plant nearby. This ignores the fact that the facility risk will be added to the lifestyle risk, not substituted for it. In addition, the facility risk is estimated from analytical models but the lifestyle risk is computed from a large database. Even local residents who cannot articulate these distinctions seem to be aware of them intuitively and reject risk analysis results.

Set Goals for the Risk Communication Program

Hazard awareness is an important first step in the process of hazard adjustment. People must be informed about the hazards to which their community is exposed and should be given this information from different perspectives. For example, people should know what a disaster would mean in terms of the public health. In addition, they need to know how likely it is that a disaster will occur where they live. In the case of hurricanes, a reasonable goal is to ensure residents understand the causes of hurricanes, the probabilities of being struck by a hurricane over the next ten years, and the threats hurricanes bring. Also, local residents should understand the risk to themselves and their families, damage to their property, and disruption to daily activities. To help people understand, a risk communication program should provide detailed maps showing areas at risk from wind, storm surge, and inland flooding. Explain the vulnerability of different buildings to these threats. For example, you can define the areas that would be affected by hurricanes, using the Saffir-Simpson Hurricane Scale (a 1-5 rating). Display these risk areas on large-scale maps. Such maps should indicate streets, rivers, political boundaries, and other local landmarks that help people identify the risk areas in which their homes and workplaces are located. Provide information about the personal consequences of hazard impact by showing drawings of different types of structures. Show mobile homes, typical single-family residences, and typical multifamily structures and the level of expected damage from each hurricane category. Develop this information carefully because recent studies found that only one- to two-thirds of coastal residents can accurately identify their hurricane risk areas (Arlikatti, Lindell, Prater and Zhang, in press; Zhang, Prater and Lindell, 2004).

To be successful, your risk communication program must foster people's sense of personal responsibility for self-protection. Remind local residents of the limits to what local government and industry can do in reducing hazard damage. Remind residents that they can prevent death or injury to themselves and their families and damage to their homes. Through risk communication, people should be made aware of the available hazard adjustments. To get people to adopt hazard adjustments, you must convince them the hazard adjustments have high efficacy and low resource requirements. This should motivate people to take protective actions. However, recognize that even the most scientifically sound and effectively implemented risk communication program will not lead a large percentage of people to take immediate protective action. Nonetheless, a long-term perspective will put environmental hazards on the political agenda and achieve important results over time (Birkland, 1997; Prater and Lindell, 2000).

4.2.2 Step 2: Perform an Operational Analysis

As an emergency manager, there are five tasks you must perform when performing an operational analysis.

Task 1: Identify and assess feasible hazard adjustments for the community and its households/businesses

The purpose of this task is to make sure people know how to protect themselves (Lindell and Perry, 2000). You can access resources such as the American Red Cross Web site (www.redcross.org/services/disaster/beprepared). At this site, you will find information about recommended household adjustments for a wide range of hazards. You can help people evaluate these in terms of resource requirements, such as financial cost, time and effort, knowledge and skill, tools and equipment, and required cooperation with others.

Task 2: Identify ways to provide incentives, sanctions, and technological innovations

To encourage people to protect themselves, you may have to punish them with sanctions, reward them with incentives, or inform them about technological innovations. Sanctions are appealing because they avoid the obvious costs associated with incentives. For example, ticketing drivers who don't wear seatbelts was a successful sanction. However, sanctions require enforcement. Your jurisdiction can reduce the costs of preparing for hazards by providing incentives such as grants, low interest rate loans, or tax credits. An alternative incentive, providing specific plans or checklists for hazard adjustments, is informational rather than financial. For example, providing plans for homeowners to bolt their houses to their foundations helps do-it-yourselfers who have a modest level of construction experience. Adding a community tool bank also makes this feasible for those who lack the necessary tools and equipment.

Task 3: Identify the available risk communication sources in the community

Sources can be categorized as authorities, news media, and peers. These sources are judged in terms of their credibility, which is a combination of expertise and trustworthiness. Perceptions of credibility vary depending upon whether a source is speaking about hazards or hazard adjustments. Official sources are generally the most credible. People look for sources to have impressive credentials, previous experience, or the respect of others (Perry and Lindell, 1990b).

Ethnic minorities trust different types of sources. Research has focused on the perceptions that Mexican Americans, African Americans, and Caucasians have of source characteristics. Authorities (particularly firefighters and members of the police department) tend to be regarded as credible by the majority of all three ethnic groups (Lindell and Perry, 1992). African Americans and Caucasians tend to be more skeptical of the mass media than Mexican Americans. In general, Mexican Americans are more likely than other groups to consider peers to be the most credible sources. There is evidence, however, that the results vary by community. This reflects historical differences in relationships between ethnic groups and authorities in these specific communities.

You must identify what groups trust which risk information source in your community. Know which minority groups live and work in the community, if the members are located in one area (and where), and how they view the risk information sources. You can gain this knowledge from census data, informants, and personal observation. Census data can identify those areas having a greater than average percentage of ethnic minorities. Informants can supplement this data and can give you an inside view on how different groups and neighborhoods view sources of information. For example, does a particular group trust the Mayor? It is important to identify the opinion leaders in each ethnic group. Contact these opinion leaders to see if they are willing to serve as additional sources for your risk communication program. The best information comes from long-term outreach programs. Meet with people and speak at neighborhood associations and civic organizations. Involve a diverse group of citizens in advisory committees. Community involvement provides you with information about how the residents regard information sources. It also enhances your visibility, fosters dialogue, and gives citizens access to accurate information.

Task 4: Identify the available risk communication channels in the community

Electronic and print media are available in most communities. Using the media is one way to communicate risks to the community. Additional ways include informal face-to-face conversations and formal meetings (Hance et al., 1988; Mileti, Fitzpatrick & Farhar, 1990). Even though you have access to all of these channels, you may be limited by your budget. To gain access to low-cost opportunities for publicity, you must establish contacts with local media. In addition, a long-term relationship with local businesses sometimes generates contributions to pay for low-cost items such as brochures and posters.

Task 5: Identify specific audience segments

It is easier and cheaper to develop one communication message for the entire community, but it won't be as effective as tailoring your message to different groups. Individuals have different concerns. To develop specific messages for different groups,

- ▲ Know the geographic and demographic characteristics of your community.
- ▲ Know where each group likes to find information. Radio stations, in particular, focus on specific audiences defined by age and ethnicity.
- ▲ Make sure messages are in the appropriate languages for different groups of non-English speakers. Some communities have dozens of different languages and dialects spoken there.
- ▲ If members of a community group tend to be fatalistic about hazards, be sure to target them with messages emphasizing hazard adjustments that are easiest to implement. Maintaining a four day supply of food and water is a good starting point.
- ▲ Ensure your messages are understood. Follow up with people face-to-face to see if they are taking the appropriate actions.

4.2.3 Step 3: Mobilize Resources

As an emergency manager, there are five tasks you must perform to mobilize resources for risk communication.

Task 1: Obtain the support of senior appointed and elected officials

You need the support of senior officials (Lindell, 1994b). They help you acquire resources and help put emergency management on the political agenda. Getting this support is also an important step toward obtaining the participation of other government agencies. If you can't get senior officials or organizations to support you, work with their staff members. Staff members can sometimes convince their bosses to support critical issues. You must successfully stress that the community is at risk and hazard mitigation can reduce disaster impacts. You should also propose emergency preparedness measures and recovery preparedness measures as effective solutions.

Task 2: Enlist the participation of other government agencies

No matter how supportive senior officials are, they have limited resources. This means you should work with other agencies to share the cost of risk communication. Ensure that each agency is aware of the risk communication programs being planned by other governmental agencies, nongovernmental organizations, and hazardous technological facilities. Gather the resources of multiple agencies within local government (Drabek, 1990; Gillespie et al., 1993; Lindell et al., 1996). Identify ways you can work together to achieve the goals of both organizations. For example, work with the police department to ensure that neighborhood watch groups are provided with information about environmental hazards.

Task 3: Enlist the participation of nongovernmental and private sector organizations

Organizations such as the American Red Cross and the Salvation Army play an important role. They help communities prepare for emergencies and recover from disasters. They routinely work with needy families. Consequently, they can identify areas with a high concentration of population that are most vulnerable to disasters. These organizations can also identify methods of assisting households to prepare for emergencies, reduce the vulnerability of their homes, or find safer places to which they can move.

In addition, water, wastewater, fuel, and electric utilities play a significant role in promoting the adoption of hazard adjustments. Most of these respond to routine emergencies such as severe storms, so they are aware of the demands disasters can place on the community. In addition, these organizations routinely send bills to the residents of their service areas. This gives you an opportunity to include notices about hazards and hazard adjustments to customers in their bills.

Task 4: Work with the mass media

The mass media, with all the television channels, Web sites, newspapers, and radio stations, reach many residents each day. Working with the media allows you to get your message out. With more people seeing your message, more people will become aware of the role of emergency management within the community. Also, reporters know their audiences and focus on them. This allows you to target messages to specific audience groups. These groups are defined by gender, age, ethnicity (and language), and socioeconomic status.

Reporters do not always consider hazard information to be newsworthy. To combat this, federal agencies such as the National Weather Service urge government officials to "declare" weeks for hazards such as tornadoes and hurricanes. You can take advantage of the publicity generated by these agencies. Work with the news media to develop the background materials reporters need in an escalating crisis, emergency response, or disaster recovery. Anticipate what types of information reporters are likely to seek during these events. Prepare fact sheets and other "boilerplate" that can be used no matter what conditions occur during an emergency.

Task 5: Work with neighborhood associations and civic organizations

Most communities have many neighborhood groups and civic organizations whose members are active in their community (Chavis and Wandersman, 1990; Florin and Wandersman, 1984). These groups work with the community and their members and can help instruct them on the hazards and hazard adjustments. Time is often available for this because many of these organizations want to meet on a regular schedule but do not have enough activities to fill their agendas. Consequently, they are often willing to host speakers whose topics will interest their members.

4.2.4 Step 4: Develop a Program

As an emergency manager, there are four tasks you must perform when developing a program.

Task 1: Staff and train a crisis communication team

Establish a crisis communication team. The team forms a critical link between experts and the population. The team must be able to communicate effectively with both groups. There should be one spokesperson on the team who understands the scientific aspects of the situation and can explain it to everyone at a level they can understand. Spokespersons with technical credentials will generally be considered credible. And spokespersons with previous disaster experience will be seen as credible as well. It also is helpful if team members receive training from public relations experts (Hance et al., 1988).

The crisis communication team should have written operating procedures that include documentation of all emergency response activities. They should also maintain an event log that records what information has been requested and released. Criteria used to guide critical decisions, such as those involving protective actions for the public, should also be documented. The team should monitor the news media and designate a rumor control center. This center should be staffed by operators who are frequently updated on the status of the incident and the response to it.

The crisis communication team should recognize that reporters describe events in terms of stories that are framed by five questions—*who, what, when, where,* and *why* (Churchill, 1997). Reporters want to know the specific causes of an event. Other questions include who was (or will be) affected. They want information on casualties, property damage, and economic disruption. They want to know what authorities have done (and will do) to respond to the situation. It frequently is difficult to answer these questions because information is lacking. The spokesperson should avoid speculation (and especially premature blame), but, rather, admit he or she does not know the answer and will find out as soon as possible.

Reporters rarely have scientific backgrounds, so technical details might be unnecessary and confusing. All technical jargon must be translated into plain English. If you help reporters do this, you will have a better chance in getting your message out. Work with local reporters to make sure the information is easily understood, but recognize that this will not solve all problems. Some major crises draw reporters from outside the U.S. Reporters from national or international media will not cover stories in exactly the same way as local reporters, but the most important information needs will be common to all categories of reporters.

Finally, the risk communication plan and procedures should be evaluated using drills and exercises. Each drill or exercise should be followed by an evaluation of the plan and procedures, as well as the staffing, training, and materials used.

Task 2: Establish procedures for maintaining an effective communication flow during an escalating crisis or emergency response

All organizations should establish procedures for coordinating information. It is critical that each organization receives all the information it needs as promptly as possible. The types of information needed in an escalating crisis depends on the circumstances. Recommendations regarding the content of incident notifications for nuclear and chemical facilities are summarized in Table 4-3. Adopt this table as a template because it is based upon extensive experience with escalating crises and disaster responses. It is *essential* that you discuss this with facility operators. You both need to understand your information capabilities and needs. Agree in advance what information will be exchanged when the need arises.

Task 3: Develop a comprehensive risk communication program

The four key factors when designing a message are

- ▲ Personal risk.
- ▲ Personal responsibility.
- ▲ Guidance for protective action.
- ▲ Sources for further information.

You should design messages to address all of these factors. Messages should include any important details but, generally, be short and concise. Too many details can overwhelm people and prevent them from listening. Information should be presented in a way that attracts attention, so people will understand and remember it more easily. Address risk perception but do not over emphasize it. You should address risk perception because probabilities are difficult for many to understand. The statement "there is a 1% probability of a damaging earthquake within the next year" might have little impact on people's behavior. However, adding probabilities over time by making the mathematically equivalent statement that there is roughly a 20% chance of an earthquake in the next twenty years makes more of a difference in risk perception (Kunreuther, 2001; Slovic et al., 1978). However, increasing the accuracy of people's risk perceptions does not help if people fail to take action to protect themselves. When

Table 4-3: Essential Incident Data
Date and time of report
Name, affiliation, and telephone number of information source
Location, type, and current status of the incident
(a) Derailment, containment failure, fire, explosion, liquid spill, gaseous release
(b) Hazardous material name, physical properties (gas, liquid, solid), environmental cues (sights, sounds, smells), and potential health effects
(c) Hazardous material release duration and quantity released
(d) Casualties and damage already incurred
Incident prognosis
(a) Potential for fire or explosion at site
(b) Potential for fire or explosion affecting residential, commercial, or industrial areas
(c) Hazardous material quantity available for release and expected release duration
(d) Locations and populations requiring protective action
(e) Types of protective actions recommended: evacuation, sheltering in-place, expedient respiratory protection, interdiction of food/water
Weather conditions (current and forecast wind speed and direction)
Chronology of important events in the development of the incident
Current status of response
(a) Facility/shipper/carrier actions: assessment, preventive, corrective, population protective actions
(b) Local/state/federal agency actions: assessment, preventive, corrective, and population protective actions

issuing warnings, you can increase the accuracy of people's risk perceptions by addressing four questions about the risk they face:

- ▲ What is the risk?
- ▲ Where is it going to happen?

- ▲ When is it going to happen?
- ▲ What will the effects be (Mileti, 1993)?

Discussing how people need to take responsibility for their own safety is important. Letting residents know that they must be self-sufficient for 72 hours increases personal responsibility for self-protection (Lindell and Perry, 2004). Some people expect the government to come to their aid right away. You must explain to them that there are limits not only to what the government can provide but also how quickly the government can provide those things. Residents become more self-sufficient when they know how to protect themselves.

Residents should be given specific instructions on what protective actions to take and, in some cases, how to implement those actions. In the case of evacuation, for example, people should be reminded of items to take with them that may not be obvious (e.g., important legal records such as birth certificates). If some evacuation routes are hazardous or congested, residents should also be given alternate routes that are available. If people don't have their own cars, they should be informed about bus pickup points.

Finally, sources for further information should be addressed because residents might need specific information that hasn't been addressed in the warning message. Some residents might need information about evacuation procedures for children at school. Others might need information about what number to call for assistance in evacuating physically handicapped members of their households.

Task 4: Plan to make effective use of informal communication networks

It is important for you to recognize that people talk to their peers throughout all phases of emergency management. Use these informal networks to increase the level of hazard adjustment adoption. However, friends, relatives, neighbors, and coworkers might not understand a message, or they might not remember the message correctly. To reduce confusion, release information through several channels. Provide many opportunities for people to hear messages so they will retain the common elements of these messages.

Task 5: Establish procedures for obtaining feedback from the news media and the public

Feedback is a critical part of any communication process. It provides receivers an opportunity to confirm that they have comprehended the message, to reconcile inconsistencies within or between messages, or to obtain information that is not available in the messages they have received. Feedback is part of informal face-to-face discussions, but opportunities for receiving feedback are more limited in public hearings. Public comments frequently are limited to a few minutes at the end of a meeting. The need for feedback is why many experts recommend informal channels of communication (e.g., Committee on Risk Perception and Communication,

1989; Covello, 1986; Hance, et al, 1988). Therefore, agency procedures that require public hearings should be supplemented by less formal meetings.

During emerging crises, reporters might unintentionally distort your message because they didn't quite understand what you were saying. This makes it important to read local newspapers, listen to radio, and view television broadcasts to see if your information is being correctly presented. In addition, you can obtain feedback from citizens via rumor control centers, using a telephone number or a Web site that has been publicized in advance.

4.2.5 Step 5: Implement the Risk Communication Program

As an emergency manager, there are five tasks you must perform when implementing the risk communication program.

Task 1: Build source credibility by increasing perceptions of expertise and trustworthiness

It is important for personnel from each agency to develop a history of effective job performance that enhances their credibility. This experience is gained during minor incidents, such as minor floods, that cause damage and disruption of normal activities. Credibility is also enhanced by effective performance in public hearings or in meetings with neighborhood associations and civic organizations. Of course, expertise is only one component of credibility; trustworthiness is also essential. Earn a community's trust by being competent, caring, honorable, and considering outrage factors when working with the public (Covello et al., 1988; Hance et al., 1988).

Task 2: Use a variety of channels to disseminate hazard information

Use not only the news media, but informal channels to communicate your message. Also use opportunities to meet with neighborhood associations and civic organizations.

Task 3: Describe community or facility hazard adjustments being planned or implemented

Inform residents of any actions being taken to reduce the probability of an incident so they understand that their risk is being reduced. Acknowledge that there is no action that can guarantee complete safety. For example, land-use and building construction practices can reduce, but not eliminate, the threat of natural hazards. The same can be said about engineered safety features in connection with technological hazards. In addition, describe actions being taken to facilitate a response to an emergency should one occur.

Task 4: Describe feasible household hazard adjustments

As described in previous sections, let people know what actions they can take to protect themselves and reduce damage to their homes.

FOR EXAMPLE

Hurricane Katrina: Those Who Were Left Behind

In New Orleans, 30% of the population lived below the poverty line before Hurricane Katrina. Many of them lived in the 9th Ward, a low lying area that was wiped out by the flooding. Some of these people had no cars of their own, or the cars they had were so unreliable that they could not be used for long distance travel. As a result, even those who wanted to evacuate could not do so.

Task 5: Evaluate program effectiveness

Set goals for the risk communication program and determine how they should be measured. It is very common for emergency managers to measure program effectiveness by reporting the number of meetings attended and the number of brochures distributed. Better measures of effectiveness include increases in the number of households with hazard insurance, family emergency plans, earthquake-prone homes with water heaters strapped to the foundations, and hurricane-prone homes with window shutters.

SELF-CHECK

- Describe why it is important to set goals for the risk communication program.
- List the key message factors in a risk communication program.
- Name four tasks in the program development step.
- Name five tasks you must perform when implementing the risk communication program.

4.3 Risk Communication during an Escalating Crisis or Emergency Response

An **escalating crisis** is a situation in which there is a significantly increased probability of an incident occurring that will threaten the public's health, safety, or property. Not everyone will agree that there is a crisis. As a practical matter, a crisis exists if authorities, the news media, or a significant proportion of residents believe there is one. The principle behind this is that "perception is reality." If the news media or local residents believe there is a crisis, then there is a crisis unless authorities can convince them otherwise. Thus, authorities must be prepared to explain specifically why they believe a situation is or is not a crisis.

4.3.1 Step 1: Classify the Situation

You can exert control over people's definition of a situation by defining a system for classifying threat levels. For example, the National Weather Service has established a classification system that consists of watches and warnings. A tornado warning is more serious than a tornado watch. The U.S. Nuclear Regulatory Commission (NRC) (1980) classifies a nuclear power plant incident as an Unusual Event, Alert, Site Area Emergency, or General Emergency. The categories in the emergency classification system correspond to meaningful differences in the levels of response by local authorities. The classification system needs to be established in advance, defined objectively, and agreed to by all responding organizations (Lindell and Perry, 1992). By establishing an emergency classification system, authorities commit themselves to taking specific actions when certain criteria are met. With this system, decisions are made based on rational scientific considerations rather than emotion or other considerations.

4.3.2 Step 2: Implement a Risk Communication Program

Once there is an emergency, authorities will act. Some of the actions will include protecting the population, protecting the environment, and assessing the situation. One of the most important incident management actions is risk communication. As an emergency manager, you must perform six tasks when implementing an emergency response program.

Task 1: Activate the crisis communication team promptly

You need to contact all appropriate authorities. Make sure all communication links are open and all sources of information and expertise are used. Monitor information from other organizations so you can identify any disagreement. If there is disagreement, prepare an explanation for it before the media contacts you about the disagreement. Your explanation will be more credible if you contact the news media than if you wait until they contact you.

Review the information in press kits and background materials. Contact personnel who are in the crisis and brief them regularly. This will help them answer questions if they are contacted by friends or the media.

Review your communication objectives (Churchill, 1997). Evaluate all press releases, press conferences, and public meetings in light of these objectives. In most environmental emergencies, the principal objective is to save lives. This objective could be expressed in evacuation orders. One of the objectives should *not* be to prevent panic, which disaster researchers have found to be extremely rare (Drabek, 1986; Lindell and Perry, 1992). Nor should authorities make fun of what they consider to be unnecessary protective actions by those who *think* they are at risk, as long as such actions do not prohibit the protection of those whom the authorities believe *are* at risk. Do not attempt to promote one protective action by criticizing another. For example, some misguided attempts have been made to promote sheltering in-place by asserting that people will get into major traffic accidents if they evacuate. Not only is this incorrect (the accident risks in evacuation appear to be no greater than those of normal driving; Lindell and Perry, 1992), but it is likely to lead those at risk to believe that there is nothing they can do to protect themselves.

Task 2: Determine the appropriate time to release sensitive information

You must determine when to alert others of the danger. Your team needs to know when to release information. There are no hard and fast rules about when information should be released. Early information often turns out to be incorrect because the facts are still coming in. However, an early release can enhance your credibility and give you more control. Being the first to break bad news allows you to put the information into an appropriate context. In addition, controlling the timing of a press release can have a significant impact on the amount of attention it receives. A press release distributed on a slow news day might receive more coverage than the same information released on a busy day or on a Friday afternoon. The disadvantage of delaying the release of information is that this can be misinterpreted as a cover-up if the information is leaked (Hance et al., 1988). It is also important to respond to reporters' questions when they are aware that something important is happening. Statements of "no comment" are interpreted as meaning authorities are withholding important information.

Task 3: Select the communication channels that are appropriate to the situation

An escalating crisis is newsworthy, so you will have little difficulty in obtaining media coverage. Initiate communication with reporters through press releases and press conferences. Press releases give you the most control over the agenda, and interviews with individual reporters provide the least control. You need to ensure reporters are accurately disseminating the information. However, this alone cannot ensure those at risk are receiving, heeding, and comprehending the information they need. You need to promote dialogue through two-way communication, preferably in small groups.

Task 4: Maintain source credibility with the news media and the public

You must obtain timely and accurate data. If the available data is incomplete, you should be honest about what is and is not known. A candid confession of ignorance might be uncomfortable at the time, but it is less dangerous to one's credibility than making up an answer that is later found out to be incorrect.

The news media have many sources of information. This is why it is important to respond to reporters when they need information for a deadline. If you do not respond, they will obtain whatever information they can from whatever sources are available at the time (Churchill, 1997). If you don't have information, it is better to explain that data is being collected, describe how it will be analyzed, and indicate when the information will be released.

Trust is a major issue. There is little trust in society and what is there can be lost easily. Television anchors tend to be among the few people other than independent scientists who are generally trusted (Kasperson, 1987). Television anchors are trusted because they are familiar, authoritative, and have developed a track record of accuracy. You may be stereotyped as a typical bureaucrat. This is the reason why it is important to work with the community on issues before crises arise and publicize your accomplishments.

Task 5: Provide timely and accurate information about the hazard to the news media and the public

Press releases should be no longer than two pages with simple short sentences in plain English. They should contain

- A dateline (date and location of release)
- The organizational source (including point of contact) for the information
- A summary lead sentence
- A brief description of any attachments

Press releases should be supplemented by fact sheets, which contain background information. There should be attachments including a biographical summary of the spokesperson and other details about the hazard and official responses (Churchill, 1997).

Be prepared to describe the process by which risks are being assessed and what the risks are. In general, it is important to presume the average member of the audience is intelligent but uninformed about environmental risks. Avoid acronyms and technical jargon. Also anticipate the possibility of confrontational tactics by the news media or some members of the public. If confronted with different interpretations, be prepared to calmly restate your scientific qualifications, support your own position, and explain the weaknesses in alternative positions.

Task 6: Evaluate performance through post-incident critiques

Organizations must learn from experience. Thus, each incident in which you must disseminate risk information to the news media or the public should be followed by a thorough critique of performance (Lindell and Perry, 1992;

FOR EXAMPLE

Being in the Know During a Hurricane

Michael Brown, head of FEMA during Hurricane Katrina, shocked people in a television interview with CNN claiming he didn't know New Orleans residents had taken shelter in the New Orleans convention center. This claim was made despite days of television reporting about the lack of medical help, food, supplies, or police protection.

National Response Team, 1987). All members of the crisis communication team should review the goals of the risk communication program, the event logs kept during the incident, and other available documentation to identify weaknesses in performance. Experiences in drills, exercises, and incidents have demonstrated the importance of focusing on the performance of the organization rather than the performance of individuals because this enhances a spirit of cooperation. Thus, each participant should be encouraged to follow up on any weaknesses by identifying potential improvements in plans, procedures, and training.

SELF-CHECK

- Define **escalating crisis.**
- List some actions that authorities should take when there is an emergency.
- Describe why television anchors are generally trusted.
- Describe what members of the crisis communication team should review when evaluating performance through post-incident critiques.

SUMMARY

We get caught up in our daily lives. How many times have you driven someplace familiar only to arrive not knowing how you got there? If this feeling is familiar, then you understand how people ignore life-threatening risks unless they think the risks are real. They'll go about their daily lives as if nothing is happening around them. As an emergency manager, it's up to you to make sure they don't become involved in a disaster wondering how that happened to them. In this lesson, you learned how critical communication is when it comes to influencing people's perceptions of danger. Apply the communication skills this lesson discussed and you just might save lives.

KEY TERMS

Adaptive plan
The answer to the question "What is the best method of protection?" Those at risk generally have at least two options—taking protective action or continuing normal activities.

Escalating crisis
A situation in which there is a significantly increased probability of an incident occurring that will threaten the public's health, safety, or property.

Information need
A need that results from the question, "What information do I need to answer my question?"

Information search plan
A plan that results from addressing the question, "Where and how can I obtain this information?"

Protection motivation
A positive response to the question of whether there will be personal consequences if disaster occurs.

Risk
The *possibility* that people or property could be hurt. Risk is defined in terms of the likelihood that an event will occur at a given location within a given time period and will inflict casualties and damage. This risk must be effectively communicated to the people who are likely to be affected.

Risk assessment
An evaluation of what will be the personal consequences if the disaster occurs.

Warning
A risk communication about an imminent event that is intended to produce an appropriate disaster response.

ASSESS YOUR UNDERSTANDING

Go to www.wiley.com/college/lindell to evaluate your knowledge of the basics of risk perception and communication.
Measure your learning by comparing pre-test and post-test results.

Summary Questions

1. Those who do not believe the threat is real are likely to continue their normal activities. True or False?
2. As perceived probability and magnitude increase, so do people's likelihood of taking protective action. True or False?
3. What is *not* one of the five basic risk communication functions to address in the continuing hazard phase?
   (a) Strategic analysis
   (b) Resource mobilization
   (c) Program development
   (d) Capability assessment
4. Most people do not believe the risks of technological facilities are greater than those of natural hazards. True or False?
5. The hazards that produce the greatest community conflict are those having a potential for inflicting significant harm on bystanders. True or False?
6. You cannot exert control over people's definition of a situation by defining a system for classifying threat levels. True or False?

Review Questions

1. What is a risk perception and why is it important?
2. What affects people's perception of risks?
3. What is a protective action? Give one example.
4. What three things need to happen for people to seek shelter?
5. What are three of the *perceived implementation barriers* inhibiting residents from taking protective action?
6. What are the components of a good, detailed adaptive plan?
7. Name the possible outcomes of communication action implementation.
8. Name the eight warning stages and actions of a communication action implementation plan.

9. What are the five basic risk communication functions you should address in the continuing hazard phase?
10. When analyzing the community, what do you need to know about the community?
11. What are the five tasks to complete when performing an operational analysis?
12. What are three things you must do to develop specific messages for different groups?
13. One of the most important incident management actions is risk communication. What are the tasks of risk communication?

Applying This Chapter

1. Local residents have ignored your community's tornado hazard because they think they are protected by a large hill west of town. What information sources can you use to tell community residents that the hill won't protect them and that they need to take action to protect themselves?
2. Your county health department has found some traces of a toxic chemical in wells located near a pesticide factory. The chemical plant manager, who has been very cooperative with local government in the past, doesn't think the chemical is coming from his plant. Should you release information about possible chemical contamination now? If not now, when (if ever)?
3. You have been asked by your city manager to evaluate the community's risk communication program. The city council doesn't think it's necessary to spend money evaluating the program. How would you convince them of the importance of evaluating the communication program?

YOU TRY IT

Risk Communication

You know a Category 5 hurricane is scheduled to make landfall in your city. How do you effectively communicate the risk?

Evacuation: Tailor Your Message

Knowing that many who did not evacuate New Orleans were poor and geographically concentrated in low-lying areas, how would you tailor your message? What concerns would the poor have that you would need to address?

Maintaining Credibility

How can you maintain credibility with the public and the media during a crisis? What steps can you take to ensure your own personal performance does not distort your message?

5

PRINCIPAL HAZARDS IN THE UNITED STATES

Causes and Effects

Starting Point

Go to www.wiley.com/college/lindell to assess your knowledge of the basic hazards in the United States.
Determine where you need to concentrate your effort.

What You'll Learn in This Chapter

▲ Types of hydrologic hazards and their characteristics
▲ Types of geophysical hazards and their characteristics
▲ Types of biological hazards and their characteristics
▲ Types of technological hazards and their characteristics
▲ The causes and effects of different types of hazards
▲ The risks of working with hazardous substances

After Studying This Chapter, You'll Be Able To

▲ Differentiate among different types of hazards
▲ Demonstrate how to prepare for different types of hazards
▲ Analyze the risks of hazardous substances
▲ Demonstrate how to work with others to reduce the risks of working with hazardous substances

Goals and Outcomes

▲ Organize hazards into categories
▲ Formulate a plan for handling hazardous substances
▲ Manage the risks associated with hazards
▲ Create relationships with others to prepare for different types of hazards

INTRODUCTION

To manage hazards, you must first understand them. Each hazard has a different cause and effect, and every type of hazard has a distinct set of characteristics. You must understand their causes and effects, as well as the distinctive characteristics of each hazard.

Environmental hazards are commonly classified as natural or technological.

- ▲ **Natural hazards** are extreme events that originate in nature. The natural hazards are commonly categorized as meteorological, hydrological, or geophysical.
- ▲ **Technological hazards** originate in human-controlled processes but are released into the air and water.

The most important technological hazards include explosives, flammable materials, toxic chemicals, radiological materials, and biological hazards. Sometimes terrorists deliberately release technological hazards to meet political objectives. Whether a hazardous release is accidental or deliberate might make a difference in the magnitude of the impact but not the types of impacts emergency managers must confront. You must know how to confront and deal with all types of hazards.

This chapter discusses various types of hazards: meteorological, hydrological, geophysical, and technological. It also examines the risks and the effects and describes how to deal with them. As is the case with other emergencies, you must know how to work with others to deal with hazards. This chapter looks at how to do this when dealing with hazards.

5.1 Meteorological Hazards

The main meteorological hazards are severe storms (including blizzards), severe summer weather, tornadoes, hurricanes, and wildfires.

5.1.1 Severe Storms

The National Weather Service (NWS) defines a **severe storm** as one that has wind speeds exceeding 58 mph, that produces a tornado, or that releases hail with a diameter of three-quarters of an inch or greater. The threats from severe storms are:

- ▲ Lightning strikes
- ▲ Downbusts and microbursts
- ▲ Hail
- ▲ Flash floods

Lightning can cause casualties. However, casualties are rare and are easily handled by local emergency medical services units. The bigger threat is that lightning strikes can initiate wildfires that threaten entire communities. This is especially true during droughts. Downbursts (up to 125 mph) and microbursts (up to 150 mph) are threats to aircraft as they take off or land. This creates a potential for mass casualty incidents. Generally, hail, even large hail, causes few casualties. Hail damage rarely causes significant disruption. Flooding can cause casualties and property damage.

Severe Winter Weather

Severe winter storms pose a greater threat than those at other times of the year. A severe winter storm is classified as a blizzard if its wind speed exceeds 38 mph and its temperature is less than 21°F. These conditions produce significant wind chill effects. Severe winter storms can:

- ▲ Immobilize travel
- ▲ Isolate residents of remote areas
- ▲ Deposit enormous amounts of snow on roofs that collapse long-span roofs of gymnasiums, theaters, and arenas
- ▲ Bring down telephone and electric power lines

The likelihood of winter storms is greatest in the northern states. However, winter storms can be extremely disruptive farther south where cities have less snow removal equipment.

Extreme Summer Weather

On the opposite extreme, heat can be a silent killer. Heat-related illnesses are:

- ▲ Heat cramp. The least serious condition, characterized by mild fluid and electrolyte imbalances.
- ▲ Heat syncope. A condition in which there is a sudden loss of consciousness that disappears when the victim lies down.
- ▲ Heat exhaustion. A condition of weakness or dizziness.
- ▲ Heat stroke. A condition in which the victim might be delirious or comatose. Unless treated effectively by rapid cooling, heat stroke can produce neurological damage and fatalities in about 15% of those affected.

Temperature and humidity are combined into a heat index of **apparent temperature.** Apparent temperatures of 80–90°F warrant *caution* because exposure and physical activity can cause fatigue. Take *extreme caution* once 90–105°F is reached because exposure and physical activity can cause heat cramp and heat exhaustion. *Danger* exists from 105–130°F because exposure and physical activity can cause heat stroke. *Extreme danger* exists at 130°F because heat stroke will occur.

Citizens in the Southwest deserts, Mississippi Valley, and southern states are the most prone to heat-related illnesses. The groups at greatest risk are outdoor workers, the old, the very young, and the ill. Heat is a problem in the cities where buildings reradiate sunlight and block the wind. People without air conditioning have the greatest exposure if they live in high crime areas and are afraid to open the windows for fans.

5.1.2 Tornadoes

Tornadoes form when cold air from the north collides with a warmer air mass. The cold air descends because of its greater weight and is replaced by rising warm air. This process initiates rotational flow inside the air mass. As the tornado forms, pressure drops inside the vortex, and the wind speed increases. The resulting high wind speed can destroy buildings, vehicles, and large trees. The resulting debris becomes part of the wind field, which adds to the tornado's destructive power. There are approximately 900 tornadoes each year in the United States. Most tornadoes strike Texas, Oklahoma, Arkansas, Missouri, and Kansas. There is also significant vulnerability in the North Central states and the Southeast from Louisiana to Florida. Tornadoes are most common during the spring, with the months of April, May, and June accounting for 50% of all tornadoes. There is also predictable daily variation, with four to eight o'clock in the afternoon being the most frequent time of impact. Tornadoes have distinct directional tendencies as well, most frequently traveling northeast (54%), east (22%), and southeast (11%). Only 8% travel north, 2% travel northwest, and 1% travel west, southwest, or south. There also is a tendency for tornadoes to follow low terrain. A tornado's forward speed (the speed of the funnel over the ground) can range from 0–60 mph but is usually about 30 mph. Tornadoes vary substantially in intensity and this attribute is characterized by the Fujita scale, which has a low end of F0 (maximum wind speed of 40 mph) and a high end of F5 (maximum wind speed of 315 mph). However, only about one-third of all tornadoes exceed F2 (111 mph). The impact area of a typical tornado is 4 miles in length but has been as much as 150 miles. Many structures in a stricken community receive only moderate or minor damage. This is because 90% of the impact area is affected by a wind speed of less than 112 mph. Only about 3% of tornadoes cause deaths. Half of these deaths are residents of mobile homes because these structures are substantially less sturdy than site-built homes.

An increased number of tornadoes have been reported during recent years. This is due partly to improved radar and spotter networks. However, tornadoes have been observed in locations where they have not previously been seen. This suggests that some long-term changes in climate are also involved. Detection is usually achieved by trained meteorologists observing characteristic clues on Doppler radar. Over the years, warning speed has been improved by the National Oceanic & Atmospheric Administration (NOAA) weather radio, which provides timely and specific warnings. Those who do not receive a warning can assess their danger from a tornado's distinct physical cues:

- ▲ Dark, heavy cumulonimbus clouds.
- ▲ Intense lightning.
- ▲ Hail and downpour of rain immediately to the left of the tornado path.
- ▲ Noise like a train or jet engine.

The most appropriate protective action to take is to go to a specially constructed safe room (FEMA, 1998). If a safe room is not available, building occupants should go to an interior room on the lowest floor. Mobile home residents should go to a community shelter. Those who are outside should seek refuge in a low spot (e.g., a small ditch or depression) if shelter is unavailable.

5.1.3 Hurricanes

As we have seen with Hurricane Katrina, a **hurricane** is the most severe type of tropical storm. The earliest stages of hurricanes are marked by thunderstorms that intensify through a series of stages: tropical wave, tropical disturbance, tropical depression, and tropical storm. Few of these storms escalate to a major hurricane. In an average year, there are 100 tropical disturbances, 10 tropical storms, 6 hurricanes, and only 2 of these hurricanes strike the US coast. Hurricanes in Categories 3–5 account for 20% of landfalls, but over 80% of damage. During the 20th century, only three Atlantic hurricanes were Category 5 at the time they made landfall. Hurricane season begins the first of June, reaches its peak during September, and usually decreases through the end of November. The 2005 hurricane season was unusual for its large number of hurricanes and its duration into the month of December.

Tropical storms draw their energy from warm seawater, so they form only when the sea surface temperature exceeds 80°F. For most Atlantic hurricanes, this takes place in tropical water off the West African coast. The storms are generated when surface water absorbs heat and evaporates, and the resulting water vapor rises to higher altitudes. When it condenses there, it releases rain and latent heat of evaporation. An easterly steering wind, which blows from east to west, pushes these storms westward across the Atlantic.

Hurricanes have a definite structure.

- ▲ The **eye of the hurricane** is an area of calm conditions that has a 10–20 mile radius, the eye is surrounded by bands of high wind and rain.

▲ The **eyewall** is the bands of high wind and rain that spiral and form a ring around the eye.

The entire hurricane, which can be as much as 600 miles in diameter, rotates counterclockwise in the Northern Hemisphere. This produces a storm surge that is located in the *right front quadrant* relative to the storm track. Hurricanes have a forward movement speed averaging about 12 mph, but any hurricane can be faster or slower. Each hurricane's speed can vary over time and even stall for an extended period of time. Atlantic hurricanes tend to track toward the west and north. Yet, they can change direction. Storm intensity weakens as it reaches the North Atlantic. This is because it derives less energy from the cooler water at high latitudes. Hurricanes also weaken when they make landfall. Landfall cuts the storm off from its source of energy and adds the friction of the interaction with the rough land surface.

Hurricanes produce four specific threats:

▲ High winds
▲ Tornadoes
▲ Inland flooding
▲ Storm surge

The strength of the wind can be seen in the third column of Table 5-1. This table shows that the pressure of the wind on vegetation and structures is proportional to the square of the wind speed. For example, as the wind speed *doubles* from 80 mph in a Category 1 hurricane to 160 mph in a Category 5 hurricane, the velocity pressure *quadruples* from less than 20 pounds per square foot (psf) to over 80 psf. Damage from high wind and the debris that is entrained in the wind field is a function of a structure's exposure. Wind exposure is highest in areas directly downwind from open water or fields. Upwind hills, woodlands, and tall buildings decrease exposure to the direct force of the wind. However, they increase exposure to flying debris. Flying debris includes tree branches and building materials that have been torn from their sources.

Storm clouds in the outer bands of a hurricane can sometimes produce tornadoes that are mostly small and short-lived. Hurricanes can also produce torrential rain at rates up to 4 inches per hour for short periods of time. One U.S. hurricane produced 23 inches of rain over a 24-hour period. Such downpours cause severe local ponding (water that falls and doesn't move because the land is so flat) and inland flooding.

Until Hurricane Katrina, hurricane disasters have resulted in relatively few casualties in the U.S. since the early part of the twentieth century. The worst hurricane disaster occurred in Galveston, Texas, in 1900 when over 6000 lives were lost in a community of about 18,000. However, coastal counties have

Table 5-1: Saffir-Simpson Hurricane Categories

Saffir-Simpson category	*Wind speed (mph)*	*Velocity pressure (psf)*	*Storm surge (feet)*	*Wind effects*
1	74–95	19.0	4–5	• Vegetation: some damage to foliage • Street signs: minimal damage • Mobile homes: some damage to unanchored structures • Other buildings: little or no damage
2	96–110	30.6	6–8	• Vegetation: much damage to foliage; some trees blown down • Street signs: extensive damage to poorly constructed signs • Mobile homes: major damage to unanchored structures • Other buildings: some damage to roof materials, doors, and windows
3	111–130	41.0	9–12	• Vegetation: major damage to foliage; large trees blown down • Street signs: almost all poorly constructed signs blown away • Mobile homes: destroyed • Other buildings: some structural damage to small buildings
4	131–155	57.2	13–18	• Street signs: all down • Other buildings: extensive damage to roof materials, doors, and windows; many residential roof failures
5	>155	81.3	>18	• Other buildings: some complete building failures

experienced explosive population growth in recent decades, which creates the potential for another catastrophic loss of life. Moreover, economic losses are increasing substantially over time. Inflation makes only a small contribution; most of the increase is due to increased population in vulnerable areas and increased wealth (per person) in those areas (Pielke and Landsea, 1998). There is extreme variation in losses by decade, due to variability in the number of storms. For example, the two decades from 1950 to 1969 experienced 33 hurricanes, whereas the equivalent period from 1970 to 1988 experienced only 6 hurricanes.

Hurricanes are rapidly detected by satellite and continually monitored by specially equipped aircraft. Storm forecast models have been developed that provide increasing accuracy in the prediction of the storm track. Nonetheless, there are forecast uncertainties about the eventual location of landfall, as well as the storm's size, intensity, forward movement speed, and rainfall. One of the biggest problems is that the time required to evacuate some urbanized areas requires warnings to be issued at a time when storm behavior remains uncertain. Many coastal cities take 36 hours or more to evacuate. As you can see from the strike probability data in Table 5-2, these cities must begin evacuating even though the storm has only a 20–25% chance of hitting them. Moreover, there is significant uncertainty about the wind speed, and thus, the inland distance that must be evacuated.

Appropriate protective actions for hurricanes are well understood. Within the storm surge/high wind field risk areas, shelter in-place is recommended only for elevated portions (i.e., above the wave crests) of reinforced concrete buildings having foundations anchored well below the scour line. The scour line is the level at which wave action erodes the soil on which the building rests. Authorities recommend that evacuation be completed before routes are flooded or tropical storm force winds (39 mph) tip motor homes and other vehicles (see Figure 5-1). Outside storm surge risk areas, shelter in-place is suitable for most permanent buildings with solid construction. However, debris sources should be controlled. Temporary shutters (or at least plastic film) should be installed on windows. Evacuation is advisable for residents of mobile homes in high wind zones.

Table 5-2: Uncertainties about Hurricane Conditions Before Landfall

Forecast period (hours)	*Absolute landfall error (nautical miles)*	*Maximum probability*	*Miss/hit ratio*	*Average wind speed error (mph)*
72	>200	10%	9 to 1	23
48	150	13–18%	7 to 1	18
36	100	20–25%	4 to 1	15
24	75	35–50%	2 to 1	12
12	50	60–80%	2/3 to 1	9

Figure 5-1

As we saw with Hurricane Katrina, hurricanes can cause massive flooding and destruction.

5.1.4 Wildfires

All fires require the three elements of the fire triangle: *fuel*, which is any substance that burns; *oxygen* that combines with the fuel; and enough *heat* to ignite fuel. The resulting combustion yields heat (sustaining the reaction) and combustion products such as toxic gases and unburned particles of fuel that are visible as smoke. Wildfires are distinguished mostly by their fuel.

- **Wildland fires** burn areas with nothing but natural vegetation for fuel.
- **Interface fires** burn into areas containing a mixture of natural vegetation and built structures.
- **Firestorms** are distinguished from other wildfires because they burn so intensely that they warrant a special category.

Firestorms actually create their own local weather and are virtually impossible to extinguish. Wildfires can occur almost anywhere in the United States but are most common in the arid West where there are extensive stands of conifer trees and brush that serve as ready fuels. Once a fire starts, the three principal variables determining its severity are fuel, weather, and topography.

Fuels differ in a number of characteristics that collectively define fuel type. These include the fuel's ignition temperature (low is more dangerous), amount of moisture (dry is more dangerous), and the amount of energy (resinous wood is more dangerous). A given geographical area can be defined by its fuel loading, which is the quantity of vegetation in tons per acre, and fuel continuity, which refers to the proximity of individual elements of fuel. Topography affects fire behavior by directing the hot air produced by the fire as well as the larger wind currents of the prevailing wind. Canyons can accelerate the wind by funneling it through narrow openings. A fire's forward movement speed doubles on a 10 degree slope and quadruples on a 20 degree slope. Weather affects fire behavior by wind speed and direction as well as temperature and humidity. Wind speed and direction have the most obvious effects on fire behavior. Strong winds push the fire front forward and carry burning embers far in advance of the main front. High temperature and low humidity promote fires by decreasing fuel moisture.

Wildland fires are a major problem in the United States. An average of about 73,000 such fires per year burn over three million acres. Approximately 13% of these wildfires are caused by lightning, but humans cause 24% of them accidentally. Humans also cause 26% of the fires deliberately. In California's 1991 Oakland Hills Fire, 25 people were killed and 150 injured. Over 3,000 homes were damaged or destroyed. Major contributors to the severity of this interface fire were the housing materials (predominantly wood siding and wood shingle roofs), vegetation planted immediately adjacent to the houses, and narrow winding roads that made access for fire fighting equipment difficult.

The US Forest Service maintains a Fire Danger Rating System that monitors changing weather and fuel conditions throughout the summer fire season. Some of the fuel data are derived from satellite observations. The weather data come

FOR EXAMPLE

Hurricane Katrina

Despite the fact that Hurricane Katrina was a Category 4 hurricane and not the Category 5 that was predicted, it was still the most expensive natural disaster in U.S. history. The impact of Katrina wiped coastal communities in Mississippi and Louisiana off the map. It also caused levee failure in New Orleans, covering 80% of the city with water. The flood water rose so quickly that hundreds were trapped in their homes and drowned. Some people who took shelter in their attics waited for days on their roofs to be rescued. Biloxi mayor A.J. Holloway, noting that his town was destroyed as well, said, "This is our tsunami."

from hundreds of weather stations. Appropriate protective actions include evacuation out of the risk area, evacuating to a safe location (e.g., an open space such as a park or baseball field having well-watered grass that will not burn), and sheltering in-place within a fire-resistant structure (e.g., a concrete building with no nearby vegetation).

SELF-CHECK

- Define natural hazards.
- List elements required for a fire.
- Describe the conditions under which tornadoes form.
- Describe how hurricanes are categorized.

5.2 Hydrological Hazards

The principal hydrological hazards of concern to environmental hazard managers are floods, storm surges, and tsunamis.

5.2.1 Floods

A **flood** is an event in which abnormally large amount of water accumulates in an area in which it is usually not found. Flooding is a widespread problem in the United States that accounts for three-quarters of all presidential disaster declarations. Flooding is determined by a hydrological cycle in which precipitation falls from clouds in the form of rain and snow (see Figure 5-2). When it reaches the ground, the precipitation either soaks into the soil or travels downhill as surface runoff. Some of the water that infiltrates the soil is taken up by plant roots and transported to the leaves where it is transpired into the atmosphere. Another portion of the ground water gradually moves down to the water table and flows underground until reaching water bodies such as wetlands, rivers, lakes, or the ocean. Surface runoff moves directly to surface storage in these water bodies. At that point, water evaporates from surface storage, returning to clouds in the atmosphere.

There are seven different types of flooding that are widely recognized.

1. **Riverine** (main stem) flooding occurs when surface runoff gradually rises to flood stage (overflows natural banks) and later falls.
2. **Flash flooding** occurs when runoff reaches its peak in less than six hours, which usually occurs in hilly areas with steep slopes and sparse vegetation.

Figure 5-2

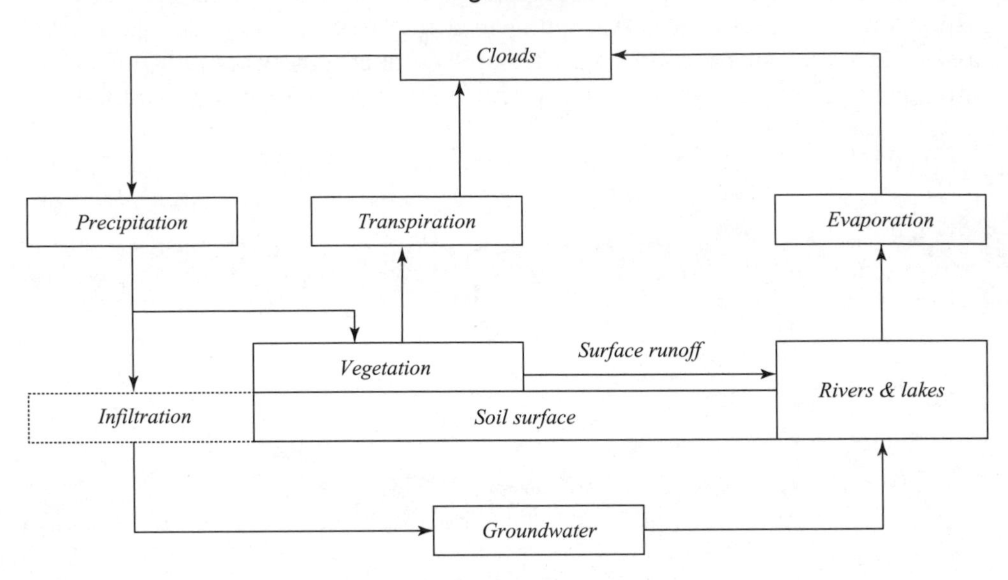

The hydrological cycle.

It also can occur in urbanized areas with rapid runoff from impermeable surfaces such as streets, parking lots, and building roofs.

3. **Alluvial fan flooding** occurs in deposits of soil and rock found at the foot of steep valley walls in arid Western regions.
4. **Ice/debris dam failures** result when an accumulation of material temporarily blocks the flow of water and raises its surface above the stream bank before giving way.
5. **Surface ponding** occurs when water accumulates in areas so flat that runoff cannot carry away the precipitation fast enough.
6. **Fluctuating lake levels** can occur over short-term, seasonal, or multi-year periods, especially in lakes that have limited outlets or are entirely landlocked.
7. **Control structure (dam or levee) failure,** has many characteristics in common with flash flooding.

Floods are measured either by discharge or stage.

- ▲ **Discharge** is defined as the volume of water per unit of time.
- ▲ **Stage,** the height of water above a defined level, is the unit needed by emergency managers because flood stage determines the level of casualties and damage.

Discharge is converted to stage by means of a rating curve (see Figure 5-3). The horizontal axis shows discharge in cubic feet per second and the vertical axis shows stage in feet above flood stage. Note that high rates of discharge produce much higher stages in a valley than on a plain because the valley walls confine the water.

Flooding is affected by a number of factors. The first of these, precipitation, must be considered at a given point and also across the entire watershed (basin). The total precipitation at a point is equal to the duration of precipitation times its intensity. This is frequently measured in inches per hour. Total precipitation over a basin is equal to precipitation summed over all points in the surface area of the basin. The precipitation's contribution to flooding is a function of temperature. Rain is immediately available whereas snow must first be melted by warm air or rain. The precipitation from a single storm might be deposited over two or more basins and the amount of rainfall in one basin might be quite different from that in the other basin (see Figure 5-4). Consequently, there might be severe flooding in a town on one river (City A) and none at all in a town on another river (City B) even if the two towns received the same amount of rainfall from a storm.

As the hydrological cycle demonstrates, flooding is also affected by surface runoff. Runoff is determined by terrain and soil cover. One important aspect of terrain is its slope. Runoff increases as slope increases. In addition to slope steepness, slope length and orientation to prevailing wind and sun are also determinants of flooding. Slope geometry is also an important consideration. Convergent slopes (shaped like a letter U) provide runoff storage in puddles, potholes, and ponds. By contrast, divergent slopes (shaped like an inverted U) provide rapid runoff dispersion. Mixed slopes have combinations of these.

Figure 5-3

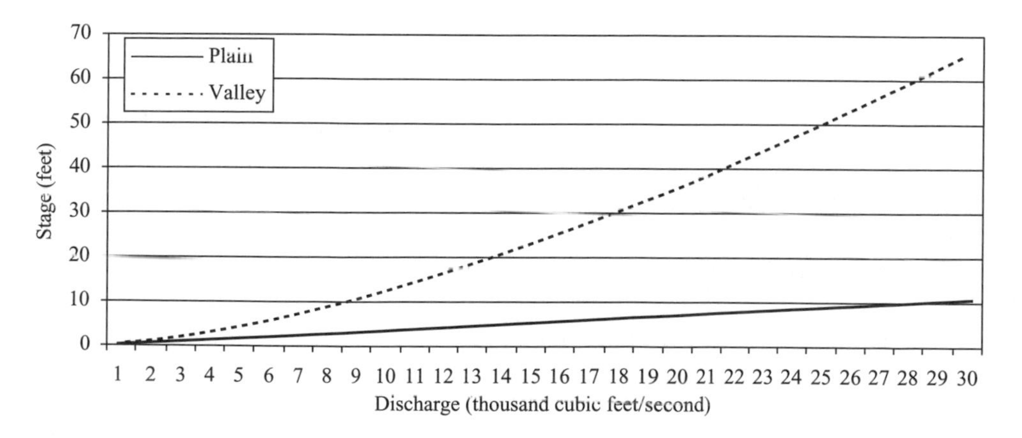

Stage Rating Curve.

Figure 5-4

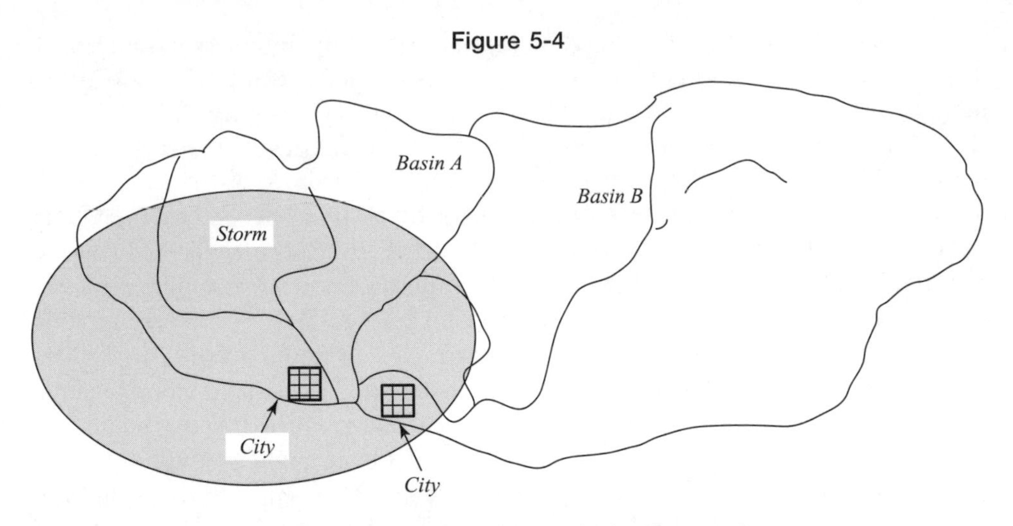

Map of the distribution of precipitation from a storm.

Slope mean and variance determine the amount of storage. A slope with a zero mean and high variance (a plain with many potholes) provides a larger amount of storage than a slope with a zero mean and low variance (a featureless plain). A slope with a positive mean and high variance (a slope with many potholes) provides a larger amount of storage than a slope with a positive mean and low variance.

Soil cover also affects flooding because dense low plant growth slows runoff and promotes infiltration. In areas with limited vegetation, surface permeability is a major determinant of flooding. Surface permeability increases with the proportion of organic matter content because this material absorbs water like a sponge. Permeability is also affected by surface texture. Clay, stone, and concrete are very impermeable because particles are small and smooth. Gravel and sand are very permeable—especially when the particles are large and have irregular shapes that prevent them from compacting. Finally, surface permeability is affected by soil saturation because even permeable surfaces resist infiltration when soil pores (the spaces between soil particles that ordinarily are filled with air) become filled with water. Groundwater flows through local transport to streams at the foot of hill slopes and through remote transport through aquifers. Rapid in- and outflow through loosely compacted valley soil increases peak flows whereas very slow in- and outflow through upland areas maintains flows between rains.

Evapotranspiration takes place two ways. First, there is direct evaporation to the atmosphere from surface storage in rivers and lakes. Second, there is uptake from soil and subsequent transpiration by plants. Transpiration draws moisture from the soil into plants' roots, up through the stem, and out through the leaves' pores. The latter mechanism is generally much higher in summer than in winter due to increased heat and plant growth, but transpiration is negligible during periods of high precipitation.

Stream channel flow is affected by channel wetting, which infiltrates the stream banks (horizontally) until they are saturated as the water rises. In addition, there is seepage because porous channel bottoms allow water to infiltrate (vertically) into groundwater. Channel geometry also influences flow because a greater channel cross-section distributes the water over a greater area, as does the length of a *reach* (distinct section of river) because longer reaches provide greater water storage. High levels of discharge to downstream reaches can also affect flooding on upstream reaches because flooded downstream reaches slow flood transit by decreasing the elevation drop of the river.

Flooding increases when upstream areas experience deforestation and overgrazing, which increase surface runoff to a moderate degree on shallow slopes and to a major degree on steep slopes as the soil erodes. The sediment is washed downstream where it can silt the channel and raise the elevation of the river bottom. These problems of agricultural development are aggravated by flood plain urbanization. Like other cities throughout the world, American cities have been located in flood plains because the water was the most efficient means of transportation until the mid-1800s. Many cities were located at the head of navigation or at transshipment points between rivers. In addition, cities have been located in flood plains because level alluvial soil is very easy to excavate for building foundations. Finally, urban development takes place in flood plains because of the aesthetic attraction of water. People enjoy seeing lakes and rivers, and pay a premium for waterfront real estate.

One consequence of urban development for flooding is that cities involve the replacement of vegetation with *hardscape*. **Hardscape** is impermeable surfaces such as building roofs, streets, and parking lots. This hardscape decreases soil infiltration, thus increasing the speed at which flood crests rise and fall. Another factor increasing flooding is intrusion into the flood plain by developers who fill intermittently flooded areas with soil to raise the elevation of the land. This decreases the channel cross-section, forcing the river to rise in other areas to compensate for the lost space.

Flood risk areas in the US are generally defined by the **100-year flood,** an event that is expected to have a 100-year recurrence interval and, thus, a 1% chance of occurrence in any given year. It is important to understand these extreme events are essentially independent, so it is possible for a community to experience two 100-year floods in the same century. Indeed, it is possible to have them in the same year even though it is improbable. A 100-year flood is an arbitrary standard of safety that reflects a compromise between the goals of providing long-term safety and developing economically valuable land. A 50, 200, or even a 500-year standard could be used instead. Community adoption of a

50-year flood standard would provide more area for residential, commercial, and industrial development. The encroachment into the flood plain would lead to more frequent damaging floods than would a 100-year flood standard. Alternatively, a community might use different standards for different types of structures. For example, it might restrict the 100-year floodplain to low intensity uses (e.g., parks), allow residential housing to be constructed within the 500-year floodplain, and restrict nursing homes, hospitals, and schools to areas outside the 500-year floodplain.

There are three different types of automated devices that support prompt detection of imminent flooding. Radar can assess the amount of rainfall at any point in a watershed. Rain gages detect rainfall amounts at predetermined points in a watershed. Stream gages detect water depth at predetermined points along a river. Manual devices can also detect floods. Spotters assess rainfall amounts, water depth, or levee integrity at specific locations. After data on the quantity and distribution of precipitation have been collected, they are used to estimate discharge volumes over time from the runoff characteristics of a given watershed at a given time. Discharge volume is used with downstream topography to predict downstream flood heights.

Timely and specific warnings of floods are provided by commercial news media as well as NOAA Weather Radio All Hazards. The most appropriate protective action for persons is to evacuate in a direction perpendicular to the river channel. Flash floods in mountain canyons can travel faster than a motor vehicle so it is safest to climb a canyon wall rather than try to drive out. It also is important to avoid crossing running water. Just two feet of fast moving water can float a car and push it downstream with 1000 pounds of force.

5.2.2 Storm Surge

Storm surge is an increased height of a body of water that exceeds the normal tide. This is most commonly associated with hurricanes but also can be caused by extratropical cyclones. As Table 5-1 indicates, the height of a storm surge increases as the hurricane category increases. This is mostly because of the storm wind pushing the water forward toward the coast. Storm surge is highest where coastal topography and bathymetry (submarine topography) have shallow slopes.

At one time, storm surge was the primary source of casualties in all countries, but inland flooding is now the primary cause of hurricane deaths in the U.S. In part, this is because analysts use sophisticated computer programs to identify the areas that are most likely to be inundated by storm surge. This allows local emergency managers to identify the areas that need to be evacuated. People can safely stay in the surge inundation zone only if they seek refuge in steel reinforced concrete structures on deep pilings. Consequently, evacuating inland to higher ground is the safest protective action.

5.2.3 Tsunamis

Tsunamis are sea waves that are usually generated by undersea earthquakes, but volcanic eruptions or landslides can also cause tsunamis. Tsunamis are rare events. Over the course of a century, 15,000 earthquakes generated only 124 tsunamis, a rate of less than 1% of all earthquakes and only .7 tsunamis per year. This low rate of tsunami generation is related to earthquake intensity. Two-thirds of all Pacific tsunamis are generated by shallow earthquakes exceeding 7.5 in magnitude.

Tsunamis can travel across thousands of miles of ocean at speeds up to 400 mph. They slow to 25 mph as they begin to break in shallow water and run up onto the land. Tsunamis are largely invisible in the open ocean because they are only 1-2 feet high. However, they have a wave length up to 60 miles and a period as great as one hour. Compare this to ordinary ocean waves that have wave heights up to 30 feet, wave lengths of about 500 feet, and a period of about 10 seconds. Tsunami waves encounter bottom friction when the water depth is less than 1/20 of their wavelength. Then the bottom of the wave front slows and is overtaken by the rest of the wave, which must rise over it. For example, when a wave reaches a depth of 330 feet, its speed is reduced from 400 mph to 60 mph. Later, reaching a depth of 154 feet reduces its speed to 44 mph. This causes the next 650 feet of the wave to overtake the wave front in a single second. As the wave continues shoreward, each succeeding segment of the wave must rise above the previous segment. It can't go down because water is not compressible and it can't go back because the rest of the wave is pressing it forward. Because the wavelength is so long and wave speed is so fast, a large volume of water can pile up to a very great height. This phenomenon is most likely where the continental shelf is very narrow. It is important to note that the initial clue to tsunami arrival might be that the water level drops, rather than rises. An initial *wave* is created if the seafloor rises suddenly, whereas an initial *trough* is created if the seafloor drops. In either case, the initial phase is followed by the alternate phase.

Tsunamis threaten shorelines worldwide. They do not occur at any specific time of day or in any specific time of year. International tsunami warning systems base their initial detection on seismic monitoring to detect major earthquakes. Then tidal gauges located throughout the Pacific basin verify the generation of a tsunami. This warning system can alert people many hours before a remote tsunami strikes. For example, the 1960 Chilean earthquake generated a tsunami that took 15 hours to reach Hawaii. However, the situation is much different for a locally generated tsunami. A tsunami generated 100 miles away from the coast will arrive in 15 minutes, so people in coastal communities should take prompt action if they feel severe earthquake shaking. In some cases, the only warning is the arrival of a trough (making it appear that the tide went out unexpectedly). That should also be a cue for

FOR EXAMPLE

Tsunami Detection

The devastating Indian Ocean tsunami of 2004 killed approximately 225,000 people. One group that survived, however, was the Moken. These people live on the Andaman Islands. The Moken know the sea intimately, so they knew a tsunami was coming when they saw the sea receding. They also watched the behavior of the animals, such as pools of dolphins swimming for deeper water. The Moken have passed down stories of tsunamis for generations. In their folklore, they talk about how the angry sea eats people. These factors led the Moken to run for higher ground, warning nearby tourists in the process.

prompt action. If you wait until you see a wave arriving, it probably will be too late.

The physical magnitude of a tsunami is extremely impressive. Wave crests can arrive at 10 to 45 minute intervals for up to six hours and the highest wave, as much as 100 feet at the shoreline, can be anywhere in the wave train. The runup zone, which is the area that is inundated by a tsunami, is a concept that is quite similar to the 100-year floodplain. Because of the complexities in wave behavior, and the offshore water depth and onshore topography, runup zones must be calculated by competent analysts using sophisticated computer programs. The physical impacts include deaths from drowning and traumatic injuries from wave impact. Property damage is caused by the same mechanisms.

As is the case with storm surge, steel reinforced concrete structures on deep pilings are required to withstand wave battering and foundation scour. Consequently, evacuation to higher ground is the most effective method of protection. People must evacuate a long distance if they live on low-lying coasts. Evacuation can also be difficult even where there are nearby hills if the primary evacuation route runs parallel to the coastline.

SELF-CHECK

- Define **discharge and stage.**
- List factors that affect flooding.
- Identify how often tsunamis occur.

5.3 Geophysical Hazards

The earth has three distinct geological components.

- ▲ The **core** consists of molten rock at the center of the earth.
- ▲ The **crust** is solid rock and other materials at the earth's surface that vary in depth from 4 miles under the oceans to 40 miles in the Himalayas.
- ▲ The **mantle** is an 1800 mile thick layer between the core and the crust. The earth's crust is defined by large plates that float on the mantle and move gradually in different directions over time.

Tectonic plates diverge, converge, or move laterally past each other. When they diverge, new material is generated from below the earth's mantle, usually at mid-ocean ridges, that flows at a rate of a few inches per year away from the source. This process produces a gradual expansion of the plate toward an adjoining plate. Thus, one plate converges with another plate and the heavier material (a seafloor) is subducted under lighter material (a continent). In the United States, this process is taking place in the Cascadian Zone along the Pacific coast of Washington, Oregon, and Northern California. Tectonic activity produces movement that causes earthquakes and tsunamis. These tectonic processes cause the most important geophysical hazards in the United States—volcanic eruptions and earthquakes.

5.3.1 Volcanic Eruptions

Volcanoes are formed when a column of magma (molten rock) rises from the earth's mantle into a magma chamber and erupts at the surface, where it is called lava. Successive eruptions, deposited in layers of lava or ash, build a mountain. Major eruptions create craters that are gradually replaced in dome-building eruptions. Cataclysmic eruptions leave only a depression where the mountain once stood. American volcanoes (recently erupted) are located in Alaska (92) and Hawaii (21), and the west coast. Oregon has 22 volcanoes, California has 20, and Washington has 8. Vulcanologists distinguish among 20 different types of volcanoes, but the two most important are shield volcanoes and stratovolcanoes. Shield volcanoes produce relatively gentle *effusive* eruptions of low-viscosity lava, resulting in shallow slopes and broad bases (e.g., Kilauea in Hawaii) whereas stratovolcanoes produce *explosive* eruptions of highly acidic lava, gas, and ash, resulting in steep slopes and narrow bases (e.g., Mt. St. Helens in Washington).

Threats from volcanoes include lightweight gases and ash that are blasted high into the air and the heavier lava and mud that travel downslope. There are simple asphyxiants (carbon dioxide and methane) that are dangerous because they displace atmospheric oxygen. There also are chemical asphyxiants (carbon monoxide, CO) that are dangerous because they prevent the oxygen that people breathe from reaching their bodies' tissues. In addition, there are corrosives (sulfur dioxide, hydrogen sulfide, hydrogen chloride, hydrogen fluoride, and sulfuric acid) and radioactive gases such as radon. Pyroclastic flows are hot gas and ash mixtures (up to 1600°F) discharged from the crater vent. Tephra consists of solid particles of rock ranging in size from talcum powder ("ash") to boulders ("bombs"). Lahars are mudflows and floods, usually from glacier snowmelt, with varying concentrations of ash. The impacts of volcanic eruption tend to be strongly directional because ashfall and gases disperse downwind. Ash flows follow blast direction and lava and lahars travel downslope through drainage basins. The forward movement speed of the hazard varies. Pyroclastic flows can move at more than 100 mph. Gas and tephra movements are determined by wind speed, usually less than 25 mph. Lava moves at walking speed (5 mph) but can travel faster (35 mph) on steep slopes. Lahars move at the speed of water flow, usually less than 25 mph but can exceed 50 mph.

The physical magnitude of the hazard also differs for each specific threat. Ash and lahars can range from six inches up to tens of meters in depth. Lava flows and ash flows are so hot that any impact is fatal. The impact area also varies by threat. Tephra deposition depends on eruption magnitude, wind speed, and particle size. After large eruptions, traces of ash have circled the globe. Lava flows, lahars, and pyroclastic flows follow localized drainage patterns. Safe locations can be found only a short distance from areas that are totally devastated. Volcano risk areas are described in Table 5-3.

The physical impacts of a volcanic eruption vary with the type of threat. Gases can cause casualties from inhalation exposure. Pyroclastic flows are more dangerous because they can cause casualties from blast, thermal exposure, and inhalation of gas and ash. In addition, they also can cause property damage from

Table 5-3: Volcano Risk Areas

Category	*Name*	*Distance**	*Threats*
1	Extreme	0–100 yards	High risk of heat, ash, lava, gases, rockfalls, and projectiles
2	High	100–300 yards	High risk of projectiles
3	Medium	300–3000 yards	Medium risk of projectiles
4	Low	2–6 miles	Low risk of projectiles
5	Safe	>6 miles	Minimal risk of projectiles

*Does not include mudflows and floods that can travel up to 100 kilometers or tsunamis that can travel thousands of kilometers.

blast, heat, and coverage by ash (even after the ash has cooled). Tephra causes property damage from excess weight collapsing roofs, shorting of electric circuits, clogged air filters in vehicles, and abrasion of machinery. Deaths and injuries can be caused by bomb impact trauma, and health effects can result from ash inhalation. Lava causes property damage from excess heat and coverage by rock (when cooled). Thermal exposure to lava can cause casualties, but these are rare because lava moves so slowly. Lahars can cause property damage from flooding and coverage by ash (when water drains off) and deaths from drowning. In addition, volcanic eruptions can cause tsunamis and wildfires as secondary hazards.

The threat of volcanic eruption can be detected by physical cues indicating rising magma. These include earthquake swarms, outgassing, ash and steam eruptions, and changes in ground slope. Protective measures include sweeping ash from building roofs and evacuating an area at least 6 miles in radius for a crater eruption and 12–18 miles in the direction of a flank/lateral eruption. People should be evacuated from floodplains threatened by lahars. There are substantial uncertainties in the timing of eruptions, so people can be forced to stay away from their homes and businesses for months after evacuating. In some cases, the expected eruption never materializes, causing severe conflict among scientists, local civil officials, and disrupted residents.

5.3.2 Earthquakes

When an **earthquake** occurs, energy is released at the hypocenter.

- ▲ The **hypocenter** is a point deep within the earth. However, the location of an earthquake is usually identified by the epicenter.
- ▲ The **epicenter** is a point on the earth's surface directly above the hypocenter.

Earthquake energy is carried by three different types of waves:

- ▲ P-waves, primary or pressure waves, travel rapidly.
- ▲ S-waves, secondary or shear waves, travel more slowly but cause more damage.
- ▲ Surface waves have low frequency and are damaging to tall buildings.

The physical magnitude of an earthquake is different from its intensity.

- ▲ **Magnitude** is measured on the Richter scale where a one-unit increase represents a 10-fold increase in seismic wave amplitude and a 30-fold increase in energy release from the source.
- ▲ **Intensity** measures the impact at a given location and can be assessed either by behavioral effects or physical measurements.

The behavioral effects of earthquakes are classified by the Modified Mercalli Intensity scale (see Table 5-4). This scale defines each category in terms of the effects of earthquake motion on people and the physical environment. Physical measurements describe seismic forces in horizontal and vertical directions either

Table 5-4: Modified Mercalli Intensity (MMI) Scale for Earthquake

Category	*Intensity*	*Type of damage*	*Max. acceleration* (mm/sec^{-2})
I	Instrumental	Detected only on seismographs	<10
II	Feeble	Some people feel it	<25
III	Slight	Felt by people resting; like a large truck rumbling by	<50
IV	Moderate	Felt by people walking; loose objects rattle on shelves	<100
V	Slightly strong	Sleepers awake; church bells ring	<250
VI	Strong	Trees sway; suspended objects swing; objects fall off shelves	<500
VII	Very strong	Mild alarm; walls crack; plaster falls	<1000
VIII	Destructive	Moving cars uncontrollable; chimneys fall and masonry fractures; poorly constructed buildings damaged	<2500
IX	Ruinous	Some houses collapse; ground cracks; pipes break open	<5000
X	Disastrous	Ground cracks profusely; many buildings destroyed; liquefaction and landslides widespread	<7500
XI	Very disastrous	Most buildings and bridges collapse; roads, railways, pipes and cables destroyed; general triggering of other hazards	<9800
XII	Catastrophic	Total destruction; trees driven from ground; ground rises and falls in waves	>9800

as the number of millimeters per second squared (mm/sec^2), or as a multiple of the force of gravity ($g = 9.8$ meters/sec^2).

The intensity of an earthquake at a given point is determined by a number of factors. First, earthquake intensity decreases with distance from the epicenter, just as dropping a rock into a still pool causes large waves at the source that decrease in size to small ripples farther away. Second, basins (loose fill surrounded by rock) focus energy waves as they reflect off the surrounding rock, just as waves reflect off the walls of a swimming pool. Third, soft soil transmits energy waves much more readily than bedrock, just as pushing a plate of Jell-O will cause the Jell-O to shake even after the plate has come to rest. The result is that earthquake intensity can be very different at two points that are equidistant, but in different directions, from the epicenter.

Within the impact area, the primary threats are ground shaking, surface faulting, and ground failure (see Figure 5-5). Ground shaking is a hazard because it

Figure 5-5

Earthquakes are a type of geophysical hazard.

creates lateral (sideways) and upward motion in structures designed only for gravity (downward) loads. In addition, nonreinforced structures respond poorly to tensile (upward stretching) and shear (lateral) forces, as do "soft-story" (e.g., buildings with pillars rather than walls on the ground floor) and asymmetric (e.g., L-shaped) structures. High-rises can demonstrate resonance, a tendency to sway in synchrony with the seismic waves, thus amplifying their effects.

Cracking in the earth's surface is less of a problem than is popularly imagined. The vulnerability of buildings to surface faulting is easily avoided by zoning regulations that prevent building construction within 50 feet of a fault line. Unfortunately, zoning restrictions are not feasible for utility networks that must cross the fault lines.

Ground failure is a loss of soil bearing strength and takes three different forms. Landsliding occurs when a marginally stable soil assumes a more natural angle of repose. The term **landslide** refers to the downward displacement of rock or soil because of gravitational forces. Landslide risk areas can be mapped by conducting geological surveys to identify areas having slopes with soil that is likely to separate when saturated or shaken. Fissuring occurs when loose fill is located close to other soils that are less prone to this behavior. Finally, soil liquefaction is caused by loss of grain-to-grain support in saturated soils. Ground failure is a threat because building foundations need stable soil to support the structure. Even partial failure of the soil under the foundation can destroy a building by causing it to tilt at a dangerous angle.

As yet, there is no definitive evidence of physical cues that provide reliable warning of an earthquake. Unusual animal behavior has been observed, but has not proved to be a reliable indicator. The Chinese successfully predicted an earthquake for the city of Haicheng in 1975 and thousands of lives were saved by evacuating the city. However, there was no forewarning of the 1976 earthquake in Tangchan. Currently, earth scientists are examining many potential predictors such as increased radon gas in wells, increased electrical conductivity and magnetic anomalies in soil, changes in ground elevation, slope, and location.

Currently, there is no method of advance detection and warning for local earthquakes because these are so close to the impact area that P-waves and S-waves arrive almost simultaneously. Protective measures can be understood by the common observation that "earthquakes don't kill people, falling buildings kill people." Thus, building occupants are advised to shelter in-place under sturdy furniture while the ground is shaking. Those who survive the collapse of a building, typically attempt to rescue those survivors who remain trapped. However, the success of this improvised response depends upon the type of building. Nonreinforced masonry buildings are much more likely to collapse. Search and rescue from these structures can be relatively easy. By contrast, steel-reinforced concrete buildings are much less likely to collapse, but search and rescue is extremely difficult unless sophisticated equipment is available to well-trained urban search and rescue (USAR) teams.

FOR EXAMPLE

Destruction of an Ancient City

In 2003, an earthquake destroyed 70% of the ancient city of Bam, Iran. Approximately 70,000 people died. The high death toll is in part due to the fact that the residents lived in nonreinforced brick homes. The earthquake disintegrated these structures and buried people under the debris. By contrast, concrete buildings would have allowed for air pockets between building slabs. Iran does not have many search and rescue teams. Those that were requested were not physically able to get there in enough time to save any survivors. One woman survived because she was under a table next to a ventilation pipe. Iran is still working to rebuild the beautiful ancient city of Bam.

Unfortunately, almost all victims will die by the time remote USAR teams arrive. Crush injuries usually kill a person within 24 hours and most USAR teams take longer than this to mobilize and travel to the incident site. Another problem with earthquakes is that destruction of infrastructure impairs emergency response. Consequently, households, businesses, and local governments must be self-sufficient for at least 72 hours until outside assistance can arrive.

SELF-CHECK

- Define **volcanoes.**
- Identify the difference between **magnitude and intensity.**
- Describe the physical cues that signal an approaching earthquake.
- Define **landslide.**

5.4 Technological Hazards

Hazardous materials (also known as **hazmat**) are regulated by several federal agencies. Many federal agencies, including the US Coast Guard and US Environmental Protection Agency (EPA), have responsibilities for emergency response to hazmat incidents. Hazmat is defined as substances that are "capable of posing unreasonable risk to health, safety, and property" (49CFR 171.8).

Until the Emergency Planning and Community Right to Know Act of 1986, the location, identity, and quantity of hazmat throughout the United States was generally undocumented. People who produce, handle, or store amounts of any of the 400 Extremely Hazardous Substances (EHSs) must notify local agencies, their State Emergency Response Commission, and the U.S. EPA. The Chemical Abstract Service (CAS) lists 1.5 million chemical formulations with 63,000 of them hazardous. There are over 600,000 shipments of hazardous materials per day. Petroleum products account for 100,000 of these shipments. Fortunately, only a small proportion of these chemicals account for most of the number of shipments and the volume of materials shipped (see Table 5-5). These hazmat shipments result in an average of 280 liquid spills or gaseous releases per year, the vast majority of which occur in transport. Of these spills and releases, 81% take place on the highway and 15% are in rail transportation. These incidents cause approximately 11 deaths and 311 injuries per year (Federal Emergency Management Agency, 1993).

You will find hazmat produced, stored, or used in diverse locations including

- ▲ facilities such as petrochemical and manufacturing plants.
- ▲ warehouses (e.g., agricultural fertilizers and pesticides).

Table 5-5: Volume of Production for Top 12 EHSs, 1970–1994

Rank in top 50	*Chemical name*	*Year* 1970	1980	1990	1994	*% increase 1970–1994*
1	Sulfuric acid	29,525	44,157	44,337	44,599	51
8	Ammonia	13,824	19,653	17,003	17,965	30
10	Chlorine	9,764	11,421	11,809	12,098	24
13	Nitric acid	7,603	9,232	7,931	8,824	16
23	Formaldehyde	2,214	2,778	3,360	4,277	93
25	Ethylene oxide	1,933	2,810	2,678	3,391	75
31	Phenol	854	1,284	1,769	2,026	137
33	Butadiene	1,551	1,400	1,544	1,713	10
34	Propylene oxide	590	884	1,483	1,888	220
36	Acrylonitrile	520	915	1,338	1,543	197
37	Vinyl acetate	402	961	1,330	1,509	275
47	Aniline	199	330	495	632	218

Adapted from Lindell and Perry (1995).

- ▲ water treatment plants (chlorine is used to purify the water).
- ▲ breweries (ammonia is used as a refrigerant).

Hazmat is transported in a variety of ways, using ships, barges, trucks, pipelines, railways, trucks, and airways. The quantities of hazmat on ships, barges, and pipelines can be as large as at many fixed site facilities. Quantities are small when transported by rail, smaller still when transported by truck, and smallest when transported through the air. Small-to-moderate releases of hazardous materials at facilities are occupational hazards. These often pose little risk to public safety because the risk area lies within the facility. However, releases of this size during hazmat transportation are a public hazard because passers-by can easily enter the risk area and become exposed. The amount that is actually released is often much smaller than the total quantity that is available in the container. Prudence, however, dictates that planners assume the plausible worst case of complete release within a short period of time. In the case of toxic gases, this is 10 minutes (U.S. Environmental Protection Agency, 1987). In addition to the quantity released, the size of the risk area depends upon the hazmat's chemical and physical properties.

The Department of Transportation groups hazmat into ten different classes. Classification of a substance into one of these categories does not mean it cannot be a member of another class. For example, hydrogen sulfide is transported as a compressed gas that is both toxic and flammable.

1. **Explosives** are compounds or mixtures that undergo a rapid chemical transformation that is faster than the speed of sound. This generates a release of large quantities of heat and gas. For example, one volume of nitroglycerin expands to 10,000 volumes when it explodes. It is this rapid increase in volume that creates the surge in pressure characteristic of a blast wave. Explosives vary in their sensitivity to heat and impact. Explosives can cause casualties and property damage. Destructive effects from the quantities of explosives found in transportation can be felt as much as a mile or more away from the incident site.

   - ▲ Class A consists of high explosives that *detonate* (up to 4 mi/sec), producing overpressure, fire, and missile hazards.
   - ▲ Class B consists of low explosives that *deflagrate* (approximately .17 mi/sec—about 4% as fast as a detonation) and are fire and missile hazards.
   - ▲ Class C consists of low explosives that are fire hazards only.

2. **Compressed gases** are divided into flammable and nonflammable gases.

   - ▲ Nonflammable gases—such as carbon dioxide, helium, and nitrogen—are usually transported in small quantities. These are a significant hazard only if the cylinder valve is broken, causing the contents to escape rapidly through the opening and the container to become a missile hazard.

- ▲ Flammable gases—acetylene, hydrogen, methane—are missile and fire hazards. Rupture of gas containers can launch fragments up to a mile, so evacuation out to this distance is advised if there is a fire. Large quantities of flammable gases, such as railcars of liquefied petroleum gas (LPG), are of significant concern because the released gas travels downwind after release until it reaches an ignition source such as the pilot light in a water heater or the electrical system of a car. At distances of one-half mile or more, the gas cloud can erupt in a fireball that flashes back toward the release point.

3. **Flammable liquids,** which evolve flammable vapors at 80°F or less, pose a threat similar to flammable gases. A volatile liquid such as gasoline rapidly produces large quantities of vapor that can travel toward an ignition source and erupt in flame when it is reached. When a flammable liquid is spilled on land, initiate a downwind evacuation of at least 300 yards. A flammable liquid that floats downstream on water could be dangerous at even greater distances. One that is toxic requires special consideration. A fire involving a flammable liquid should initiate an evacuation of 800 yards in all directions.
4. **Flammable solids** self-ignite through friction, absorption of moisture, or spontaneous chemical changes. Flammable solids are less dangerous than flammable gases or liquids, because they do not disperse over wide areas. A large spill requires a downwind evacuation of 100 yards. A fire should stimulate consideration of an evacuation of 800 yards in all directions.
5. **Oxidizers and organic peroxides** include halogens (chlorine and fluorine), peroxides (hydrogen peroxide and benzoyl peroxide), and hypochlorites. These chemicals destroy metals and organic substances and enhance the ignition of combustibles (a spill of liquid oxygen can cause the ignition of asphalt roads on a hot summer day). Oxidizers and organic peroxides do not burn but are hazardous because they promote combustion and some are shock sensitive. A large spill should prompt a downwind evacuation of 500 yards and a fire should initiate an evacuation of 800 yards in all directions.
6. **Toxic chemicals** are classified in a number of ways.

- ▲ Class A includes gas or vapors, a small amount of which is an inhalation hazard.
- ▲ Class B consists of liquids or solids that are ingestion or absorption hazards. Toxic materials are a major hazard because of the effects they can produce when inhaled into the lungs, ingested into the stomach, or

absorbed through the skin by direct contact. Inhalation is the greatest concern because high concentrations achieved during acute exposure can kill a person instantly. However, prolonged ingestion can cause cancers in those who are exposed. It can also cause genetic defects in their offspring. Chemical contamination of victims poses problems for volunteers and professionals providing first aid and transporting victims to hospitals. These chemicals vary substantially in their volatility and toxicity. Evacuation distances following a spill or fire must be determined from the Table of Protective Action Distances in the *2004 Emergency Response Guidebook* (Department of Transportation, 2000, see www.dot.gov). http://hazmat.dot.gov/pubs/erg/erg2004.pdf, page 7.

7. **Infectious substances** have rarely been a significant threat because there are few shipments of these substances, and they are transported in small quantities with restrictive packaging and marking. However, infectious substances have the potential for being used in terrorist attacks.
8. **Radioactive materials** are substances that undergo spontaneous decay, emitting radiation in the process. The types and quantities of materials transported in the United States have very small impact areas. With the exception of nuclear power plants, releases of radioactive materials are likely to involve small quantities. Nonetheless, even a few grams of a lost radiographic source for industrial or medical X rays can generate a high level of concern. We have also recently recognized the terrorist threat of a "dirty bomb" that uses a conventional explosive to scatter radioactive material over a wide area. A large spill should prompt a downwind evacuation of 100 yards. A fire initiates an evacuation of 300 yards in all directions.
9. **Corrosives,** which are substances that destroy living tissue at the point of contact, can be either acidic or alkaline. Examples of acidic substances include hydrochloric acid and sulfuric acid. Examples of alkaline substances (caustics) include sodium hydroxide, potassium hydroxide, and ammonia. In addition to producing chemical burns of human and animal tissues, corrosives also degrade metals and plastics. The substances in this class that are most frequently used and transported are not highly volatile, so the geographical area affected by a spill is likely to be no greater than 100 yards unless the container is involved in a fire or the hazmat enters a waterway (e.g., via storm sewers). These chemicals vary substantially in volatility and toxicity. Evacuation distances following a spill or fire must be determined from the Table of Protective Action Distances in the *2004 Emergency Response Guidebook* (page 7).
10. **Miscellaneous dangerous goods,** as the name of this category suggests, is composed of a diverse set of materials such as air bags, certain

vegetable oils, polychlorinated biphenyls (PCBs), and white asbestos. These materials are low-to-moderate fire or health hazards to people within 10–25 yards.

5.4.1 Fires

There is an important distinction between gases and liquids. A **gas** expands to fill the available volume in a space. A **liquid** spreads to cover the available area on a surface. Any liquid contains some molecules that are in a gaseous state; this is called **vapor.** All liquids generate increasing amounts of vapor as the temperature increases and the pressure decreases. Conversely, at a given temperature and pressure, the amount of vapor in a liquid varies from one substance to another. There are three temperatures of each flammable liquid that are important because they determine the production of vapor. In turn, vapor generation is important because *it is the vapor that burns, not the liquid.* The three important temperatures of a liquid substance are its boiling point, flash point, and ignition temperature.

- ▲ Boiling point is the temperature at which its vapor pressure is equal to atmospheric pressure. Vapor production is negligible when a fuel is below its boiling point but increases significantly after it exceeds this temperature. Liquids with lower boiling points are more dangerous than those with higher boiling points.
- ▲ Flash point is the temperature at which the liquid gives off enough vapor to flash momentarily when ignited by a spark or flame. A liquid is defined as combustible if it has a flash point above 100°F. It is flammable if it has a flash point below 100°F. Liquids with lower flash points are more dangerous than those with higher flash points.
- ▲ Ignition temperature is the minimum temperature at which a substance becomes so hot that its vapor ignites even in the absence of an external spark. Liquids with lower ignition temperatures are more dangerous than those with higher ignition temperatures.

Gases and vapors have flammable limits that are defined by the concentration (percent by volume in air) at which ignition can occur in open air or an explosion can occur in a confined space. The lower flammable (explosive) limit (LFL/LEL) is the minimum concentration at which ignition will occur. Below that limit the fuel/air mixture is "too lean" to burn. The upper flammable (explosive) limit (UFL/UEL) is the maximum concentration at which ignition will occur. Above that limit the fuel/air mixture is "too rich" to burn.

The most dangerous flammable substances have a low ignition temperature, low LEL, and wide flammable range. Indeed, gasoline is widely used precisely because of these characteristics. It has a low flash point (–45 to –36°F), a low LFL (1.4 to 1.5%), and a reasonably wide range (6%). By contrast, peanut oil is useful in cooking because it has the opposite characteristics.

An important hazard of flammable liquids is a boiling liquid expanding vapor explosion (BLEVE), which occurs when a container fails at the same time the temperature of the contained liquid exceeds its boiling point at normal atmospheric pressure. BLEVEs involve flammable or combustible compressed gases that are not classified as "explosive substances," but can produce fireballs as large as 1000 feet in diameter and launch container fragments up to a half mile from the source of the explosion.

5.4.2 Toxic Industrial Chemical Releases

The release of a toxic chemical from its storage container will usually produce a cloud that is very similar to a plume of smoke that travels downwind from a campfire. In fact, the plume of smoke from a fire is just the visible particles of unburned fuel from the fire. However, airborne dispersion of toxic industrial chemicals can be much more dangerous because they can produce serious inhalation exposures at distances up to 10 miles. The spread of a toxic chemical release can be defined by a dispersion model that includes

- The hazmat's chemical and physical characteristics, including
  - Quantity.
  - Volatility (higher volatility means more chemical becomes airborne per unit of time).
  - Buoyancy (whether it tends to flow into low spots because it is heavier than air).
  - Toxicity (the biological effect due to cumulative dose or peak concentration).
  - Physical state—whether it is a solid, liquid, or a gas at ambient temperature and pressure.
- The hazmat's release characteristics, including
  - Release rate (in pounds per minute); a higher release rate puts a larger volume of chemical into the air, and increases its concentration.
  - Size (surface area) of the spilled pool if the substance is a liquid.
  - Temperature (a temperature higher than that of the surrounding environment increases the evaporation rate of a liquid).
  - Pressure (a pressure higher than that of the surrounding environment increases the dispersion rate of a gas).
- The topographic conditions in the release area.
- The meteorological conditions at the time of the release.

Topographical conditions relevant to spills include the slope of the ground and the presence of depressions. Slopes allow a liquid to rapidly move away from the source of the spill. Depressions decrease the size of a liquid pool which reduces the size of the pool's surface area and the rate at which vapor is generated from it. For this reason, dikes are erected around chemical tanks to confine spills in case the tanks leak. Hazmat responders build temporary dikes around spills for the same reason. Topographical characteristics also affect the dispersion of a chemical release in the atmosphere. Hills and valleys channel the wind direction and can increase wind speed at constriction points. For example, where a valley narrows, wind speed increases due to a "funnel" effect. Forests and buildings are rough surfaces that increase turbulence in the wind field, causing greater vertical mixing. This mixing decreases the chemical concentration at ground level where it is most dangerous to residents. By contrast, large water bodies have smooth surfaces that minimize turbulence. This allows a chemical release to maintain a dangerous concentration at ground level.

Wind speed, wind direction, and atmospheric stability class affect the dispersion of a chemical. Figure 5-6 shows a release dispersing uniformly in all directions when there is no wind (Panel A). The circle corresponds to the *level of concern* (LOC), which is the "concentration of an EHS [Extremely Hazardous Substance] in air above which there may be serious irreversible health effects or death as a result of a single exposure for a relatively short period of time" (U.S. Environmental Protection Agency, 1987, pp. 2-13). The nearby town lies outside the vulnerable zone, so it does not need to take protective action. However, Panel B describes a situation in which there is a strong wind, so the affected area is much smaller in the upwind and cross-wind directions. However, the vulnerable zone is much larger in the downwind direction, so the nearby town lies inside the vulnerable zone and would need to take protective action.

As Table 5-6 indicates, the atmospheric stability class can vary from Class A through Class F. Class A, the most unstable condition, occurs during strong sunlight and light wind. This dilutes the released chemical by causing vertical mixing into a large volume of air. Class F identifies the most stable condition, which takes place during clear nighttime hours when there is a light wind. This condition has

Figure 5-6

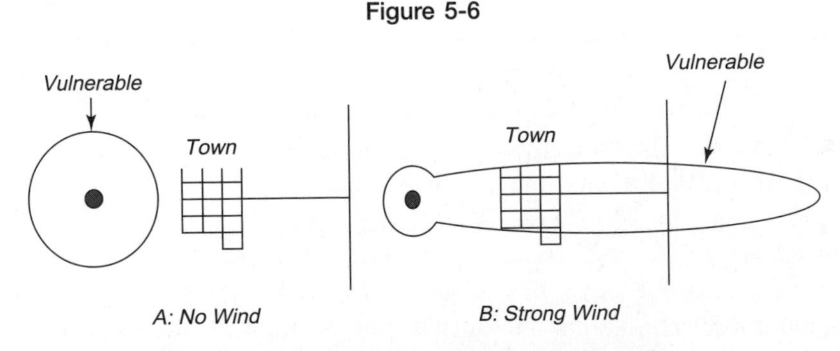

Effects of Wind Speed on Chemical Dispersion.

Table 5-6: Atmospheric Stability Classes

	Strength of sunlight			*Nighttime conditions*	
Surface wind speed (mph)	*Strong*	*Moderate*	*Slight*	*Overcast ≥50%*	*Overcast <50%*
<4.5	A	A–B	B	-	-
4.5–6.7	A–B	B	C	E	F
6.7–11.2	B	B–C	C	D	E
11.2–13.4	C	C–D	D	D	D
>13.4	C	D	D	D	D

little vertical mixing, so the released chemical remains highly concentrated at ground level. In some cases, meteorological characteristics remain stable for days at a time but, in other cases, change from one hour to the next. To be on the safe side, you should assume the worst meteorological conditions during a hazmat release (variable wind direction and high atmospheric stability) unless competent meteorological authorities tell you otherwise.

- A: Extremely Unstable Conditions
- B: Moderately Unstable Conditions
- C: Slightly Unstable Conditions
- D: Neutral Conditions (heavy overcast day or night)
- E: Slightly Stable Conditions
- F: Moderately Stable Conditions

Your main concern is the protection of the population at risk. The risk to this target population varies with distance from the source of the release. However, distance is not the only factor that should be of concern. In addition, the density of the population should be considered. There are also likely to be differences in susceptibility within the risk area population because individuals differ in their dose-response relationships. One relevant variable is age, with the youngest and oldest tending to be the most susceptible. Another relevant variable is physical condition. Those with compromised immune systems are the most susceptible.

Toxic chemicals differ in their exposure pathways—inhalation, ingestion, and absorption. Inhalation is the route by which entry into the lungs is achieved. This is a major concern because toxic materials can pass rapidly through lungs to bloodstream and on to specific organs within minutes of exposure. Ingestion is of less immediate concern because entry through the mouth into the digestive system (stomach and intestines) is a slower route into the bloodstream and on to specific organs. Depending on the chemical's concentration and toxicity, ingestion exposures could be tolerated for days or months. Authorities might choose to prevent ingestion exposures by withholding contaminated food from the market or recommending that those in the risk area drink boiled or bottled water. Absorption involves entry directly through the pores of the skin (or through the eyes), so it tends to be a greater concern for first responders than local residents.

The harmful effects of toxic chemicals are caused by alteration of cellular functions. These effects can be either acute or chronic. Acute effects occur during the time period from 0–48 hours. Irritants cause chemical burns (dehydration and exothermic reactions with cell tissue). There are two types of asphyxiants: simple asphyxiants and chemical asphyxiants. Simple asphyxiants, such as carbon dioxide, displace oxygen within a confined space or are heavier than oxygen so they displace it in low-lying areas such as ditches. By contrast, chemical asphyxiants prevent the body from using the oxygen even if it is available in the atmosphere. For example, carbon monoxide combines with the hemoglobin in red blood cells more readily than does oxygen so the carbon monoxide prevents the body from obtaining the available oxygen in the air. Anesthetics/narcotics depress the central nervous system. In extreme cases, they suppress autonomic responses such as breathing and heart function.

Chronic, or long-term, effects can be general cell toxins, known as cytotoxins, or have organ-specific toxic effects. The word *toxin* is preceded by a prefix referring to the specific system affected.

- ▲ *Hemotoxins* affect the circulatory system.
- ▲ *Hepatotoxins* affect the liver.
- ▲ *Nephrotoxins* affect the kidney.
- ▲ *Neurotoxins* affect the nervous system.

Chemicals that cause cancer are referred to as **carcinogens.** Mutagens cause mutations in those directly exposed. Teratogens cause mutations to the genetic material of those directly exposed and, thus, mutations in their offspring. The severity of any toxic effect is determined by the chemical's

- ▲ Rate and extent of absorption into the bloodstream.
- ▲ Rate and extent of transformation into breakdown products.
- ▲ Rate and extent of excretion of the chemical from the body.

Research on toxic chemicals has led to the development of dose limits. Some important concepts in defining dose limits are the LD-50, which is the dose (usually of a liquid or solid) that is lethal to half of those exposed and the LC-50, which is the concentration (usually of a gas) that is lethal to half of those exposed. Based upon these dose levels, dose limits are administrative quantities that should not be exceeded.

5.4.3 Weaponized Toxic Chemicals

Terrorists might attempt to use weaponized toxic agents, which are toxic chemicals that require smaller doses to achieve a significant effect (e.g., disability or death). One consequence of the more advanced toxic agents is that they can affect victims through absorption in secondary *contamination*. That is, chemical residues on a victim's skin or clothing can affect those who handle that individual. Indeed, any object on which the chemical is deposited becomes an avenue of secondary contamination (World Health Organization, 2004). A list of the most likely weaponized toxic agents is presented in Table 5-7. Some of these agents, such as anthrax and botulism, are produced by biological processes that affect victims through the production of toxins. Consequently, the World Health Organization (WHO) considers these to be chemical rather than biological weapons.

Your police or fire department might detect a terrorist attack involving a toxic chemical agent when it responds to a report of mass casualties. This would be especially likely if many people in the same place at the same time displayed symptoms including headaches, nausea, breathing difficulty, convulsions, or sudden death. The appropriate response to this situation is the same as in any other hazmat incident. You need to control access to the incident site, decontaminate the victims, and

Table 5-7: Weaponized Toxic Agents

Agent	*Example*
Tear gases/other sensory irritants	Oleoresin capsicum ("pepper spray")
Choking agents (lung irritants)	Phosgene
Blood gases	Hydrogen cyanide
Vesicants (blister gases)	Mustard gas
Nerve gases	O-Isopropyl Methylphosphonofluoridate (Sarin gas)
Toxins	Clostrinium botulinum ("botulism")
Bacteria and rickettsiae	Bacillus anthracis ("anthrax")
Viruses	Equine encephalitis

transport them to medical care. Also be aware of the assistance that is available from local poison control centers. The capabilities needed to respond effectively to an attack using toxic chemicals are the same as those needed for an industrial accident involving these materials. Unfortunately, few communities have this capability. Even communities with many chemical facilities rarely have hospitals with the capability to handle mass casualties from toxic chemical exposure.

In the event of a terrorist attack, the incident site is considered a crime scene by law enforcement authorities. Consequently, you must learn about the basic procedures these authorities follow. These procedures include collecting evidence, maintaining a chain of custody over that evidence, and controlling access to the incident scene. Access to the scene should be carefully coordinated to avoid a conflict between victim rescue and law enforcement procedures for crime scene security.

5.4.4 Radiological Material Releases

Radioactive materials are used for a variety of purposes. Small quantities are sources of radiation for medical and industrial diagnostic purposes. Large quantities of other radiological materials produce the steam needed to drive electric generators at power plants. Enriched uranium fuel fissions when struck by a free neutron (see Figure 5-7). Some energy is released as heat, which is used to produce the steam that drives electric turbines. There also is ionizing radiation that can take the form of alpha, beta, or gamma radiation. Alpha radiation can travel only a very short distance. It is easily blocked by a sheet of paper, so it is most likely to be dangerous when inhaled. Beta radiation can travel a moderate distance but can be blocked by a sheet of aluminum foil. Gamma radiation can travel a long distance and can be blocked only by very dense substances such as stone, concrete, and lead. The free neutrons continue a sustained chain reaction.

Figure 5-7

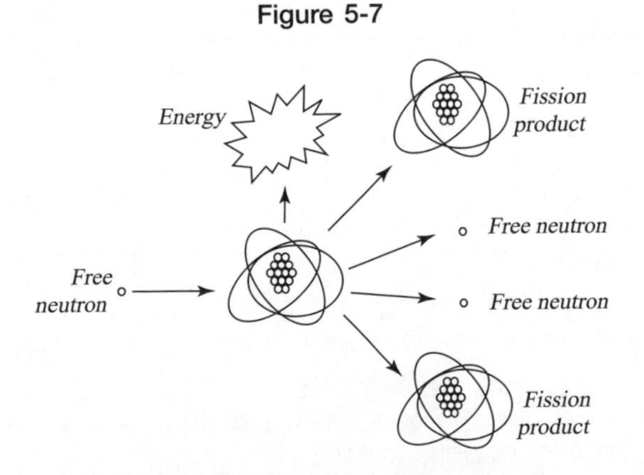

The atomic fission reaction

The fission products are waste products that must be stored in a permanent repository.

The radioactive fuel for a nuclear power plant is contained in fuel pins that are welded shut and inserted into long rods. These fuel rods are inserted into a reactor vessel that uses cooling water to carry away the heat from the fission reaction. For safety, the reactor vessel is located in a containment building that is constructed with thick walls of steel-reinforced concrete. It can withstand high internal pressures or the pressure of an airplane crash. However, it has many penetrations for water pipes, steam pipes, and instrumentation and control cables. These penetrations are sealed during normal operations, but the seals might be damaged during an accident. Such damage to the seals could allow radioactive material to escape into the environment.

During a severe accident involving a major loss of cooling water, the fuel will:

- ▲ Melt through the steel cladding.
- ▲ Melt through the reactor vessel.
- ▲ Escape the containment building.

This process could produce a release as soon as 45–90 minutes after the cooling water leaks out. If the core melts, the danger to offsite locations depends on containment. Early health effects are likely if there is early total containment failure. They are also possible if there is early major containment leakage. Otherwise early health effects are unlikely. Unfortunately, containment failure is not predictable (McKenna, 2001).

A radioactive release involves a mix of **radionuclides,** which are radioactive substances that vary in atomic weight. The mix of radionuclides involved in a given release is called the **source term.** Source term characteristics are defined by three classes of radionuclides—particulates, radioiodine, and noble gases. Particulates include strontium, which is chemically similar to calcium. This chemical similarity makes strontium dangerous because it tends to be deposited in bones where it irradiates the marrow that produces blood cells. Radioiodine (Iodine-131) is dangerous because it substitutes for nonradioactive iodine in the thyroid and causes thyroid cancer. Noble gases such as krypton don't react chemically with anything. However, they are easily inhaled and produce radiation exposures while they remain in the lungs.

The source term is also defined by its volatility. Higher volatility means more of the radionuclide becomes airborne and stays airborne. The amount of radioactivity released is measured by the number of disintegrations per unit time (curies). These disintegrations are what a Geiger counter measures. The amount of radioactivity is measured in curies of an individual radionuclide.

As is the case with other hazardous materials, breathing air contaminated with radioactive materials can cause inhalation exposure. Similarly, eating food or drinking liquids that are contaminated with radioactive materials can cause ingestion exposure. Unlike other hazardous materials, radiological materials release energy—especially high energy gamma radiation. Thus, radiological materials can produce exposures through direct irradiation from a plume of radioactive material that is floating overhead but never makes contact with those below. This distinction between contamination and irradiation is important to remember. Irradiation involves the transmission of energy to a target that absorbs it. Contamination occurs when radioactive particles are deposited in a location within the body where they provide continuing irradiation. Emergency responders helping people who have been irradiated cannot be harmed by that irradiation. By contrast, emergency responders helping people who have been contaminated by chemical or radiological materials can pick up the contamination from the victims.

Measuring radiation doses is more complicated than measuring toxic chemical doses. A Roentgen is a measure of exposure to ionizing radiation. A rad is a measure of absorbed dose, and a rem ("Roentgen equivalent man") is a measure of committed dose equivalent. (The current international unit, one Sievert, equals 100 rem). The term **committed** refers to the fact that contamination by radioactive material on the skin or absorbed into the body will continue to administer a dose until it decays or is removed. The term equivalent refers to the differences in the biological effects of alpha, beta, and gamma radiation. Weighting factors are used to make adjustments for the effects of the different types of radiation. For planning purposes, one rem is approximately equal to one rad.

The health risks of exposure to radiation are indentified as early fatalities, prodromal effects, and delayed effects. Early fatalities occur within a period of days or weeks. Prodromal effects are early symptoms of more serious health effects. Delayed effects are cancers that might take decades to develop. Genetic disorders do not reveal their effects until the next generation is born. Prodromal effects manifest themselves in less than 2% of the population at a dose of 50 rad. However, 50% exhibit prodromal symptoms at 150 rad and 98% would show these symptoms at 250 rad. The delayed effects of radiation exposure are fatal cancers, nonfatal cancers, and genetic disorders. Organ differences in dose-response arise because rapidly dividing cells, found in the intestines and hair follicles, are especially susceptible. There also are differences in dose-response. For example, fetuses are extremely susceptible because all of their cells are dividing rapidly. Other population segments include those at risk of any environmental insult: the very old, the very young, and those with compromised immune systems.

Protective actions for radiological emergencies are based upon three fundamental factors—time, distance, and shielding. Evacuation reduces the amount of time exposed and increases distance from the source. Sheltering in-place can provide shielding if this is done within dense materials that absorb energy and are airtight. To determine when protective action should be initiated, the EPA has developed early phase protective action guides (PAGs), which are specific criteria for initiating population protective action in radiological emergencies (Conklin and Edwards, 2001). Note that the whole body dose listed in Table 5-8 for initiating evacuation (1 rem) is only a small fraction of the exposure level that would be expected to produce prodromal effects in the most susceptible 2% of the general population.

The U.S. Nuclear Regulatory Commission (NRC) has conducted extensive analyses to identify the most effective protective action to be taken under various conditions. These analyses have examined the relative effectiveness of continuing normal activity, sheltering in a home basement or large building, or evacuating before or after plume arrival at three distances (one, three, and five miles) from a nuclear power plant. According to McKenna (2000), even a late evacuation is better than home shelter, large building shelter is better than late evacuation, and early evacuation is best of all. The decrease in risk as a function of distance is quite large. Those who are five miles from the source have a 50% lower chance of exceeding 200 rem.

5.4.5 Biological Hazards

Most biological agents that might be used in deliberate terrorist attacks also exist as natural hazards. These biological agents exist at low levels of prevalence in human populations. They also exist in animal populations and they can spread to humans. Indeed, one-quarter of the world's deaths in 1998 were caused by infectious diseases. The major consequence of most biological agents is the magnification of their effects by infection, unlike chemical agents that generally *dissipate* over time and distance. Biological agents magnify their effects by multiplying within the target group, but chemical agents cannot do this.

Table 5-8: EPA Protective Action Guides

Organ	*EPA PAGs*[a] *(rem/Sv)*	*Protective Action*[b]
Whole body	1–5 (.01–.05)	Evacuation
Thyroid	25 (.25)	Stable Iodine (KI)

a) Dose inhalation from and external exposure from plume and ground deposition.
b) Actions should be taken to avert PAG dose.
c) Evacuation is considered to be the most effective protective action for nuclear power plant accidents at American sites.

Terrorists can spread disease by contaminating the food or water supply, causing people who drink or eat contaminated food and water to become ill. In this scenario, the terrorists would be using our own food distribution system against us. However, USDA and state departments of agriculture monitor the food distribution system and look for signs of trouble. In addition, the state emergency management agencies provide support when needed. For example, state and federal officials worked together in recent cases of mad cow disease.

A terrorist can also disperse biological agents in aerosol clouds of liquid droplets or solid particles to cause inhalation exposure. An aerosol could be released in the open or through a building's heating, ventilation, and air conditioning (HVAC) system. Dispersal through the HVAC system produces more casualties because of the greater concentration of the agent. The effectiveness of the dispersion depends on the agent's particle size and weight (smaller, lighter particles travel farther). Wind and other weather conditions can affect the dispersion of the agent and even dilute it so it is no longer effective. Nonetheless, if one person becomes ill and rapidly spreads the disease, there could be an epidemic.

It can be difficult to detect if someone has been infected with a biological agent because symptoms of biological agents resemble symptoms of the common cold and flu. Moreover, symptoms usually do not appear until long after exposure occurs. This makes it possible for victims to travel a significant distance from the site of the attack before they show symptoms. If the victims are infected with a contagious agent, they might cause widespread secondary outbreaks before authorities are aware that an attack has occurred. A biological agent is most likely to be identified by noting a significant increase in people appearing with such symptoms. Healthcare providers in emergency rooms and clinics are likely to be the first people to notice such an increase.

We do have sensors that can identify some biological agents in the early stages before people develop symptoms. However, these sensors are very expensive so they are currently deployed only at the most critical facilities. This makes it important for emergency managers to establish a close relationship with their local health departments. This way, they can coordinate effectively with others to identify the agent, treat the victims, and decontaminate the incident site.

If there is an outbreak, there are two actions to take: isolate and quarantine. Isolation prevents the ill patients from infecting others. It is associated with special treatment to remedy the disease. Quarantine involves those who might have been exposed to a biological agent but do not currently exhibit symptoms. They might not become ill and, indeed, might not even have the disease. However, it is critical to prevent them from infecting others if they do have the disease. Quarantine is somewhat similar to sheltering in-place from toxic chemical hazards. The difference is that people being quarantined are asked (or legally required) to remain indoors to protect others from themselves (because they are the

> **FOR EXAMPLE**
>
> **Three Mile Island**
>
> The 1979 accident at the Three Mile Island nuclear power plant illustrates the difficulty of evacuating only the areas downwind from a release. Wind speed and direction changed repeatedly during the first day of the accident. Consequently, any recommendation to evacuate the area downwind from the plant would have referred to a different geographic area at different times during the day. This would have made evacuation difficult because of the many hours needed to evacuate a given downwind sector. Consequently, the evacuation of one downwind sector would have been in progress when the order to initiate an evacuation in a different direction was initiated. This could have caused traffic jams if a major release had occurred while evacuations were in progress.

hazard) rather than to protect themselves from an external hazard. Vaccines can also protect people against some biological agents. However, vaccines are not available for all biohazards and the available quantities are limited.

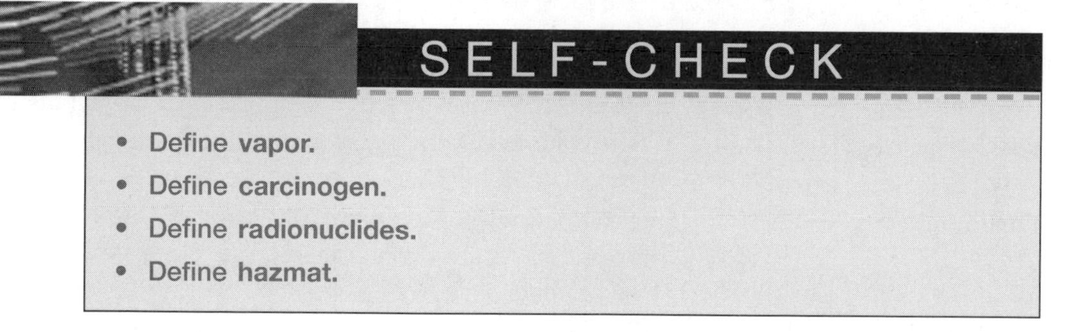

SUMMARY

To manage hazards, you need to understand them. As this chapter illustrates, each type of hazard has a different cause and effect. Each hazard also has a distinct set of characteristics and potential dangers. Not only do you need to know how to identify hazards, but you must also understand what to do to prepare for one, inform others of them, and aid others in the event a hazard threatens your community. If there is forewarning of an impending disaster, evacuation is often a good solution. If there is no warning, you must mitigate ahead of time or bear the consequences.

KEY TERMS

100-year flood An arbitrary standard of safety that reflects a compromise between the goals of providing long-term safety and developing economically valuable land.

Apparent temperature The combination of temperature and humidity into a heat index.

Carcinogens Chemicals that cause cancer.

Committed The fact that contamination by radioactive material on the skin or absorbed into the body will continue to administer a dose until it decays or is removed.

Compressed gases Gases that are cooled to a liquid state so they occupy a small enough volume to be transported at a reasonable economic cost.

Core Molten rock at the center of the earth.

Corrosives Substances that destroy living tissue at the point of contact because they are either acidic or alkaline.

Crust Solid rock and other materials at the earth's surface that is defined by large plates floating on the mantle and moving gradually in different directions over time.

Discharge The volume of water passing a specific point per unit of time.

Earthquake A sudden release of energy that has been built up as two tectonic plates attempt to move past each other.

Epicenter A point on the earth's surface directly above the hypocenter.

Explosives Compounds or mixtures that undergo a rapid chemical transformation that is faster than the speed of sound.

Eye of the hurricane The area of calm conditions that has a 10 to 20 mile radius. The eye is surrounded by bands of high wind and rain that spiral and form a ring around the eye.

Eyewall The spiral that forms a ring around the eye of a hurricane.

Firestorms	Fires that are distinguished from other wildfires because they burn so intensely that they create their own local weather and are virtually impossible to extinguish.
Flammable liquids	Liquids that evolve flammable vapors at 80°F or less, thus posing a threat similar to flammable gases.
Flammable solids	Solids that self-ignite through friction, absorption of moisture, or spontaneous chemical changes.
Flood	An event in which abnormally large amount of water accumulates in an area in which it is usually not found.
Gas	A substance that expands to fill the available volume in a space.
Hardscape	Impermeable surfaces, such as building roofs, streets, and parking lots.
Hazmat	Hazardous materials that may pose unreasonable risk to health, safety, and property."
Hurricane	The most severe type of tropical storm.
Hypocenter	A point deep within the earth from which an earthquake's energy is released.
Intensity	The measure of energy release at a given impact location, which can be assessed either by behavioral effects or physical measurements.
Interface fires	Fires that burn into areas containing a mixture of natural vegetation and built structures.
Landslide	The downward displacement of rock or soil because of gravitational forces.
Liquid	A substance that spreads to cover the available area on a surface.
Magnitude	The measure of energy release at the source. Earthquake magnitude is measured on the Richter scale where a one-unit increase represents a 10-fold increase in seismic wave amplitude and a 30-fold increase in energy release from the source.
Mantle	An 1800 mile thick layer between the core and the crust.

Miscellaneous dangerous goods A diverse set of materials such as air bags, certain vegetable oils, PCBs, and white asbestos.

Natural hazards Extreme events that originate in nature. Natural hazards are commonly categorized as meteorological, hydrological, or geophysical.

Oxidizers and organic peroxides Chemicals that include halogens (chlorine and fluorine), peroxides (hydrogen peroxide and benzoyl peroxide), and hypochlorites. These chemicals destroy metals and organic substances and enhance the ignition of combustibles.

Radioactive materials Substances that undergo spontaneous decay, emitting radiation in the process.

Radionuclides Radioactive substances that vary in atomic weight.

Severe storms A storm whose wind speed exceeds 58 mph, that produces a tornado, or that releases hail with a diameter of three-quarters of an inch or greater.

Source term The mix of chemicals or radionuclides involved in a given release.

Stage The height of water above a defined level that is used by emergency managers to predict the level of flood casualties and damage.

Storm surge An increased height of a body of water that exceeds the normal tide.

Technological hazards Hazards that originate in human-controlled processes but are released into the air and water. The most important technological hazards are explosives, flammable materials, toxic chemicals, radiological materials, and biological hazards.

Tornadoes Windstorms that form when cold air from the north collides with a warmer air mass.

Tsunamis Sea waves that are usually generated by undersea earthquakes. Tsunamis can also be caused by volcanic eruptions or landslides.

Vapor The molecules that are in a gaseous state of a substance that is a liquid at normal temperature and pressure.

Volcanoes Geological structures that transport a column of molten rock from the earth's mantle to the surface.

Wildland fires Fires that burn areas with nothing but natural vegetation for fuel.

ASSESS YOUR UNDERSTANDING

Go to www.wiley.com/college/lindell to evaluate your knowledge of the basic hazards in the United States.

Measure your learning by comparing pre-test and post-test results.

Summary Questions

1. The impact of winter storms is greatest in the northern states because winter storms are not disruptive in the south. True or False?
2. What apparent temperature range warrants extreme caution?
   (a) 80–90 degrees
   (b) 90–105 degrees
   (c) 105–130 degrees
   (d) 70–80 degrees
3. The three elements of the fire triangle are heat, oxygen, and air. True or False?
4. Runoff increases as slope increases. True or False?
5. Which of the following is not impermeable?
   (a) Clay
   (b) Stone
   (c) Concrete
   (d) Gravel
6. Volcanic eruptions can cause tsunamis and wildfires as secondary hazards. True or False?
7. All liquids generate increasing amounts of vapor as the temperature decreases and the pressure increases. True or False?
8. Molecules in a gaseous state that are generated by a substance that is in a liquid state at normal temperature and pressure are referred to by which of the following names?
   (a) Vapor
   (b) Gas
   (c) Air
   (d) Steam
9. What is not a factor in determining the protective actions for radiological emergencies?
   (a) Time
   (b) Distance
   (c) Frequency
   (d) Shielding

Review Questions

1. What are the threats of a severe storm?
2. What are three heat-related illnesses?
3. What are three physical cues of a tornado?
4. What are three threats of hurricanes?
5. Name five of the seven flood types.
6. What are the three distinct geological components of the earth?
7. What is the difference between shield volcanoes and stratovolcanoes?
8. What are the three different types of waves in earthquake energy?
9. What is the distinction between gases and liquids?
10. Name the three important temperatures of a liquid substance.
11. If there is a biological hazard outbreak, what two actions must take place?
12. What are the health effects of exposure to radiation?
13. What is the severity of any toxic effect determined by?
14. How does terrorism differ from a natural or technological hazard?
15. Why is gasoline a more dangerous flammable substance than peanut oil?
16. What could your jurisdiction do to reduce damage and casualties from earthquakes?

Applying This Chapter

1. Before Hurricane Katrina struck, local officials issued evacuation orders for parts of Louisiana and Mississippi. Millions left, but thousands of people chose to stay. Some did not believe the hurricane would be as powerful as it was because they had heard prior dire predictions before that did not come true. Others simply did not have the money to evacuate. Thousands went to the shelters in New Orleans that quickly became overcrowded. Two shelters, the Superdome and the Convention Center, were sufficient for the first few days after the storm but conditions quickly became intolerable. After having witnessed Hurricane Katrina, how would you evacuate your own city for a severe hurricane? How would you set up the shelters?
2. You are the emergency manager for a coastal town that has seen hurricane activity in the past. You have been asked to prepare a presentation on flooding. What would you include in your plan? What types of flooding are important to discuss? Describe the presentation you would put together.

3. You are the emergency manager for a small Midwestern town. One of your jobs is to create a public safety campaign that informs residents about biological hazards and how to deal with biological threats. What steps would you take to inform residents about these types of hazards?
4. You are the emergency manager for a small Midwestern town that is prone to tornadoes. One of your jobs is to create a public safety campaign that informs residents about the physical signs of an approaching tornado and what to do when they witness the physical signs. Which residents and structures are most at risk? Why?

YOU TRY IT

Tsunami Evacuation

You are the emergency manager for Hilo, Hawaii. How important would tsunami detection be to you and why? What are some of the methods used for detection? How soon would you need to evacuate the shoreline if a tsunami was detected?

Surviving an Earthquake

You are the emergency manager for San Francisco, a city that last had an earthquake in 1989. One of your jobs is to create a public safety campaign that informs residents of precautions to take when there is an earthquake. What steps do you tell residents to take if there is an earthquake?

Toxic Chemical Release

The 1984 release of methyl isocyanate from a pesticide factory in Bhopal, India caused thousands of deaths and tens of thousands of injuries. What are the risks to first responders in a toxic chemical release? What are the risks to the general population? Do you think conditions in the U.S. today are different from those in India two decades ago? How are they different?

6

HAZARD, VULNERABILITY, AND RISK ANALYSIS

Focusing Efforts

Starting Point

Go to www.wiley.com/college/lindell to assess your knowledge of the basics of risk analysis.
Determine where you need to concentrate your effort.

What You'll Learn in This Chapter

- ▲ Physical, structural, and social vulnerabilities
- ▲ The three preimpact conditions to be concerned with
- ▲ Characteristics of a hazard and how they affect losses
- ▲ Effects of a hazard and how to guard against them
- ▲ Emergency management interventions
- ▲ Community hazard and vulnerability analyses

After Studying This Chapter, You'll Be Able To

- ▲ Analyze physical, structural, and social vulnerabilities
- ▲ Recognize the characteristics of hazards
- ▲ Differentiate among groups that are most vulnerable to hazards
- ▲ Demonstrate the exposures and effects of secondary hazards
- ▲ Examine emergency management interventions
- ▲ Analyze hazards and vulnerabilities

Goals and Outcomes

- ▲ Evaluate community vulnerabilities
- ▲ Create a chemical inventory
- ▲ Assess and prepare for hazards
- ▲ Design a preimpact disaster recovery plan
- ▲ Plan and compose emergency management interventions
- ▲ Select what hazard and vulnerability information to include on your Web site

INTRODUCTION

A disaster occurs when an extreme event exceeds a community's ability to cope with that event. By anyone's standards, Hurricane Katrina was a disaster for the city of New Orleans and other coastal communities of eastern Louisiana and Mississippi. The affected communities were unable to respond and recover without outside help.

Understanding how disasters affect communities is important for four reasons.

1. You can determine what makes your community vulnerable.
2. You can identify which groups are likely to be more affected than others.
3. You can identify the specific characteristics of an incident that determine the level of disaster impact.
4. You can determine how to protect your community.

This chapter looks at the ways disasters affect communities and explains how to analyze these disaster impacts.

6.1 Community Vulnerability to Disasters

A community's vulnerability to disasters can be explained by models proposed by Cutter (1996), Lindell and Prater (2003), and Prater, Peacock, Lindell, Zhang and Lu (2004). Specifically, Figure 6-1 illustrates that the effects of a disaster are determined by three preimpact conditions—hazard exposure, physical vulnerability, and

Figure 6-1

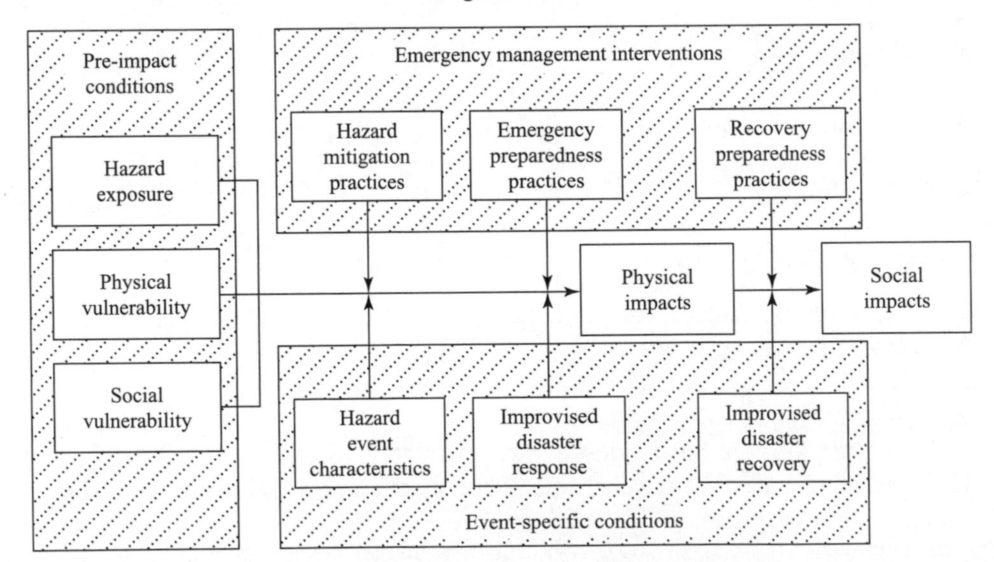

Conceptual model of disaster impacts.

FOR EXAMPLE

Impact Area

A disaster's impact area can be limited to a small geographic area such as when tornadoes touch down in an area that is limited to a few miles. Or disasters can have widespread impact. One could argue that the impact of Hurricane Katrina was felt throughout the country as those displaced by the hurricane were taken to cities far away from the physical destruction. Evacuees, for example, were sent to shelters in cities such as Boston and Portland.

social vulnerability. There are also three event-specific conditions—hazard event characteristics, improvised disaster responses, and improvised disaster recovery. Two of the event-specific conditions, hazard event characteristics and improvised disaster responses, combine with the preimpact conditions to produce a disaster's physical impacts. The physical impacts, in turn, combine with improvised disaster recovery to produce the disaster's social impacts. Communities can engage in three types of emergency management interventions to ameliorate disaster impacts. Physical impacts can be reduced by hazard mitigation practices and emergency preparedness practices, whereas social impacts can be reduced by recovery preparedness practices.

SELF-CHECK

- Describe pre-impact conditions according to the disaster impacts model.
- Describe emergency management interventions according to the disaster impacts model.
- List the variables that determine the physical impacts of disasters.
- List three types of conditions that determine social impacts of a disaster.

6.2 Preimpact Conditions

The specific impacts of a disaster are determined by three preimpact conditions:

▲ **Hazard exposure** (living, working, or being in places that can be affected by hazard impacts).

- ▲ **Physical vulnerability** (human, agricultural, or structural susceptibility to damage or injury from disasters).
- ▲ **Social vulnerability** (lack of psychological, social, economic, and political resources to cope with disaster impacts).

6.2.1 Hazard Exposure

Each part of the United States is prone to specific hazards. For example, residents who live on the Atlantic and Gulf coasts are exposed to hurricanes. People who live in the Pacific Northwest and Hawaii are exposed to volcanoes. People who live near chemical plants are exposed to explosions and hazardous materials releases. You can try to assess your community's hazards, but it is difficult. We have data on the weather and water levels for only the past hundred years, which limits our knowledge of floods, hurricanes, and tornadoes. It is difficult to estimate chemical and nuclear reactor accidents, because each facility is unique. It is even more difficult to calculate the likelihood of a terrorist attack, as that hazard depends on the behavior of other people.

6.2.2 Physical Vulnerability

There are three types of physical vulnerability:

- ▲ **Human vulnerability.** Humans are vulnerable to extremes of temperature, pressure, and chemical exposures (see Figure 6-2). These environmental conditions can cause death, injury, and illness. Specific segments of the affected population respond to hazards differently. That is, given the same level of exposure, some people will die, others will be severely injured, others slightly injured, and the rest will survive unscathed. Typically, the most susceptible to any environmental stressor are the very young, the very old, and those with weakened immune systems.
- ▲ **Agricultural vulnerability.** Like humans, agricultural plants and animals can also be hurt by hazards. Like humans, there are differences among individuals within each plant and animal population. However, agricultural vulnerability is more complex than human vulnerability because there are more species. Each species has its own response.
- ▲ **Structural vulnerability.** Buildings are damaged or destroyed by hazards. The design and materials used in construction determines the level of vulnerability. The construction of most buildings is governed by building codes intended to protect the building occupants from structural collapse. However, the buildings do not necessarily provide protection from extreme wind, seismic, or hydraulic loads. Nor do they provide a complete barrier to toxic air pollutants.

Figure 6-2

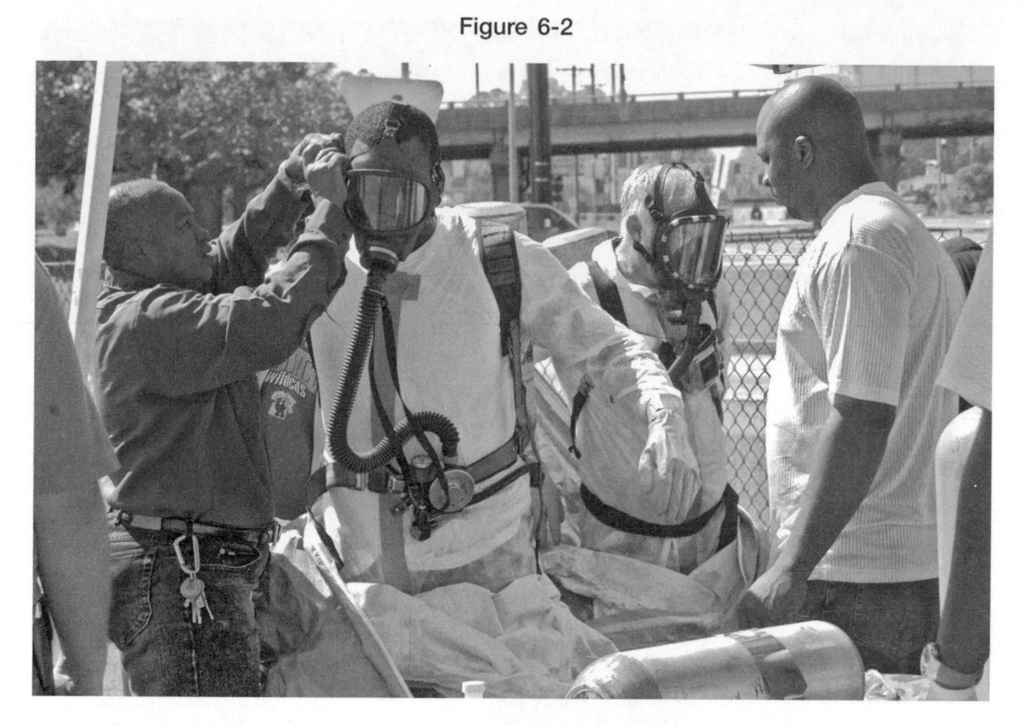

People are vulnerable to hazardous chemicals and gasses and have to wear gas masks to protect themselves.

6.2.3 Social Vulnerability

Social vulnerability refers to "the characteristics of a person or groups and their situation that influence their capacity to anticipate, cope with, resist and recover from the impacts of a natural hazard" (Wisner, Blakie, Canon, and Davis, 2004). Simply put, some people have the psychological, social, economic, and political resources to cope with disaster impacts and others don't. You have to identify groups that are the most socially vulnerable.

FOR EXAMPLE

The Collapse of the World Trade Center

Unfortunately, we all saw that the World Trade Center was vulnerable to terrorists. The buildings were not designed to withstand passenger jets crashing into them at high speed. The impact and fire consumed weakened the buildings' structural frameworks and caused them to collapse.

- Define **human vulnerability.**
- Define **agricultural vulnerability.**
- Define **structural vulnerability.**
- Define **social vulnerability.**

6.3 Event-Specific Conditions

Three event-specific conditions affect a disaster's impact on the community: hazard event characteristics, improvised disaster responses, and improvised disaster recovery.

6.3.1 Characteristics of the Hazard

A given hazard agent may initiate a number of different threats. For example, hurricanes can cause casualties and damage through wind, rain, storm surge, and inland flooding (Bryant, 1991). Volcanoes produce ash fall, explosive eruptions, lava flows, mudflows and floods, and forest fires. Each hazard has six significant characteristics:

1. The speed of onset, which affects how much warning you have.
2. The availability of perceptual cues (wind, rain, or ground movement), which also affects how much warning you have.
3. The intensity of the hazard, which is defined by the amount of energy or hazardous materials released.
4. The scope of the hazard, which is the size of the geographical area and the number of people and businesses affected by disaster impacts.
5. The duration of impact (the length of time the disaster impacts persist), which might be many years in the case of some hazardous materials.
6. The probability of occurrence. Hazards that are likely to occur are more likely to mobilize communities to engage in hazard mitigation and emergency preparedness measures that reduce their vulnerability (Prater and Lindell, 2000).

6.3.2 Improvised Disaster Response

Disaster victims are often portrayed as dazed, panicked, or disorganized. However, people actually adapt when disasters strike. It may take a while for people to adapt, because **normalcy bias** delays people's realization that an improbable event is, in fact, occurring to them. Delays also occur because people often want more information before they take actions to protect themselves. Finally, it takes time for people to develop social organizations that can cope with the unfamiliar situation caused by a disaster. Victims are also often portrayed as only concerned about their own safety and property. However, many victims save others and watch over the other people's property. Residents in nearby areas come to the disaster area to offer assistance. When existing organizations seem incapable of meeting the needs, they expand to take on new members, extend to take on new tasks, or new organizations emerge (Dynes, 1974).

6.3.3 Improvised Disaster Recovery

After there is no longer a threat to lives or property, communities must begin the long process of disaster recovery. Immediate tasks in this process include:

- ▲ Damage assessment.
- ▲ Debris clearance.
- ▲ Reconstruction of infrastructure (electric power, fuel, water, wastewater, telecommunications, and transportation networks).
- ▲ Reconstruction of buildings in the residential, commercial, and industrial sectors.

Individuals and organizations provide funds to help provide disaster assistance. The victims themselves might have financial assets (e.g., savings and insurance) as well as tangible assets (e.g., property) that are undamaged. Low-income victims tend to have small savings. They are also more likely to be victims of insurance redlining and, thus, have been forced into contracts with insurance companies that go bankrupt after the disaster. Therefore, even those who plan

FOR EXAMPLE

School Closing

One of the many casualties of Hurricane Katrina was Tulane University. With the school year about to start, Tulane University had to close its doors due to the destruction in New Orleans. Tulane students had to find other universities to attend. Tulane professors had to relocate and find other short-term teaching jobs. Communication between officials, professors, and students was difficult as well. Losing their server and electricity, Tulane officials had to put together a temporary Web site and use groups on Yahoo! to communicate.

ahead for disaster recovery can find themselves without anything (Peacock and Girard, 1997). Victims can bring in additional money through overtime employment. Charities and local government can donate money and other gifts. The government can also provide assistance through tax deductions or deferrals.

SELF-CHECK

- Define normalcy bias.
- Name four characteristics of a hazard.
- Name three sources of funding for disaster recovery.
- Name the steps a community must take immediately to recover from a disaster.

6.4 Effects of a Disaster

The physical impact of a disaster is measured in deaths, injuries, and property damage. These losses are the most obvious, easily measured, and first reported by the media. The social impact, which includes psychosocial, demographic, economic, and political effects, develop over a long period of time and can be difficult to assess. Despite the difficulty in measuring these social impacts, it is important to monitor them, and even to predict them if possible. Social impacts can cause significant problems for the long-term functioning of specific types of households and businesses in an affected community.

6.4.1 Physical Impact

The physical impact can be measured in casualties and damages.

Casualties

In ranking the disasters, hurricanes caused the most fatalities (Noji, 1997). Worldwide data from 1947–1980 shows:

- ▲ Hurricanes produced 499,000 deaths.
- ▲ Earthquakes produced 450,000 deaths.
- ▲ Floods caused 194,000 deaths.

As the aftermath of the Indian Ocean tsunami illustrated, it can be difficult to determine how many deaths and injuries are caused by a disaster. In some cases it is impossible to determine how many are missing and, if so, whether this is due to death or relocation. Estimates of injuries are similarly problematic. Even when bodies can be counted, there are problems because the disaster may be only a contributing factor to death. For example, someone with a chronic heart problem may have a heart attack while lifting debris after an earthquake.

Damage

Losses of buildings, animals, and crops are rising very rapidly in the United States (Mileti, 1999). Such losses usually result from physical damage or destruction. They can also be caused by chemical contamination, or loss of the land itself to erosion. It usually is the case that collapsing buildings cause damage to personal possessions as well as casualties. This suggests that strengthening the structure will protect the contents and occupants. In most cases this is true, but some hazard agents can damage building contents without affecting the structure itself. For example, earthquakes can destroy the contents of seismically resistant buildings whose contents are not securely fastened. Thus, risk area residents may need to adopt additional hazard adjustments to protect contents and occupants even if they already have structural protection.

One of the most significant structural impacts of a disaster is the loss of people's homes. Losing a home can start a long process of recovery that typically passes through four stages (Quarantelli, 1982a).

1. The first stage is *emergency shelter.* This is any location, including a car or tent, that provides protection from normal wind, rain, and temperature extremes.
2. The next step is *temporary shelter.* This includes food and sleeping facilities that are sought from friends and relatives or are found in commercial lodging. Mass care facilities are acceptable but are generally the least preferred option.
3. The third step is *temporary housing.* This allows victims to reestablish household routines in nonpreferred locations or structures.
4. The last step is *permanent housing* in preferred locations and homes.

Because the poor have fewer resources, they take longer to go through the stages of housing, sometimes remaining in severely damaged homes (Girard and Peacock, 1997). In other cases, they are forced to accept as permanent what was intended as temporary housing (Peacock, Killian, and Bates, 1987). There may still be low-income households in temporary housing even after high-income households all have relocated to permanent housing (Berke, Kartez and Wenger, 1993; Rubin, Sapperstein and Barbee, 1985).

Other important damage includes damage to the land. We understand what damage to expect from some hazards but not all. For example, ashfall from the

1980 Mt. St. Helens eruption was expected to devastate crops and livestock, but this did not happen (Warrick et al., 1981).

6.4.2 Social Impacts

Long-term social effects of disasters tend to be minimal in the United States. Most disasters affect only a small part of the country's total area. They also tend to strike undeveloped areas more frequently than urban areas. This is because there is more undeveloped land than developed land. Even when there is a major disaster, government and charities direct recovery resources to the impact area and prevent bad long-term effects. For example, Hurricane Andrew inflicted $26.5 billion in losses to the Miami area, but this was only 0.4% of the U.S. Gross Domestic Product (Charvériat, 2000).

Psychosocial Impact

Disasters can cause a wide range of negative psychological responses (Bolin, 1985; Gerrity and Flynn, 1997; Houts, Cleary and Hu, 1988; Perry and Lindell, 1978). These responses include:

- Fatigue
- Nausea
- Confusion and an inability to concentrate
- Anxiety
- Depression and grief
- Sleep and appetite changes
- Ritualistic behavior
- Substance abuse

In most cases, these responses are mild and brief. They are the result of "normal people, responding normally, to a very abnormal situation" (Gerrity and Flynn, 1997, p. 108). Few disaster victims require psychiatric help. Most do benefit more from crisis counseling. However, some need special attention. This includes children, elderly, the mentally ill, racial and ethnic minorities, and families of those who have died. Emergency workers also need attention. They work long hours, have witnessed horrors, and belong to organizations in which discussion of emotions may be seen as a sign of weakness (Rubin, 1991).

Instead, the majority of those struck by disasters adapt by saving their own lives and those of their closest associates. Despite the stories of looting, there is actually an increase in prosocial behaviors such as donating money and items during disasters. There is a decrease in crime (Drabek, 1986; Mileti, Drabek, and Haas, 1975; Siegel, Bourque, and Shoaf, 1999). In some cases, people even risk their own lives to save the lives of strangers (Tierney, Lindell, and Perry, 2001).

Demographic Impacts

The population of a disaster impact area usually changes very little in the United States, because the number of deaths due to disasters has been relatively small. For example, the 6000 deaths in the 1900 Galveston hurricane were 17% of the city's population. There were more than a thousand deaths from Hurricane Katrina in New Orleans, but this was less than 1% of that city's population. The largest demographic impacts of disasters are due to immigration and emigration. People temporarily move to the area to take construction jobs to rebuild the city. People leave the city to find temporary housing. In many cases, people who leave return. However, some areas are not rebuilt and may become "ghost towns" (Comerio, 1998). Other reasons why people choose to leave include:

- ▲ Fear another disaster will hit.
- ▲ Loss of jobs or community services.
- ▲ Increased neighborhood or community conflict.

Economic Impact

Property damage can be measured by the cost of repair or replacement (see Figure 6-3). These losses are difficult to measure because not all the information is recorded. For insured property, the insurers record the amount of the deductible and the reimbursed loss. However, uninsured losses are not recorded. Although they can be estimated, the estimates are not very accurate. Some assets are not replaced and their loss causes a reduction in consumption or a reduction in investment. Other assets are replaced either through donations or purchases. In the latter case, the cost of replacement must come from some source of recovery funding. Some options for recovery financing include obtaining tax deductions or deferrals, unemployment benefits, loans, grants, insurance payoffs, or additional employment. Other sources include using savings, selling property, or moving to an area with better housing or employment or less risk.

In addition to direct economic losses, there are indirect losses that arise from business interruption. An earthquake in the community might have left a company's buildings, equipment, and raw materials undamaged. However, if electric power has been lost, workers will not be able to operate the machinery and produce the goods the company sells to stay in business. Business interruption can also be caused by the loss of other infrastructure such as fuel, water, sewer, telecommunications, and transportation. It also can be caused by the loss of workers if they must take time off to take care of their families and rebuild their homes

Local governments also have major economic impacts. They need money for activities such as debris removal, restoring services, and rebuilding stricken areas. Despite these increased costs, there are decreased revenues. These include loss or deferral of sales taxes, business taxes, property taxes, personal income taxes, and user fees.

Figure 6-3

Rebuilding homes is part of the reconstruction that takes place after a disaster.

Political Impact

The disaster recovery period is difficult because the community must make many changes in a very short time. Unless local agencies have developed a disaster recovery plan before the disaster strikes, it will be difficult for the community to solve all of the recovery problems at the same time. This creates conflict as one group's attempts to solve its problems create problems for others. For example, victims often try to rebuild in the same spot because they like their neighborhoods. However, neighbors may become alarmed if victims place mobile homes on their own lots while waiting for permanent housing. Conflicts arise because such housing usually is considered to be a blight on the neighborhood. Neighbors are afraid that the "temporary" housing will become permanent. Neighbors also are pitted against each other when developers attempt to build multi-family units on lots previously zoned for single-family dwellings. Such rezoning attempts threaten the market value of owner-occupied homes.

Attempts to change government policies can arise when individuals sharing a grievance join in their complaints about the recovery process. Victims may try to influence policy by forming a group. Existing political groups can expand their membership or extend their domains to include disaster-related grievances

FOR EXAMPLE

Suddenly Homeless

Much of the media coverage of Hurricane Katrina focused on New Orleans, but several towns were affected too. For example, Biloxi and Gulfport were practically wiped off the map with most of the homes destroyed by the hurricane. Many people lost everything they owned and ended up seeking shelter with friends, families, or through a church or charity.

(Dynes, 1974). New groups can emerge to influence lawmakers. Usually, groups pressure government to provide more money for disaster recovery. However, they may oppose candidates' re-elections or even seek to recall politicians from office (Olson and Drury, 1997; Prater and Lindell, 2000; Shefner, 1999).

SELF-CHECK

- Describe the demographic impacts a disaster can have in a community.
- Describe the economic impact a disaster can have in a community.
- Name three reasons why people leave a town after a disaster strikes.
- Describe the political impacts a disaster can have in a community.

6.5 Emergency Management Interventions

After every disaster, it is natural for people to want to know how the losses could have been prevented or reduced. There are three sets of actions that can reduce losses. Hazard mitigation and emergency preparedness practices directly reduce a disaster's physical impacts and indirectly reduce its social impacts. Recovery preparedness practices directly reduce a disaster's social impacts.

6.5.1 Hazard Mitigation Practices

One way to reduce disaster damage is to adopt **hazard mitigation practices**, which can be defined as actions that protect passively at the time of impact.

That is, they do not require people to take action when the disaster strikes. Hazard mitigation involves:

- ▲ Hazard source control intervening at the point of hazard generation to reduce the probability or magnitude of an event. This includes installing special couplers on railroad tank cars to prevent them from being punctured.
- ▲ Community protection works, such as dams and levees, confining or diverting materials flows.
- ▲ Land-use practices reducing or eliminating development on land that has high hazard exposure.
- ▲ Building construction practices using strong materials and hazard-resistant design, such as window shutters that protect against wind pressure and debris impacts.
- ▲ Building contents protection preventing damage to furniture and equipment such as furnaces, air conditioners, washers, dryers.

6.5.2 Emergency Preparedness Practices

Another way to reduce a disaster's physical impacts is to adopt **emergency preparedness practices,** which can be defined as preimpact actions that provide the human and material resources needed to support active responses at the time of hazard impact (Lindell and Perry, 2000).

The first step in emergency preparedness is to use the community hazard vulnerability analysis to identify the emergency response demands that must be met by performing four basic emergency response functions—emergency assessment, hazard operations, population protection, and incident management.

- ▲ *Emergency assessment actions,* such as projecting hurricane wind speed, define the potential disaster impacts.
- ▲ *Hazard operations,* such as sandbagging around structures, are short-term actions that protect property.
- ▲ *Population protection actions,* such as warning and evacuation, protect people from impact.
- ▲ *Incident management actions,* such as communication among responding agencies, activate and coordinate the emergency response.

Each of these functions must be assigned to emergency response organizations that develop plans and procedures for responding quickly and effectively. The organizations must also obtain the resources they need to perform the functions. Finally, they must also have continued training, drills, and exercises (Daines, 1991).

6.5.3 Recovery Preparedness Practices

Recovery from a major disaster takes much longer and involves much more conflict than people expect. Recovery is faster and more effective when it is based on a plan that has been developed before a disaster strikes (Geis, 1996; Olson, Olson, and Gawronski, 1998; Schwab et al., 1998; Wilson, 1991; Wu and Lindell, 2004).

To design a preimpact recovery plan, you should

- ▲ Define a disaster recovery organization that includes major stakeholders from land-use and building construction agencies, business groups, and neighborhood associations.
- ▲ Identify the location of temporary housing. This is a difficult issue and usually causes conflict. Resolving this before a disaster can speed up the recovery.
- ▲ Determine how to perform essential tasks. These include damage assessment, condemnation, debris removal, rezoning, restoring services, temporary repair permits, and permit processing. All of these tasks must be completed before the impact area can be rebuilt (Schwab et al., 1998).
- ▲ Address the licensing and monitoring of contractors and retail price controls to ensure victims are not exploited. Also address the administrative powers and resources available. Local government will be overwhelmed by all of the work that needs to be done immediately after a disaster, so agencies should make arrangements to borrow staff from other jurisdictions and to use trained volunteers such as local engineers, architects, and planners.
- ▲ Determine how recovery tasks will be carried out at historical sites, which involve special issues such as constraints on the demolition of damaged structures and the materials used during reconstruction (Spennemann and Look, 1998).
- ▲ Recognize the recovery period as a unique time to enact policies for hazard mitigation and incorporate this objective into the recovery planning process.

FOR EXAMPLE

Who is Doing What?

In the aftermath of Hurricane Katrina, there was a lot of confusion among federal, state, and local officials over which group was responsible for various tasks. The governor believed FEMA should be removing the bodies of the deceased victims, while FEMA stated it was the responsibility of the state. The mayor of New Orleans wanted the National Guard to forcefully remove people from their homes, while the federal government believed that was a local law enforcement issue.

SELF-CHECK

- Define hazard mitigation practices.
- Define emergency preparedness practices.
- Explain why hazard mitigation practices and emergency preparedness are important.
- Outline the key steps of a preimpact recovery plan.

6.6 Conducting Community Hazard and Vulnerability Analyses

We will now discuss how you can use the information in the previous sections to guide assessments of your communities' exposure to specific hazards and vulnerability to physical and social impacts.

6.6.1 Mapping Natural Hazard Exposure

Communities vary in their exposure to environmental hazards. An important part of your job is to identify the hazards that threaten your community. There are many useful sources of information you can use to do this. One source is the set of maps contained in FEMA's (1997) **Multi Hazard Identification and Risk Assessment.** This source describes exposure to most natural hazards and some technological hazards. The maps of natural hazard exposures in the book can be supplemented by visiting several Web sites, including those belonging to FEMA (www.fema.gov), the U.S. Geological Survey (www.usgs.gov), and the National Weather Service (www.nws.noaa.gov). Although these maps provide a good start toward assessing the potential impacts of disasters, they have three limitations:

1. Many of these maps are designed to compare the relative risk of large areas. This information does not tell you which areas *within* your community are most likely to be struck by a disaster. For example, a coastal county might be exposed to hurricanes but only a small area is exposed to significant damage. Smaller scale maps are needed to assess exposure of different areas to storm surge, inland flooding, and high wind.
2. These maps vary in the amount of information they provide. For example, hurricane maps identify areas that will be hit by Category 1-5 hurricanes. However, these maps do not provide you with the probability of each category of hurricane striking your jurisdiction. By contrast, you can use U.S. Geological Survey earthquake maps to find out the probability that an earthquake will exceed a given intensity. These maps tell you which areas in your city contain buildings that are most likely to collapse.

Within each of these areas, each building's probability of collapse must be assessed by structural engineers.

3. Third, these maps often lack the information you need to assess the relative risk of different hazards. In deciding how to allocate your scarce resources, you need to know what is the likelihood of a flood *in comparison to* a tornado, an earthquake, or a toxic chemical release.

In many cases, you can only categorize disaster impact as high, medium, or low (FEMA, 1996). This provides only a rough estimate for determining which hazards need the most attention. In some cases, these rough estimates are not a problem because all disasters have the same demands. For example, earthquakes, hurricanes, and toxic chemical spills require reliable communications among agencies and with the public. Of course, some emergency management measures are hazard specific. Mitigation measures for earthquakes, hurricanes, and toxic chemical spills are quite different from each other. In these cases, ranking your community's hazards can help you allocate resources to get the biggest reduction in likely casualties and damage.

6.6.2 Mapping Hazmat Exposures

Facilities and vehicles that contain hazardous materials pose a risk to their communities. Accidents or terrorist attacks could release these toxic materials into the environment. To assess the risks, you should create an inventory of which facilities use which chemicals.

To create a chemical inventory,

- Identify dangerous chemicals.
- Identity their locations.
- Identify their quantities at those locations.
- Identify the highway, railway, water, and air routes used to transport chemicals.
- Use this information to assess the threats these chemicals pose to the facility, its workers, its neighbors, and the environment.

You can also identify the areas, known as **vulnerable zones (VZs),** that are likely to be affected by chemical releases. A VZ can be calculated once you know

- A chemical's toxicity.
- A chemical's quantity available for release.
- The type of spill (liquid or gaseous).
- The release duration (e.g., 10 minutes).
- The weather conditions (wind speed and atmospheric stability).
- The terrain (urban or rural).

There are software programs that can help you use this information to determine the size of the affected area. To learn about two of the programs, visit www.epa.gov/ceppo/cameo or yosemite.epa.gov/oswer/ceppoweb.nsf/content/rmp-comp.htm.

You should also be familiar with the ***Technical Guidance for Hazards Analysis*** and the ***North American Emergency Response Guidebook.*** The *Technical Guidance for Hazards Analysis* lists extremely hazardous substances (EHSs) and describes a simple manual method for calculating VZs. The *Emergency Response Guidebook* contains a helpful table that lists the chemicals commonly found in transportation. It also lists each chemical's identification number. It informs you which one of the 172 emergency response guides in the book provides the information needed to respond to a spill. It also helps you to determine how far from the spill location to shelter in-place or evacuate residents. One limitation of the *Emergency Response Guidebook* is that it classifies releases only as small or large so the protective action distances are only approximate. The procedures in the *Technical Guidance for Hazards Analysis* or the *Handbook of Chemical Hazard Analysis Procedures* provide greater precision in estimating the release rate and, thus, the protective action distances.

Nuclear power plants pose special risks. The Nuclear Regulatory Commission requires each plant to establish a 10-mile radius inhalation emergency planning zone (EPZ). An EPZ is the area in which authorities might evacuate residents or shelter them in-place to avoid inhalation exposure and direct radiation. In addition, there is a 50-mile radius ingestion pathway EPZ in which authorities should monitor water, milk, and food for contamination.

6.6.3 Mapping Exposure to Secondary Hazards

Some disasters can initiate others. One way of identifying areas exposed to multiple hazards is to use a Geographical Information System (GIS) to overlay the areas subject to these different hazards. The most common secondary hazards are listed in Table 6-1.

6.6.4 Assessing Physical Vulnerability

You must identify the buildings that are located in the areas exposed to hazards. Buildings can be vulnerable to environmental hazards because of inadequate designs, inadequate construction materials, or both. You can often determine a building's structural vulnerability by its age and type—residential, commercial, or industrial. It is very important to identify facilities that have special needs. For example, one of the disturbing stories after Hurricane Katrina was that some nursing homes were evacuated late and some were not evacuated at all. These facilities have special needs because their patients could not evacuate on their own. Other examples of facilities with special needs are listed in Table 6-2. Characteristics for these facilities are listed in Table 6-3.

Table 6-1: Secondary Hazards

Primary hazard	*Secondary hazards*
Severe storms	Floods, tornadoes, landslides
Extreme summer weather	Wildfires
Tornadoes	Toxic chemical or radiological materials releases
Hurricane wind	Toxic chemical or radiological materials releases
Wildfires	Landslides (on hillsides in later rains)
Floods	Toxic chemical or radiological materials releases
Storm surge	Toxic chemical or radiological materials releases
Tsunamis	Toxic chemical or radiological materials releases
Volcanic eruptions	Floods, wildfires, tsunami
Earthquakes	Fires, floods (dam failures), tsunami, landslides, toxic chemical or radiological materials releases
Landslides	Tsunami

There are three questions you must ask when assessing a building.

1. Does the building have the strength to withstand hazard impacts such as wind, shaking, or water? If not, the concern is the structure itself.
2. Can the building protect its contents? For example, buildings that survive ground shaking without damage can transmit the motion to light fixtures, cabinets, and furniture—possibly damaging these items and harming the occupants.
3. Can the building protect its occupants? This is especially important with hazardous materials because they can infiltrate into a structure and kill the occupants without damaging the building.

The importance of each question varies from one hazard to another. For flooding and storm surge, the building must resist water that is threatening the contents and occupants. In other cases, the building must resist wind loads (tornadoes and hurricanes), blast forces (explosions and volcanic eruptions), and ground shaking (earthquakes) to protect the structure, contents, and occupants. For chemical, radiological, and volcanic ash threats, it is the tightness of construction that prevents contaminated air from entering the building. In the case of exposure to radioactive material, dense construction materials can shield occupants from plume radiation and surface contamination.

Table 6-2: Special Facilities

Health related	*Religious*
Hospitals	Churches/synagogues
Nursing homes	Evangelical group centers
Halfway houses (drug, alcohol, mental retardation)	*High density residential*
Mental institutions	Hotels/motels
Penal	Apartment/condominium complexes
Jails	Mobile home parks
Prisons	Dormitories (college, military)
Detention camps	Convents/monasteries
Reformatories	*Transportation*
Assembly and Athletic	Rivers/lakes
Auditoriums	Dam locks/toll booths
Theaters	Ferry/railroad/bus terminals
Exhibition halls	*Commercial*
Gymnasiums	Shopping centers
Athletic stadiums or fields	Central business districts
Amusement and Recreation	Commercial/industrial parks
Beaches	*Educational*
Camp/conference centers	Day care centers
Amusement parks/ fairgrounds/race courses	Preschools/kindergartens
Campgrounds/recreational vehicle parks	Elementary/secondary schools
Parks/lakes/rivers	Vocational/business/specialty schools
Golf courses	Colleges/universities
Ski resorts	
Community recreation centers	

Adapted from Lindell and Perry (1992).

Table 6-3: Characteristics of Special Facilities

Characteristics of users	*Special considerations*
Mobility of users	Ambulatory Require close supervision Nonambulatory Require life support
Permanent residence of users	Facility residents Residents of hazard impact area but not of the facility (e.g., prison guards) Transients
Periods of use	Days of week/hours of day Special events
User density	Concentrated Dispersed
Sheltering in place	Highly effective Moderately effective Minimally or not effective
Transportation support	Would use own vehicles Require buses or other high occupancy vehicles Require ambulances

Adapted from Lindell and Perry (1992).

Human Vulnerability to Inhalation Exposure

The most immediate health hazard from radiological material or toxic chemicals is inhaling them (U.S. EPA, 1987). If such hazmat is released, people must shelter in-place. A shelter has to be closed tightly enough to keep out the hazmat but also have enough oxygen to keep the occupants alive until the danger has passed. Unfortunately most buildings are leaky, allowing contaminated air to enter. This happens even with the doors and windows closed. Contaminated air can leak in through the furnace flue, ventilation fans, and leakage sites near the ceiling. It also comes through a large number of small openings including cracks around windows and doors, electrical outlets, and gaps between building walls and foundations. The rate at which indoor and outdoor air are exchanged is measured in air changes per hour. The rate of air exchange increases with the amount of leakage area, the wind speed, and the temperature difference between the indoor and outdoor air.

The best way to assess a building's suitability for sheltering in-place is to identify the presence of a vapor barrier in the walls and ceiling of a structure.

This is most common in houses in cold climates built after 1960 (Wilson, 1989). Check if most of the homes were built before or after 1960. If many were built after 1960, then sheltering in-place is more effective than if the homes are older. The presence of storm windows and doors can also be a useful indicator. Consult with your local utilities department to determine what information is available regarding the air exchange rates for different structures. Special facilities, such as hospitals, should be examined on a case-by-case basis to determine their air exchange rates.

Human Vulnerability to Radiological Materials

Radioactive material released from a nuclear power plant, during a transportation accident or from a "dirty bomb," can produce external radiation from the radioactive plume and ground contamination. Dense building materials such as concrete, brick and stone provide the best protection (Burson and Profio, 1977). In particular,

- ▲ Sheltering in a wood frame dwelling provides little more protection from cloud and ground exposure than does "sheltering" in a vehicle.
- ▲ Sheltering on the ground floor of a masonry home with no basement or in the basement of a wood frame home gives considerably higher levels of protection. One would be exposed to 50% of the cloud and less than 20% of exposure to ground contamination.
- ▲ The basement of a masonry house is even more effective. Someone sheltering in such a facility would be exposed to 40% of the cloud and 5% of the ground contamination.
- ▲ A large office building is the most effective shelter of all, reducing cloud exposure to 20% and ground contamination exposure to 1%.

According to analyses conducted by the Nuclear Regulatory Commission, even a late evacuation is better than home shelter, large building shelter is better than late evacuation, and early evacuation is best of all (McKenna, 2000). Those who are within five miles of a nuclear power plant have high vulnerability if they remain in their homes during a release. Only large building shelter provides as much safety as early evacuation.

Assessing Agricultural and Livestock Vulnerability

It is rarely the emergency manager's job to determine how crops and livestock will react to hazards. If agriculture is a significant part of the local economy, you should consult agricultural experts such as the U.S. Department of Agriculture staff and county extension agents. Animal and plant species will react differently to extreme environmental conditions. For example, fruit orchards can be devastated by wind speeds that have no impact whatsoever on rangeland. Also, the damage to many crops depends on the stage in growth cycle. Some crops, for example, will not be bothered by wind damage until just before harvest.

6.6.5 Assessing and Mapping Social Vulnerability

Social vulnerability arises from the potential for extreme events to cause changes in people's behavior. People can vary in their potential for injury to themselves and their families. They also vary in the potential for destruction of their homes and workplaces, as well as the destruction of the transportation systems and locations for shopping and recreation they use in their daily activities.

Assessing Psychosocial Vulnerability

The news media provided a riveting account of people's behavior in New Orleans after Hurricane Katrina. Despite the stress of going days without food, water, and medical attention, most people at the Superdome and convention center were able to adapt to the situation. However, some people have inadequate *problem-focused* coping skills to solve such problems. Instead, they use *emotion-focused* coping strategies, such as alcohol and drug abuse, to get through psychologically painful situations. According to Ozer and Weiss (2004), the people who are most likely to develop serious psychological conditions such as posttraumatic stress disorder (PTSD) are those who have:

- Predisposing characteristics (e.g., very low intelligence, previous psychological trauma).
- Severe personal impacts.
- Continuing stress.
- Low social support after the traumatic event.

The groups most likely to have high psychosocial vulnerability are the very young (who have limited verbal skills), the very old (who tend to be more isolated), and those with preexisting psychological problems.

Assessing Demographic Vulnerability

Older, wealthier homeowners are likely to be tied to their community (Turner, Nigg, and Heller-Paz, 1986) and seek permanent housing there even if their homes have been destroyed. Similarly, ethnic minorities are likely to have large families that make them stay. However, communities will experience out-migration of families if people cannot find temporary housing, especially if the local economy was already in decline before the disaster. There will be in-migration for the construction labor force, but these will mostly be single males.

Assessing Economic Vulnerability

Wealth is a major part of economic vulnerability. Wealth is composed of tangible assets and financial assets. Tangible assets such as buildings, equipment, furniture, and vehicles may be destroyed during a disaster. Financial assets, such as bank accounts and stocks and bonds, are recorded electronically and will survive a disaster. Households and businesses have both types of assets but many households' wealth is in their home equity. Consequently, most of their wealth is vulnerable to disaster impact.

Businesses need labor, customers, suppliers, and distributors to operate. If many of the people in the community lose their homes, there will not be anyone to fulfill these roles. In addition, businesses cannot operate without infrastructure. Managers estimate they can operate for four hours without telephones, 48 hours without water or sewer, and 120 hours without fuel (Nigg, 1995). If businesses do not have infrastructure, they must close even if the disaster did not damage their property.

Measures of household income are available in census files. Available census data on businesses are more limited. The U.S. Census Bureau's Web site (specifically censtats.census.gov) provides zipcode-level data on the number of businesses in each economic sector broken down by number of employees. These data can be overlaid onto risk areas for different hazards, such as 500-year floodplains, hurricane surge zones, or earthquake seismic zones, to develop estimates of the community's economic vulnerability to disaster impact.

Assessing Political Vulnerability

Political vulnerability is created when people consider the response and recovery to be ineffective or, worse yet, deliberately intended to neglect them. Government agencies that are believed to lack legitimacy, expertise, and adequate information for making decisions about the allocation of public resources prove vulnerable in the aftermath of disaster. There currently are no direct measures of political vulnerability that are available.

Predicting Neighborhood Social Vulnerability

Social vulnerability is not randomly distributed. There are predictors that clearly point to parts of the population that are the most likely to be vulnerable. These factors include gender, age, education, income, and ethnicity. Table 6-4 lists a sample set of social vulnerability indicators that are available from census data (Prater et al., 2004). The neighborhoods that have the highest percentages of these vulnerable groups are ones that should concern emergency managers the most.

Neighborhoods need special attention if they are high in hazard exposure, structural vulnerability, and social vulnerability. This is because households who have the social vulnerability often occupy the areas most likely to be hit by a disaster and also occupy the oldest, poorest maintained buildings.

We can identify areas called vulnerability hotspots using a GIS. You can combine data on hazard exposure, structural lifeline vulnerability, and social vulnerability in a vulnerability hotspot analysis, as illustrated in Figure 6-4.

Table 6-4: Indicators of Social Vulnerability

Vulnerable Groups	*Vulnerability indicators*
Female Headed Households	Percent female headed households
Elderly	Percent individuals over 65 Percent of elderly households
Low income/high poverty	Percent of households below poverty level Percent of households below HUD standards
Renters	Percent of households residing in rental housing Percent of households residing in rental housing by type of dwelling units
Ethnic/racial/language minorities	Percent of individual from African-American, Hispanic, and other minorities Percent of non-English speakers
Children/youth	Percent of population in selected age groupings Percent of households with dependency ratios above a specified level
Social vulnerability hot spot analysis	Areas with combined social vulnerabilities.

Figure 6-4

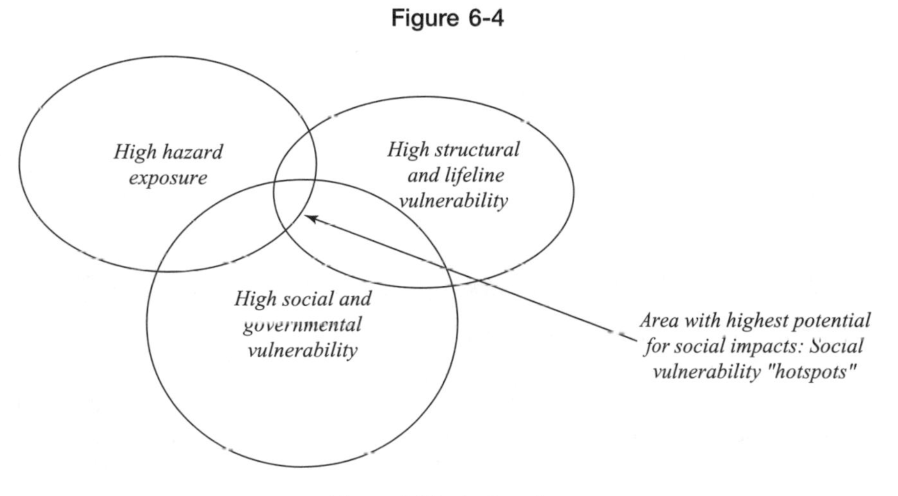

Vunerability hotspots.

6.6.6 Vulnerability Dynamics

In protecting communities, it is helpful to understand why people put themselves at risk.

- People move to land that can be used for agriculture, transportation, and recreation.
- There is a lack of accountability for investment decisions. Developers are at risk for only a short period of time. Then they pass an investment on to others (homeowners, insurers, mortgage holders) who will bear the financial impact of a disaster.
- Many risk area residents are new arrivals who are unaware of their hazard exposure. Even long-term residents sometimes have little or no information. This is because information is sometimes suppressed by those with a major stake in the community's economic development (Meltsner, 1979).
- Many people ignore low probability events because they think of these events as occurring far in the future. People also believe that the events will not happen to them (Weinstein, 1980).
- Politicians tend to ignore consequences that they expect to occur only after their term of office is over.
- People move into hazardous areas because the housing that is available in safer areas is too expensive or too far away from work, schools, or families and friends.
- People suffer more when they have physical and financial assets located *only* in the risk area. Low-income households and small businesses often have so few physical or financial assets that they cannot afford to locate some of them in safer areas.
- Hazard insurance suffers from **adverse selection,** which means that only those who are at the greatest risk are likely to purchase it (Kunreuther, 1998).
- Many risk area occupants overestimate the effectiveness of hazard adjustments, such as dams and levees (Harding and Parker, 1974).

6.6.7 Conducting Hazard Vulnerability Analysis with HAZUS-MH

Hazards US-Multi Hazard (HAZUS-MH) is a computer program that predicts losses from earthquakes, floods, and hurricane winds. The program estimates casualties, damage, and economic losses. HAZUS-MH information can be used for hazard mitigation, emergency preparedness, and recovery preparedness planning. This information can also be used to prepare for a disaster. HAZUS-MH information is also useful for emergency response and disaster recovery.

HAZUS-MH supports three levels of analysis:

1. Level 1 uses national average data to produce approximate results. A Level 1 is an initial screen that identifies the communities at highest risk.
2. A Level 2 takes refined data and hazard maps to produce more accurate estimates. Input for a Level 2 is obtained from local emergency managers, urban and regional planners, and GIS professionals.
3. A Level 3 uses community-specific information to produce the most accurate loss estimates. Input for a Level 3 is obtained from structural and geotechnical engineers, as well as other experts to examine threats such as dam breaks and levee failure.

This program has separate models for earthquakes, floods, and hurricane winds. Further information about HAZUS-MH is available at www.fema.gov/hazus.

6.6.8 Analyzing and Disseminating Hazard Vulnerability Data

Powerful computers are extremely useful for identifying areas at risk and projecting losses. In addition, desktop computers also provide you with access to internet sources. Federal and state agencies have generated many hazard analysis documents, maps, and databases that are in digital form and are available to place on various Web sites. State emergency management agency (SEMA) Web sites can be linked to your site. This allows users to immediately access information that might take them months to obtain if they were to request paper copy. Finally, information posted on a Web site can be updated frequently and is less expensive than printing and distributing paper reports.

Analyzing Hazard Vulnerability Data

Despite the great promise of computers in analyzing hazard, these tools are not often used. Lindell and Perry (2001) reported that only 59% of local emergency planning committees (LEPCs) they studied had calculated their communities' vulnerable zones. Of those, only 36% had used computer models such as CAMEO (National Safety Council, 1995) or ARCHIE (FEMA, no date). In addition, there were differences among types of computer use. Some LEPCs used databases for data on local chemical hazards and emergency response resources more than others did. Many LEMAs have a long way to go if they want to use technology to its fullest advantage.

Dissemination of Hazard Vulnerability Data Via Web Sites

Most SEMAs provide some information on their Web sites (Hwang, Sanderson, and Lindell, 2001). The most commonly addressed hazards on these sites are hurricane, earthquake, flood, fire, tornado, hazardous material, storm, terrorism, drought, and radiological material. However, many states that are also at risk from these hazards did not have any information on their sites. Also, there are other hazards that were not addressed that should receive attention. Less common hazards such as tsunami, structure failures, landslides, and avalanche are neglected even though they have the potential for significant impacts. A recent survey of Texas emergency managers and land-use planners found more than one-third of them use few sources of hazard analysis information. Nearly one-third use no hazard analysis information at all (Lindell, Sanderson & Hwang, 2002). Of those who do use hazard analysis information, two-thirds of the materials used are printed documents. Only one-third of the materials used were obtained from internet sources. Some reasons for the lack of information on Web sites include the following:

- ▲ Some LEMAs lack the resources to deliver information through internet sources.
- ▲ Some LEMAs may think a hazard analysis Web site is unnecessary because people already know about most common hazards such as storms and floods.
- ▲ Some LEMAs may not believe that Web sites are an effective method of disseminating hazard analysis information.
- ▲ Some LEMAs may provide little hazard information on their Web sites because they already distribute this information through other media.
- ▲ LEMAs might overlook hazard agents with which they have little or no recent disaster experience.

The demand for web-based hazard information will increase. According to FEMA's Web site (www.fema.gov), internet users visited FEMA's hurricane-related Web sites more than 1.25 million times in the week after Hurricane Bertha hit the U.S. in 1996. Recognized authorities need to play a leading role in distributing information (Fischer, 1998). SEMA Web sites can help you to easily collect information. Your LEMA's Web sites can add value by providing information that is specific to your community. In both cases, LEMA Web sites can help residents recognize their exposure to hazards. This can motivate them to adopt hazard adjustments that reduces their vulnerability to these threats.

RECOMMENDATIONS FOR HAZARD ANALYSIS WEB SITES

- ▲ Recognize that your Web site will be considered the most reliable and accurate source. You must make sure the information is accurate.
- ▲ Coordinate the information provided through your Web site with the guidelines contained in *Talking about Disaster: Guide for Standard*

Messages, available through the American Red Cross Web site (www.redcross.org/disaster/safety/guide.html).

- Address all main hazards to which your community is vulnerable and also provide information about the likelihood of other events so that people can judge which ones deserve the greatest priority. Display this information on maps.
- Provide nontechnical information about hazards so users can understand how a disaster will affect their communities. Important information includes the speed of onset, scope and duration of impact, and the magnitude of different types of consequences such as casualties (deaths and injuries), property damage, and economic impacts (disruption to industrial, commercial, agricultural, and governmental activity).
- Provide information about actions people can take to protect themselves, their families, and their property. Describe the specific steps required to perform any unfamiliar actions.
- Provide links to other emergency-related information, such as situation reports about current incidents and information available from other emergency-related organizations.
- Keep text clear and succinct. Use large and legible fonts and simple color design schemes so the information is easy to read.
- Provide enough figures and pictures to explain the text and maintain interest. Avoid too many graphics because this can cause the information to download slowly. Users will become frustrated.
- Ensure your Web site is compliant with the Americans with Disabilities Act, which requires pictures and graphs to be described in words and that your site be navigable without a mouse.
- Make it easy for viewers to download information by attaching documents in PDF or major word processor (e.g., Word Perfect® or MS-Word®) format.
- Include contact information with postal and email addresses, telephone numbers, and fax numbers of persons from whom users can obtain additional information or to whom they can offer suggestions.
- Verify that your server can handle many users during an emergency.

6.6.9 Assessing Risk

A major problem in applying the methods of Hazard vulnerability analysis is risk and uncertainty. FEMA's (1997) *Multi-Hazard Identification and Risk Analysis,* together with information from state and federal agency Web sites, can identify the hazards to which a community is exposed. This is a good start, but it is not enough. You should consider the probability rather than just the possibility of a disaster. An event that has a 50% chance of occurrence needs more attention than one that has only a 1% chance of occurrence. However, there are often problems with probability estimates.

- ▲ The probability that an event of a given magnitude will occur within a specified time period is uncertain.
- ▲ The probabilities have not been calculated for many technological hazards such as the probability of a toxic chemical spill during transport through a given jurisdiction.
- ▲ The probabilities are difficult to estimate because the historical record is too short. This is a very difficult problem when trying to determine the probability of very infrequent events, such as being struck by a Category 5 hurricane.
- ▲ The estimated probabilities are unstable because the system is changing. For example, the true probability of a major flood is often higher than its nominal value because upstream development is increasing the probability of downstream flooding. However, the extent to which the probability has increased is unknown.
- ▲ The event probability may be confidential. For example, chemical facilities won't share their risk analyses with anyone but public safety and emergency management personnel for security reasons.

For all these reasons, you cannot currently obtain precise information about hazard vulnerability. Moreover, such information is also unlikely to be available in the near future. This might seem to be a negative view of the usefulness of HVAs but it is not. Rather, it simply recognizes the current limitations of HVA technology and the resources you can devote to this activity.

However, you can still do a good job without extremely precise data. This is because you only need enough information to decide how much money to spend on different activities. That is, you need enough HVA data to decide how to allocate resources among hazard mitigation, emergency response preparedness, and disaster recovery preparedness. After you assess the relative threat from different types of hazards, remember that these different hazards might have the same disaster demands. For example, hurricanes and inland floods might have different probabilities of occurrence but they require similar emergency responses such as evacuation. Consequently, some investments will help you prepare for multiple hazards. In general, all hazards require effective incident management and most require population protection.

You also need HVA data to lobby for more money. Emergency management must compete with other community needs such as education, health care, and

FOR EXAMPLE

Another Tool for Vulnerability Assessment

The NOAA Coastal Services Center (CSC) also provides guidance on conducting community vulnerability assessment on their Web site (www.csc.noaa.gov/products/nchaz/startup.htm). There are several exercises you can go through to get a complete assessment for your community. The CSC Web site also contains an overview of LIDAR (Light Detection and Ranging) beach mapping to obtain highly accurate elevation data. It also describes a damage assessment tool for rapid postimpact that allows personnel to retrieve parcel data in a GIS database and integrate it with FEMA damage assessment forms. Finally, the Web site also explains how remote sensing can be used to provide broad area views of the impact area after a disaster strikes.

transportation. In this case, the HVA should be specific enough for you to persuade others that an increase will benefit the community. This is not a trivial problem. Emergency management must lobby for money to solve *future* problems. Educators, health care professionals, and transportation planners tend to be more successful because they request money to solve *current* problems. When you try to decide how much time and money to allocate for HVA, remember that HVA is not an end in itself. You should spend enough time and money on HVA to be sure you are spending the rest of your budget on the emergency management activities that will produce the greatest reduction in your community's hazard vulnerability.

SELF-CHECK

- Describe the Multi Hazard Identification and Risk Assessment.
- Describe what HAZUS-MH is used for.
- Define vulnerable zones and explain how they can be calculated.
- List five recommendations for designing a hazard analysis Web site.

SUMMARY

As an emergency manager, you must be aware of the factors that make a community vulnerable to a disaster. The more vulnerable a community is, the higher the losses are likely to be. You also must identify key groups of people that are most likely to be affected. Finally, you must understand the characteristics of the disaster that determine the level of impact it has on a community. With these details in mind, you can determine how to best protect your community using mitigation practices and emergency response practices.

KEY TERMS

Adverse selection The tendency for hazard insurance to be purchased mostly by those who are at the greatest risk of filing a claim for losses.

Agricultural vulnerability The vulnerabilities of all species of plants and animals.

Emergency preparedness practices Preimpact actions that provide the human and material resources needed to support active responses at the time of hazard impact.

Hazard exposure Living, working, or otherwise being in places that can be affected by hazard impacts.

Hazard mitigation practices Actions that protect passively at the time of impact.

Hazards US-Multi Hazard (HAZUS-MH) A computer program that predicts losses from earthquakes, floods, and hurricane winds. The program estimates casualties, damage, and economic losses.

Human vulnerability People's susceptibility to death, injury, or illness from extreme levels of environmental hazards.

Multi Hazard Identification and Risk Assessment A FEMA manual that describes exposure to many natural and technological hazards.

Normalcy bias People's tendency to delay recognition that an improbable event is occurring and affecting them.

North American Emergency Response Guidebook A manual that lists the chemicals commonly found in transportation. It details which one of its 172 emergency response guides provides the information needed to respond to a spill. It also helps you to determine how far from the spill location to shelter in-place or evacuate residents.

Physical vulnerability Human, agricultural, or structural susceptibility to damage or injury from disasters.

Social vulnerability Lack of psychological, social, economic, and political resources to cope with disaster impacts.

Structural vulnerability The susceptibility of a structure, such as a building, to be damaged or destroyed by environmental events.

Technical Guidance for Hazards Analysis A guide that lists extremely hazardous substances and describes a simple method for calculating VZs.

Vulnerable zone (VZ) The area surrounding a given source in which a chemical release is likely to produce death, injury, or illness.

ASSESS YOUR UNDERSTANDING

Go to www.wiley.com/college/lindell to evaluate your knowledge of the basics of risk analysis.
Measure your learning by comparing pre-test and post-test results.

Summary Questions

1. Which of the following does not determine the physical impact of a disaster?
   (a) hazard exposure
   (b) physical vulnerability
   (c) adverse selection
   (d) social vulnerability
2. Typically, the people who are most susceptible to any environmental stressor will be the very young, the very old, and those with weakened immune systems. True or False?
3. Human vulnerability is more complex than agricultural vulnerability. True or False?
4. The availability of perceptual cues (such as wind, rain, or ground movement) also affects how much warning you have that a disaster is about to occur. True or False?
5. Now that personal computers are so widespread, almost all communities have completely analyzed their environmental hazards. True or False?
6. Which of the following delays people's realization that an improbable event is, in fact, occurring to them?
   (a) personal conflict
   (b) normalcy bias
   (c) stress
   (d) influence of others
7. The physical impact of a disaster is measured by all of the following except which one?
   (a) media coverage of event
   (b) injuries
   (c) property damage
   (d) deaths
8. Statistics show that what causes the most casualties?
   (a) earthquakes
   (b) hurricanes

(c) fires

(d) floods

9. Which of the following is a reason why it is difficult to assess the probabilities of extreme natural and technological events?
   (a) The probabilities have not been calculated for many technological hazards.
   (b) The probabilities are difficult to estimate because the historical record is too short.
   (c) The estimated probabilities are unstable because the system is changing.
   (d) All of the above.
10. Recovery is faster and more effective when it is based on a plan that has been developed after a disaster strikes. True or False?
11. Who are the people *least* likely to develop serious psychological conditions such as posttraumatic stress disorder (PTSD)?
    (a) people with predisposing characteristics
    (b) people who suffer severe personal impacts
    (c) people with high social support after the traumatic event
    (d) people who experience continuing stress
12. Dense building materials such as concrete, brick, and stone provide the best protection if radioactive material is released from a nearby nuclear power plant. True or False?
13. The rate at which indoor and outdoor air are exchanged is measured in air changes per second. True or False?

Review Questions

1. What three preimpact conditions determine the impact of a disaster?
2. What are three event specific conditions that determine the impact of a disaster?
3. What are the three types of physical vulnerability?
4. What is social vulnerability?
5. What are the six characteristics of a hazard?
6. After there is no longer a threat to lives or property, communities must begin the long process of disaster recovery. What are three immediate tasks community members should take?
7. Why is it important to monitor social impacts of disasters?
8. How do you measure the physical impact of a disaster?
9. What are five psychological effects of disasters?

10. Name three methods of hazard mitigation.
11. What levels of analysis does HAZUS-MH support?
12. Name five reasons people put themselves at risk.
13. Name five vulnerable groups and indicators of the vulnerabilities.
14. Give three examples of facilities with special needs.
15. Give an example of a secondary hazard for an earthquake.
16. What are some tools/resources for a hazard vulnerability analysis?

Applying This Chapter

1. You are an emergency manager for a large Midwestern city and a new high-rise office building is going to be built. What types of hazards might its occupants be exposed to? Which of these hazards is the building likely to be designed to protect against? What types of people are likely to work in the building and how might they differ in their social vulnerability?
2. You are the emergency manager for New Orleans and you have people who have lost their homes and their jobs. People who have lost almost everything are coming to you and asking where they can obtain financial help. What information do you give them?
3. You are the emergency manager for a small Midwestern town, and you have been asked to describe the impact a disaster could have on the community. You are putting together a presentation on the types of physical and social impacts of disasters. You also have been tasked to describe some mitigation practices the community can take to minimize disaster impacts. What do you include in the presentation?
4. There are chemical facilities located just outside your community. You need to create a chemical inventory and develop a plan in case there is a release from one of the facilities. How do you convince community leaders to provide the funds you need to develop a chemical inventory and calculate the vulnerable zones?
5. You are working with a Web site developer to create a hazard analysis Web site for a suburb of Chicago. The designer needs your input on what content to place on the Web site. Create a presentation that shows the designer what needs to be included on the Web site.
6. You are an emergency manager for an east coast city. You have been asked to produce a report on the advantages and disadvantages of different hazard/vulnerability analysis tools. What do you include in the report? Create a presentation showing what you would include.

YOU TRY IT

Seeking Shelter

For people who are not able to stay with friends or relatives, there are four stages of housing they go through. Take the Hurricane Katrina disaster and give an example of each of the stages.

Assigning Tasks

When a disaster strikes, there can be confusion over the role of each organization. How do you ensure that everyone understands what their responsibilities are?

Analyzing the Threat

You are the emergency manager for a coastal town that not only could face a hurricane but is also close to a nuclear power plant. If either a hurricane or nuclear power plant accident were to happen, you would want to evacuate the population. How do you convince the local and national stakeholders that you need to invest in evacuation systems and capability?

7

HAZARD MITIGATION

Reducing Risk

Starting Point

Go to www.wiley.com/college/lindell to assess your knowledge of the basics of reducing risk.
Determine where you need to concentrate your effort.

What You'll Learn in This Chapter

- ▲ The legal framework for hazard mitigation
- ▲ Mitigation strategies
- ▲ Hazard mitigation measures

After Studying This Chapter, You'll Be Able To

- ▲ Analyze and use building construction practices
- ▲ Demonstrate land-use practices
- ▲ Apply mitigation strategies to different hazards

Goals and Outcomes

- ▲ Manage and reduce the risks of natural hazards
- ▲ Assess the risks of technological hazards and manage them by reducing them
- ▲ Evaluate the five categories of mitigation strategies and how they apply to different hazards

INTRODUCTION

Previously in this book, we mentioned that it is your job as an emergency manager to prepare for the many bad things that can happen to your community. Had emergency managers been better prepared in New Orleans, things would have been different after Hurricane Katrina arrived.

Not only was there public outcry over the emergency response to Hurricane Katrina, but there was also concern over why measures weren't taken to prevent losses from a major hurricane. Critics argued that the levees should have been strengthened before the hurricane. Measures like these that are taken to reduce or prevent losses are called hazard mitigation. Hazard mitigation is most effective when it takes place before disasters. However, communities should also integrate hazard mitigation into their disaster recovery. Some policy makers are trying to include mitigation in the rebuilding of New Orleans so the next hurricane will cause fewer losses. One way to define hazard mitigation is as preimpact actions that provide *passive* protection at the time of disaster impact so there is less need for emergency response actions.

This chapter looks at the topic of hazard mitigation, including the legal framework surrounding hazard mitigation, strategies used during hazard mitigation, and the application of hazard mitigation.

7.1 Legal Framework for Hazard Mitigation

The federal government cannot intervene directly in local land-use or building construction practices. However, it wants to change these practices because it pays for much of the high cost of disaster recovery (Mileti, 1999). Years ago, the federal government tried to reduce losses by reducing hazard exposure. In the case of floods, this led to a program of dams and levees. Unfortunately, flood losses continued to increase, so the federal government has more recently tried to intervene indirectly. States must update hazard mitigation plans within six months of a presidential disaster declaration as a condition for receiving federal disaster assistance.

Hazard mitigation survey teams comprising FEMA, state, and local representatives are now formed after disasters to identify community mitigation needs and opportunities. Whenever a presidential disaster declaration is made, a federal hazard mitigation officer (FHMO) is appointed to manage hazard mitigation programs. The FHMO participates in the preliminary damage assessment, helps assess local mitigation issues, develops a mitigation strategy, and also evaluates state mitigation programs for the regional analysis and recommendation. FEMA and the affected state establish a written agreement that defines the duties and responsibilities that the federal, state, and local governments assume after a disaster. These and other requirements have increased the amount of effort that state and local governments have put into hazard mitigation.

Local governments often feel that federal and state mandates are overly restrictive and do not provide enough funding. Local governments, as the direct regulators of land-use and building construction practices, are politically vulnerable to blame for withholding land from development and requiring hazard mitigation measures that drive up local development costs. States have attempted to support the local governments and meet federal requirements in many different ways. These ways include mandates that local jurisdictions apply traditional land-use planning tools such as zoning and subdivision regulations. However, states have also encouraged local governments to include hazard mitigation objectives in their everyday investment policies to reduce community hazard vulnerability.

As the cost of disasters has risen, some insurers have stopped writing policies in some hazard prone areas. The insurance industry, however, has begun to promote mitigation for households. The Institute for Business and Home Safety (IBHS), an insurance industry coordinating organization, has been a leader in this effort, through its *Showcase Communities* program (www.ibhs.org). IBHS also provides materials on disaster planning to promote business continuity after disasters through its *Open for Business* program.

Hazard mitigation faces important legal challenges in the United States. Several "regulatory takings" cases have been heard in the Supreme Court, the most famous of which was *Lucas v. South Carolina Coastal Council* (1992). These cases have sought to clarify the conditions under which jurisdictions can regulate the use of private property in order to accomplish a public purpose. Government has traditionally held the power of **eminent domain.** The government can force private owners to sell their property to the government at a fair market value if the property is to be used for a public purpose.

FOR EXAMPLE

Eminent Domain

The principles of eminent domain were recently modified by a Supreme Court decision that endorsed a broadened definition of "public benefit" (*Kelo vs. City of New London,* 2005). However, this does not change government's obligation to continue to meet other established conditions when it "takes" private property. Governments must provide adequate compensation for the economic value of any property they acquire through eminent domain. Moreover, widespread outrage about the Supreme Court's decision makes it likely that state legislatures will respond by tightening legal restrictions on the use of eminent domain to condemn private property.

SELF-CHECK

- **Define hazard mitigation.**
- **Describe what FHMO stands for.**
- **Identify the IBHS program that provides materials for disaster planning and promoting business continuity.**
- **Define eminent domain and identify who has this power.**

7.2 Mitigation Strategies

Mitigation strategies have been classified in many different ways. One of the most common is the distinction between structural and nonstructural mitigation. The most common examples of structural mitigation are dams, levees, seawalls, and other permanent barriers that prevent floodwater from reaching protected areas. Nonstructural mitigation includes activities as diverse as reducing chemical quantities stored at water treatment plants, purchasing undeveloped floodplains and dedicating them to open space, installing window shutters for buildings located on hurricane-prone coastlines, and bolting water heaters to walls in earthquake zones. The problem with classifying mitigation strategies as structural and nonstructural is that these terms are vague.

To reduce confusion, this chapter uses a system developed by FEMA (1986) to classify hazard mitigation strategies in terms of five categories—hazard source control, community protection works, land-use practices, building construction practices, and building contents protection.

7.2.1 Hazard Source Control

For some hazards, it is possible to control the source of danger. Technological hazards can be prevented. For example, fires can occur only when there is fuel, oxygen, and an ignition source. Source control for structural fires can be achieved by confining a fuel to prevent it from mixing with oxygen. You can also prevent fires by keeping any fuel/air mixture that does develop away from an ignition source. Source control for chemical releases can be achieved by using nontoxic chemicals. You can also reduce chemical quantities, and maintain equipment to prevent leaks from tanks, pipes, and valves (Ashford et al., 1993).

Hazard source control does not work for natural hazards but, there are some exceptions. Wildfire hazard can be controlled by limiting fuel loads in woodlands and controlling ignition sources. Flood hazard can be controlled by maintaining ground cover that decreases runoff by causing rainfall to infiltrate the soil.

7.2.2 Community Protection Works

Community protection works are most commonly used to divert floodwater past communities that are located in flood plains. They also can be used to provide protection from other types of water flows such as tsunami and hurricane storm surge. Finally, community protection works can protect against two types of geophysical hazards: landslides and volcanic lava flows, and some industrial hazards.

The four major types of flood control works are:

1. Stream channelization
2. Dams
3. Levees
4. Floodwalls

Channelization is the process of deepening and straightening stream channels. Deepening a channel prevents flooding by increasing the volume of water that the channel can carry. Straightening a channel allows the water to move downstream faster by shortening the distance it must travel.

Dams are elevated barriers sited *across* a streambed that increase surface storage of floodwater in reservoirs upstream from them. These structures can be made of concrete, earth, or earth with a rock core that provides additional strength. Dams have floodgates and spillways that allow their operators to release water from the reservoir. The water level in most reservoirs is managed to achieve four conflicting objectives. Electric power generation, water recreation, and irrigation are best achieved by full reservoirs. Flood control is achieved by empty reservoirs. The solution is to fill partially, leaving capacity for floods. However, severe upstream flooding might require an emergency release to protect the dam if the reservoir is full. In a severe storm, a reservoir can fill with massive amounts of water that must be released quickly. In some cases, this can cause downstream flooding that is just as bad as if the dam had never been built.

Downstream flooding can also occur if a dam fails catastrophically, and, of course, the extent of the flooding depends on the size of the reservoir. Dam safety is an issue during routine operation because 20 major dam collapses occur each year worldwide and, 83% of these collapses occur in earth- or rock-filled dams. Most of the risk comes from dams that were constructed with little or no engineering supervision.

Detention basins are similar to dams, but are earth- or rock-filled structures with dry reservoirs that fill only during floods. Unlike dams, whose floodgates and spillways are actively managed, detention basins have passive outflow. This is achieved by placing a pipe through the dam at ground level. The water in the pipe can flow through the dam unimpeded, but any amount that exceeds the capacity of the pipe backs up into the reservoir. This effectively regulates downstream discharge so it cannot exceed a predetermined rate of flow. Because the reservoir bottoms are dry most of the time, they can be used for open space. Ball fields, hiking trails, and even parking lots and picnic shelters are excellent uses for detention basins because these facilities have such low development intensity that they can be repaired quickly and cheaply after flooding subsides.

Levees are elevated barriers placed *along* the streambed that limit stream flow to the floodway. To be effective, a levee must be built on soil that provides a stable foundation. It is constructed of impervious soil, such as clay, that is compacted to prevent it from settling (Army Corps of Engineers Committee on Floodproofing, 1993). Its surfaces are usually planted with grass or other low vegetation to prevent erosion. Additional protection from erosion during flooding can be obtained by using concrete or stone to line the exposed surface. People can easily raise the height of the earthen levees. It is common to see television coverage of volunteers stacking sandbags to raise a levee as a river level rises. However, sandbagging operations conducted during a flood are emergency response actions rather than preimpact mitigation actions.

Levees have a number of design, construction, and maintenance problems. Many have been constructed by local levee districts that have limited budgets. As a result, they are poorly maintained. Levees have four basic types of failure mechanisms (see Figure 7-1):

1. **Wave action** (A) causes levee failure by attacking the face of the levee and scouring away the material from which it is constructed.
2. **Overtopping** (B) occurs when the water level exceeds the height of the levee. Once this happens, the flow of water over the top of the levee begins to erode a path that allows increasing amounts of water to flow through the opening. This can quickly flood the area behind the levee.
3. **Piping** (C) occurs when an animal burrow, rotted tree root, or other disturbance in the levee creates a long circular tunnel through or nearly through the levee. Once the water reaches the "pipe," it has an open path toward the landward face of the levee and can fail the levee.
4. **Seepage erosion** (D) occurs when the height of the water in the river puts pressure on water that has seeped into the riverbed, under the levee, and into the soil on the landward side of the levee. The resulting flow of water can eventually cause boils of muddy water that erode a path for the water to flow underneath and then behind the levee.

Levees force silt to accumulate on the river bottom rather than being deposited onto the surrounding floodplain. Thus, levees can increase flood hazard while simultaneously reducing soil fertility.

Floodwalls are built of strong materials such as concrete. They are more expensive than levees, but they are also stronger. In addition, they can be built nearly vertically. This makes floodwalls attractive in urban areas where space is

Figure 7-1

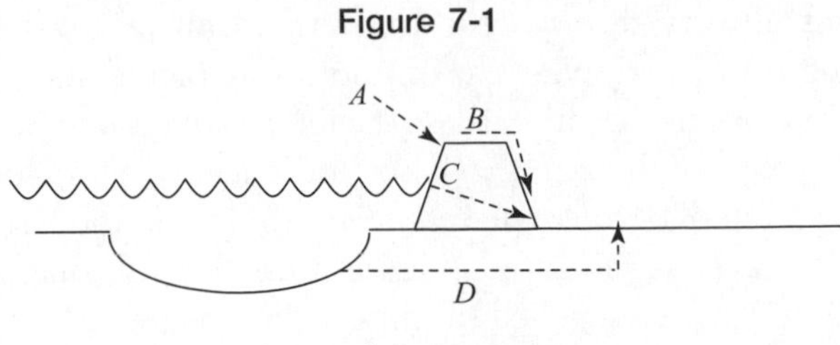

Levee failure modes.

limited and land values are high. Floodwalls must be constructed on stable soil to prevent settling or collapse. In addition, construction on impervious soil avoids seepage erosion under the floodwall. Although levees and floodwalls are typically constructed large enough to protect entire neighborhoods or communities, they also can be built small enough to protect individual structures.

Dams, detention basins, and levees have saved many lives. Nonetheless, it is important to recognize that these protection works can only protect against events up to a given magnitude—called the *design basis event* (DBE). In the United States, the DBE is typically the 100 year flood—a flood with a 1% chance of occurrence in any given year. Dams, levees, and detention basins structures will not protect against events that are more extreme than this. The biggest problem with community protection works is that people overestimate their effectiveness. In some cases, people assume that protection works will eliminate *all* flood hazards. Then they are unprepared when a flood exceeding the DBE strikes their community.

Landslide controls are designed to reduce shear stress, increase shear resistance, or a combination of these two (Alexander, 1993). Reducing shear stress decreases the pressure pushing one soil layer over the top of another. Increasing shear resistance enhances the ability of the two soil layers to remain in place. Landslide protection stabilizes slopes by hardening the soil surface to prevent water infiltration, installing drain fields to remove water from the soil, and building retaining structures such as buttresses, retaining walls, or tie-rods. You can reduce the risk of landslides by avoiding construction on unstable slopes during periods of heavy rainfall. Landslide risk can also be reduced by minimizing the loads these slopes carry. For example, limiting development prevents the weight of additional houses and roads from causing further slides.

Industrial hazard controls are used to confine hazardous material flows. For example, dikes can be constructed around storage tanks to confine any liquid releases that might occur. Such protection works are especially common around petroleum storage tanks. Risk analyses are used to determine the likely volume of any releases that might occur under different failure modes. The expected release volume and release rate can then be used to determine the required storage capacity within the dike.

7.2.3 Land-Use Practices

Land-use practices are defined by the ways people use the land. These include woodlands; livestock grazing; farmland; residential, commercial, and industrial structures; and infrastructure facilities. The local government can influence land-use practices through the use of risk communication, incentives, and sanctions.

- Risk communication changes land-use practices voluntarily because property owners develop a more accurate understanding of hazard vulnerability and hazard adjustments (Lindell and Perry, 2004).
- Incentives change land-use practices voluntarily because property owners are free to choose whether to accept them. One incentive for foregoing development of a hazard-prone property is the sale of development rights. If the incentive adequately compensates land owners for their opportunity cost (the profit they think they would have made from developing the land), they are likely to accept it. If the incentive is too small, they can proceed with development.
- Regulations change land-use practices involuntarily because property owners are subject to legal penalties such as fines and possibly jail sentences for violating the regulations.

Local governments have a variety of land-use management practices available to them that can be used to reduce hazard exposure. Chief among these is land-use planning, a well-recognized tool for natural hazard mitigation (Burby, 1998; Godschalk, Kaiser, and Berke, 1998). Under the Constitution, states delegate the land-use planning function to local governments under the police power to protect public health, safety, and welfare. States vary in the levels of planning they require, so local reliance on land-use regulation varies tremendously. Some states require local governments to develop comprehensive plans that include land-use elements and environmental hazards elements. Other states allow, but do not require, comprehensive or land-use plans, so fewer local jurisdictions in these states have such plans. Among the other tools available to local governments for hazard mitigation are zoning, subdivision regulations, capital improvements programs, acquisition of property or development rights, and fiscal policies. Local governments should explore these practices.

Acquisition of Land and Development Rights

In some cases, a community might purchase land outright to stop development in hazardous areas. Such land is often located along rivers, lakes, and seashores. This makes it a good location for public uses such as hiking and bicycle trails, golf courses, or picnic areas. You can also purchase development rights rather than the property itself. Purchasing only the development rights allows landowners to continue to use their property. An example of this is farmers who want to continue to raise crops and livestock. In exchange, the community ensures the land is not used to build residential, commercial, or industrial structures that increase physical vulnerability to hazards.

Capital Improvements Programs

A **capital improvements program (CIP)** is used to plan community infrastructure and critical facilities. They require a significant investment of public capital, so CIPs assess the need for these facilities. CIPs describe what type of facilities will be built, where they will be built, and how they will be financed. The amount of infrastructure that local government controls is small in some communities. However, local government can ensure the facilities it does control will be built in locations that limit hazard exposure. They will also be built according to construction practices that limit physical vulnerability.

Fiscal Policy

Local governments have discretion in their use of fiscal policy and taxation to distribute the public costs of private development on hazardous property. Impact fees for infrastructure such as streets and sewers offset the increased cost of construction and maintenance in hazardous areas. Cities can also offer tax incentives by reducing the property tax burden on undeveloped land in hazard areas. This reduces the incentive to develop such land in more intensive (and lucrative) uses. This is done by assessing property taxes at the current use value rather than at the market value. The owners are required to sign a written agreement to keep the property at its current level of development. If they later decide to develop the land, they must pay all of the taxes that were previously forgiven.

Hazard Mitigation Plans

Hazard mitigation plans can be part of a comprehensive plan. Comprehensive plans are based on an assessment of the community's:

- Geography and history
- Economy
- Demographic trends
- Transportation
- Housing
- Historic preservation
- Environmental protection
- Land-use practices

This data are used to define strategies for achieving the kind of development that residents desire. These strategies also determine the location of infrastructure and critical facilities. One advantage of a comprehensive plan is the increased likelihood that hazard mitigation measures will be area-wide rather than site-specific (Milliman and Roberts, 1986).

In other cases, you might have a stand-alone mitigation plan. Some communities have no comprehensive plan, or have weak and outdated plans, yet they face significant hazards. All local governments must have mitigation plans to receive federal assistance for disaster recovery. This provides an opportunity to connect hazard mitigation to land-use policy through a planning process.

A mitigation plan should be based on the hazard vulnerability analysis. The hazard vulnerability analysis HVA provides a framework for setting goals for reducing hazard vulnerability. The information can be given to public officials to help them devise effective land-use policies (Burby et al., 1985; Petak, 1984). Such hazard information can also be given to private individuals so they can make better decisions about developing their land. Local officials can publicize this information as well.

Moreover, the plan should establish specific policies for meeting these objectives. It should define a method for evaluating progress toward the stated goals on a regular schedule. The planning process is just as important as the document itself. An effective planning process involves all affected sectors of the community in discovering what needs to be done and how to go about doing it.

Zoning

Zoning was originally designed to separate "incompatible land uses." For example, heavy industry, with its noise and pollution, is kept away from residential neighborhoods. Zoning also reduces the likelihood that explosions or hazmat releases affect residential neighborhoods or special facilities such as hospitals and schools. Zoning can also keep residential and commercial property away from earthquake fault lines and floodplains.

Subdivision Regulations

Subdivision regulations control the way undeveloped land is converted into smaller parcels for building. These regulations can be used in many ways. They can mandate elevating properties above the base flood elevation or limiting development density in environmentally sensitive areas such as wetlands. Subdivision regulations can also require setbacks from faults, slopes, or floodplains. In areas prone to wildfire, regulations can require the construction of *defensible space* by clearing a 30-foot perimeter around any structure.

7.2.4 Building Construction Practices

Property owners can change their construction practices voluntarily because of risk communication or incentives. They can also change involuntarily because of building code requirements.

Building Components

Buildings can be classified as residential, commercial, and industrial. Residential structures can be classified as single-family, apartment buildings, and mobile homes. Commercial structures can be classified as low rise or high-rise. The typical residential or commercial building consists of three structural systems—the foundation, the walls, and the roof. In addition, it has a number of ancillary systems that include electric; plumbing; and heating, ventilation, and air conditioning (HVAC) systems.

A building's foundation rests on soil that is either naturally or artificially compacted so the building's weight will not cause it to sink into the ground. The foundation is usually either a thick slab of concrete poured directly onto the ground to support the floor or a basement installed in an excavated pit. A basement also has a concrete slab but, because it is below ground level (referred to as *below grade*), it has walls that are either made of a vertical concrete slab or (in some older houses) concrete blocks. In some cases, as discussed later, a building is constructed on pilings that are sunk deep into the ground.

Most residential structures are built with walls constructed in the shape of a frame made out of 2 × 4 inch wooden studs. As Figure 7-2 illustrates, each wall has a long horizontal stud, called a sill plate, that runs the length of the wall's bottom. The wall has another long horizontal stud located at the top, called a top plate. Builders customarily space vertical studs at 16-inch intervals to connect the sill plate and top plate. They leave rough openings in the walls so they can install windows and doorframes after the structural frame has been completed.

The wall frame is usually constructed lying flat on the ground because this makes it easy to assemble. Once all four walls have been assembled, they are tilted up and nailed together at the corners of the building. The wall system is

Figure 7-2

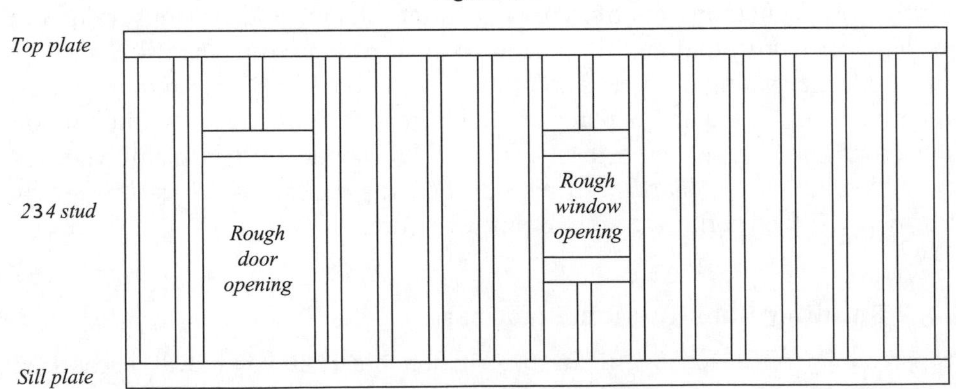

Wall frame for a small residential structure.

nailed to the foundation to secure it in place and 4 × 8 foot sheets of plywood sheathing are nailed to the wall frame. After a vapor barrier of plastic sheet is tacked over the entire surface of the sheathing to make it impermeable to water, an exterior veneer is attached. This veneer can be plywood siding, clapboard, or stucco.

It is also common to see houses with walls that appear to be built entirely of brick, but are actually brick veneer over a wooden frame. A true brick wall is made of many horizontal layers (called *courses*) of bricks that are held in place by mortar. Although this is not visible from the outside, a brick wall actually consists of two adjacent *wythes* (parallel walls). These make the wall thick enough to support the weight of the upper floors and roof. These brick walls are strong enough to support loads in *compression*. However, brick walls perform poorly when they are subjected to forces that are *shear* (pushing them sideways), *torsional* (twisting them), or *tensile* (stretching them apart by forcing them upward). Ordinary brick walls are called unreinforced masonry (URM) because they have no steel reinforcing bars to help them resist these forces. URM buildings are very likely to collapse in earthquakes.

In a small multistory building, the floors are created by nailing 2 × 6 inch floor joists at 16-inch intervals to connect two of the opposite walls of the building. Once the joists have been installed, they are covered with a rough particle board or plywood subfloor. They are then covered with a better grade of plywood that has a smoother surface. Multiple layers of subfloor are used to make the floor stiff enough to resist sagging under the weight of occupants, furniture, and appliances. The visible floor is installed over the subfloor.

The roof is usually constructed according to one of three designs—a flat roof, a gable end roof, or a hip roof. The name, *flat roof,* would seem to imply that this type of roof is exactly horizontal. However, such roofs usually have a slight slope toward the back of the building and a low parapet (6–24 inch high wall) that surrounds the roof on the front and sides. This method of construction confines the rain and forces it run off toward a gutter in the back. Panel a of Figure 7-3 illustrates a small home with a flat roof.

A gable end roof is built so it has two angled surfaces that are joined at the roof ridge. A long 2 × 6 inch board runs the length of the roof. Each side slopes downward toward the walls on opposite sides of the house (see panel b). The fact that the two surfaces of a gable end roof connect to only two of the four walls would leave open triangles over the other two end walls unless there is something to fill them. This triangle, which has the top of the wall as the base, the two sections of roof as the two sides, and the ridge as the peak, is the gable end. Of course, the gable is filled in with a structural frame, sheathed, and covered with veneer just like the rest of the wall. A hip roof is a logical extension of a gable end roof that has four angled roof surfaces instead of two. Each of the four surfaces slopes toward one of the four walls (see panel c). The principal

Figure 7-3

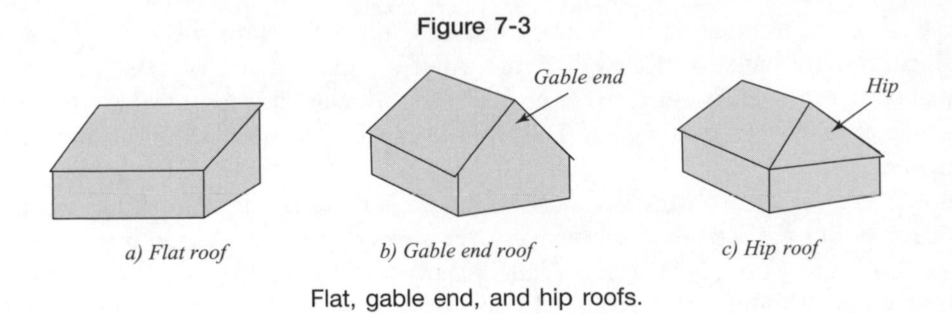

Flat, gable end, and hip roofs.

difference of the hip roof from the gable end roof is that the "vertical gable" has been tilted away from the vertical. Because it is now exposed to rainfall, it is covered with roof decking and shingles rather than wall veneer. Hip roofs resist wind damage better than gable end roofs. Gable end roofs resist wind damage better than flat roofs.

Each of these three roof designs requires a different form of bracing. A flat roof is essentially just a floor braced with horizontal joists that has an impermeable surface installed to prevent rain from entering. This surface is often a *built up roof* that has 4 × 8 foot sheets of plywood decking nailed to the joists. Layers of tarpaper and hot asphalt are applied to provide the necessary weather protection. Gable end and hip roofs require a different form of support because their surfaces are sloped. As indicated in Figure 7-4, this support consists of rafters that extend from the walls up to the roof ridge. Sometimes buildings that have very wide spans from one wall connection to the other have trusses rather than rafters. **Trusses** (shown in the figure by the structural members in dotted lines) are engineered systems that are internally braced to provide maximum strength at minimum weight. Trusses are manufactured offsite and hoisted into place by cranes. Gable end and hip roofs also have 4 × 8 foot sheets of plywood decking that are nailed to the rafters. Moreover, gable end and hip roofs usually have a single layer of tarpaper tacked to the decking and shingles nailed through the tarpaper into the decking.

Large multifamily and commercial buildings use strong materials. Because large buildings are much heavier, they have much sturdier foundations. The foundations are sometimes on concrete pilings that extend down to bedrock. These structures support the weight of the upper stories on an internal frame made from either steel girders or concrete columns that contain steel reinforcing bars. Moreover, these large buildings typically have floors that are made from poured concrete. They also have flat roofs with penthouses where the machinery is located that is needed to run the elevators.

Figure 7-4

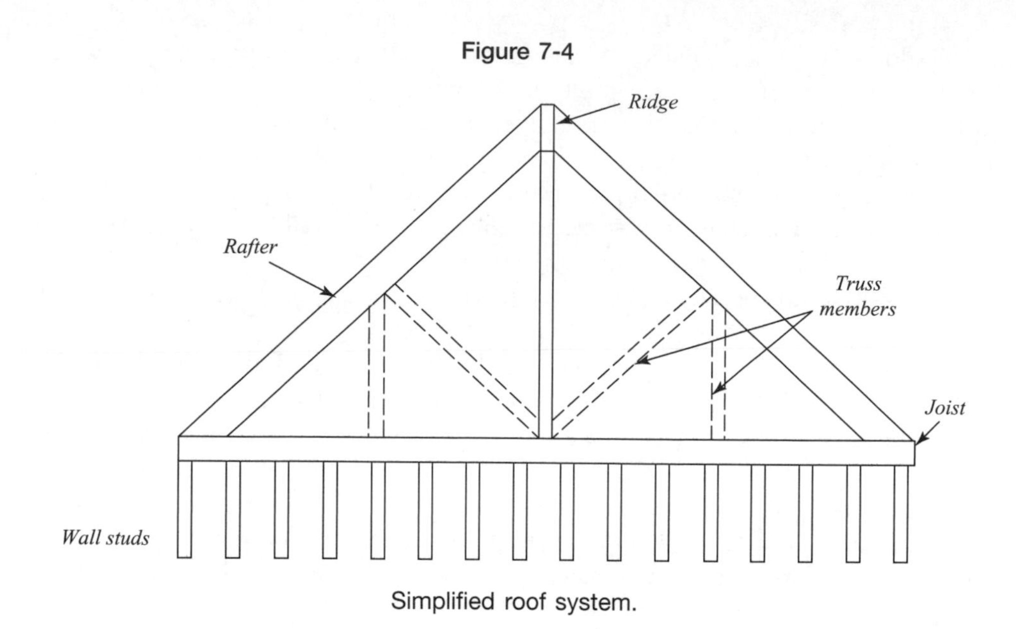

Simplified roof system.

Structural Protection from Hydrological Hazards

There are four major methods for providing protection from hydrological hazards. The most cost-effective method is to raise the house so the lowest floor is elevated above flood level. You can elevate on continuous foundation walls. You can also elevate on an open foundation.

When **elevating on continuous foundation walls,** a contractor uses a set of jacks to raise the house slightly higher than the base flood (usually the 100 year flood). The contractor builds the existing foundation walls up to the desired level. Then the contractor lowers the structure back onto the new (higher) foundation. This method increases the height of the basement walls and provides secure storage. However, it is only suitable in locations where the risks of high velocity flow and wave action are low. If flood depth is expected to be greater than about one foot, openings in the foundation must be provided to allow water to flow in and out. This prevents hydrostatic pressure, the pressure of water outside the foundation, from pushing the walls in.

When **elevating on open foundations,** the foundation supports the structure only at critical points. A foundation allows high velocity water flow and breaking waves to pass under with minimal resistance. Both methods of elevation must have foundations that are deep. This is to avoid having water currents scour the ground beneath them. Both methods of elevation are easier and cheaper for small wooden structures than for larger brick, steel, or reinforced concrete structures.

Another method for providing structural protection from hydrological hazards is *dry floodproofing*. This seals the structure so floodwater cannot enter. Walls are sealed with an impermeable coating. Shields protect penetrations such as windows and doors. Backflow valves are installed in sewer drains so water cannot enter the structure. A simple backflow valve has a rubber ball that usually rests in the bottom of the drainpipe. It floats up to block the drain when floodwater backs up through the sewer system into the basement. Dry floodproofing is not recommended where the expected flood depth is three feet or more because hydrostatic pressure can collapse URM walls and buoyancy can fracture a slab floor, or even cause the structure to "float."

A third method' of structural protection is *wet floodproofing*. This allows water to enter empty portions of the structure during flooding. Equipment is moved to a higher location or protected in place by a floodwall. In addition, vulnerable materials that cannot be moved are replaced by flood resistant materials.

A fourth method of structural protection from hydrological hazards is *relocation*. This involves moving the structure to higher ground out of the floodplain. Relocation requires jacking up the structure, supporting it on steel beams while it is being transported, and lowering it onto another foundation at the new site. Relocating wood structures is cheaper than relocating brick, steel, or reinforced concrete structures.

Protection from tsunami or hurricane surge hazards is more difficult. This is because the two coastal hazards have breaking waves. In the case of hurricane storm surge, the waves are usually small enough that they mostly threaten wood structures. However, buildings located on open coastline should be built on pilings sunk deep enough to prevent the structure from toppling if topsoil or sand is scoured away.

Tsunamis are so powerful they can threaten even sturdy masonry structures. One tsunami so thoroughly destroyed the Scotch Cap lighthouse located on an Alaskan shoreline that everything but the foundation was washed away. Structures located directly on the shoreline of tsunami-prone coasts risk total destruction. Structures that are located farther back from the shoreline, but still in the runup zone, are safe if they are built of steel-reinforced concrete. They must also have floors that are elevated above anticipated wave height. In most cases, the high cost of such construction is only reasonable for public facilities that are intended for use as shelters during a tsunami warning.

Structural Protection from Wildfire Hazard

Wildfires threaten structures with airborne firebrands that can land on roofs, radiant heat from burning trees and brush, and convective heat from direct flame contact. Property owners should not build on sites that are at the top of a slope or near a canyon draw because these will be vulnerable to fire. In addition, they should build with a 30-foot setback to eliminate vegetation near the structure. Moreover, a structure should be built with its exterior walls or roof constructed of nonflammable materials. Property owners should replace wood siding with metal, concrete, or masonry. Similarly, they should replace wood or asphalt shingles on the roof with metal or tile. They should replace wooden doors with metal ones. They should also install tempered glass to protect the windows.

Structural Protection from Wind Hazard

The pressure of the wind on structures is proportional to the square of the wind speed. Thus, a 150 mph wind speed produces four times as much pressure as a 75 mph wind. However, wind speed, and thus wind pressure, varies over the different surfaces of a building. Structures are vulnerable to wind forces in different ways at their walls, roof, and penetrations such as windows and doors. The windward wall has positive pressure that tends to push it inward. The leeward wall and side walls experience negative pressure that tends to pull them outward. Similarly, the part of the roof that is windward of the ridge experiences positive pressure that tends to push it downward, whereas the part of the roof that is leeward of the ridge experiences negative pressure that tends to pull it upward. Penetrations such as windows and doors are known as "soft spots" because they are the most likely parts of the building to fail from wind pressure or flying debris that has been entrained in the wind field. When windows or doors do fail, the wind pressurizes the house from the inside. This internal pressure forces the walls out and the roof up. The internal pressure adds to the external pressures and initiates a catastrophic structural failure. Structures fail because wind causes lateral (sideways) and upward forces on structures that are designed only to resist the downward force of gravity. These forces sometimes break the wooden studs in the wall and roof systems. However, their more common effect is to pull out the nails that hold the structure together. Many nailed connections are weak because they are driven into the wood at a 45-degree angle. This provides a weaker connection than nails driven at a 90-degree angle.

Structures can be significantly protected from wind damage. The best solution is to install stronger than normal connections among the foundation, walls, rafters or roof trusses, and roof decking. The process begins by embedding the bottom ends of bolts in a concrete foundation while it is still wet. After the wall sections have been built, holes are drilled into the sole plates so the threaded ends of the bolts can pass through. The nuts are then tightened to secure the sole plate to the foundation. This makes a very strong connection.

In turn, the roof system is connected to the wall system by *hurricane straps.* These metal straps have holes at each end through which carpenters pound nails at a 90-degree angle. One end of the hurricane strap is nailed to a rafter (or roof truss) and the other end is nailed to a wall stud. Installing hurricane straps at many locations around the perimeter of the roof makes a strong connection between the roof and walls. The roof decking, usually a layer of plywood, should be attached to the rafters by screws instead of nails. Screws bind the two pieces of wood more tightly together than nails do.

A good installation provides wind resistant shingles and also strengthens gable end walls. Installing permanent shutters for all windows can further strengthen the structure. The most convenient shutters slide on tracks. They can be raised or lowered using an electric motor. Other permanent designs include the familiar bifold shutters having two leaves, one attached to each side of the window. These can be closed and latched quickly when a storm threatens. Alternatively, inexpensive temporary shutters can be cut from plywood sheets and screwed (not nailed) to the window frame. It also is possible to replace the glass with wind (and debris) resistant materials such as Plexiglas.

Many of these wind-resistant mitigation measures can be performed as retrofits to existing structures. The cost is higher than in new construction because retrofitting requires removing exterior wall sheathing, performing the retrofit, and reinstalling the sheathing. Moreover, installing wind-resistant mitigation measures in new construction has the added advantage of being able to use wind-resistant designs as well as wind-resistant construction materials and methods. For example, new construction can use hip roofs instead of flat roofs. New construction can also avoid large roof overhangs that tend to catch the wind and pull the roof off the structure. New construction can avoid doublewide garage doors, which fail even at relatively low wind speeds. This is because wind pressure against the wide surface causes the glider wheels to pop out of their tracks. Garage door failure is a major problem for attached garages. Failure of the garage door increases the wind pressure on the door from the garage into the house. This door is not designed for wind resistance, so it becomes the next building component to fail. Kitchen door failure allows wind to pressurize the inside of the house which is usually the last step before the structure collapses.

The methods used to protect buildings from hurricane wind can also be used to protect them from tornado wind forces. The core of a tornado produces a devastating wind speed that can destroy everything down to ground level. However, buildings on the edge of the storm will have reduced damage if they are constructed using the mitigation principles described for hurricane wind. In addition, building occupants can be protected by constructing a safe room that will withstand even the most severe tornadoes (FEMA, 1998c).

Structural Protection from Seismic Hazards

Many of the basic principles for protecting buildings from wind hazard also apply to seismic protection. Seismic shaking can exert lateral and upward forces that are similar to those caused by wind. An important difference is that seismic forces come in periodic waves, so buildings sway rhythmically. However, this makes little difference in the mitigation measures used to reduce structural vulnerability.

One of the basic sources of seismic vulnerability in building design is irregularity in the building configuration. As a result, T-shaped, L-shaped, and U-shaped buildings are much more vulnerable than rectangular buildings. This is because each leg of the building reacts in a different way to the seismic forces. The stress is concentrated at the hinges where adjacent legs join. Moreover, "soft-story" buildings that are supported by piers at the ground level are vulnerable. Such designs are found in large public and commercial buildings. Apartment buildings with second story living quarters located over carports or garages at ground level are also vulnerable.

Another source of seismic vulnerability comes from the construction materials, (especially URM). Buildings constructed from brick, stone, or concrete masonry units respond poorly to earthquakes because they are too rigid to bend as the seismic waves rock them back and forth. URM buildings are common in some areas of the country such as the New Madrid Seismic Zone in the central United States URM buildings can be retrofitted by adding reinforced concrete walls, steel braces, wall ribs, or buttresses. It is common for URM structures to have flat roofs with parapets. Unfortunately, these nonstructural architectural elements are unstable. They must be secured or removed to prevent them from falling on passers-by on the sidewalks below. Unlike URM buildings, wooden structures are strong and flexible. Nonetheless, it is often wise to strengthen older wooden houses by bolting the wall system to the foundation and strapping the wall system to the roof system.

Window protection is also important in earthquakes because broken glass is a major source of injury to building occupants. This can be avoided by replacing conventional glass with Plexiglas, which will not break. Alternatively, property owners can apply plastic film over conventional glass to stabilize it. The glass will shatter during seismic shaking but the film will hold the pieces in place within the window frame. The glass will need to be replaced for aesthetic reasons, but will continue to maintain the building envelope, thus avoiding injuries to building occupants until it is replaced.

Structural Protection from Airborne Hazmat

Buildings can provide protection from inhalation of hazmat. People can find protection against these airborne pollutants if they can take refuge within a temporary "safe haven" of clean air. Most structures allow contaminated air to leak in at a rate that can be measured in air changes per hour (ACH). Another way to think of air exchange in terms of *turnover time,* which is the reciprocal of ACH, or

$$t_B = \frac{1}{ACH}$$

An infiltration rate of 1.0 ACH does *not* imply that all the clean air will be gone in one hour (Wilson, 1987). Rather, the proportion of contaminated air gradually rises until 63% of the original air has been replaced by contaminated air at the end of 1.0 t_B hours and 95% of the original air has been replaced by the end of 3.0 t_B hours. Thus, for the case of 1.0 ACH, it will take over three hours until the indoor air becomes almost completely contaminated. This result is important because it indicates that in-place sheltering is more effective than most people believe. The reason for the difference between the results can best be illustrated by examining the difference between the apparent mechanism of air exchange and the actual mechanism. It would only take one hour to replace the clean air (the incorrect result) if the contaminated air somehow "pushed out" the clean air. This is not what happens. Rather, the contaminated air that infiltrates into the structure *mixes* with the clean air. It does not "push it out." Exfiltration of a mixture of clean air and contaminated air will take longer to exhaust the clean air in a structure than will exfiltration of clean air alone. Consequently, sheltering in-place is at least three times as effective in reducing inhalation exposure as it first appears to be.

While this *time lag* effect is important, it is not the only mechanism by which sheltering in-place can reduce adverse health effects. A *damping* effect reduces the fluctuations in plume concentrations. These fluctuations arise from irregularities in meteorological conditions and local terrain. The indoor peak concentrations are much smaller than the outdoor peak concentrations. This is important for chemicals whose main health threat is its peak concentration rather than its cumulative dose.

If those in the hazard impact area continue to shelter in-place after the plume has passed outdoors, they continue to be exposed to the contaminated air indoors. Thus, once the plume has passed, the roles of the indoor and outdoor air are reversed. Now it is the clean air infiltrating into the structure that mixes with the contaminated air that is already there. Exfiltration of a mixture of clean air and contaminated air will take longer to exhaust air in a structure than of contaminated air alone. Consequently, sheltering in-place after plume passage prolongs inhalation exposure. This results in a cumulative exposure that is identical to the exposure that would have been received by remaining outdoors during the entire period of plume passage. This problem can be avoided by providing an "all clear" signal that lets those in the hazard impact area know when it is safe to come out. That is also the time they should open their doors and windows and ventilate the building to remove the rest of the contaminated air.

Building Codes

Local building codes are based on model building codes that are established by nongovernmental organizations. Building codes are influenced by engineering analyses, but political and economic factors also play a role. Model building codes do not become legal requirements until they are adopted by local communities (Nordenson, 1993). During adoption, codes are modified to accommodate the local environment. Political and economic factors play an important role at this level too (Jirsa, 1993).

Code adoption is only the first step in reducing vulnerability. The second step is to train building contractors and construction workers. They must know how to meet the new code requirements and why the codes were established. Code enforcement is essential. There are many examples of building failures in disasters caused by inadequate design, construction materials, and construction methods that violated the local building codes. Codes only work if the local adopting agencies are willing to enforce them. The extent to which building officials enforce codes by checking plans and inspecting construction depends upon local laws and budget support (Nordenson, 1993).

Building codes are in a continuous state of reconsideration even after they have been adopted. Changes are often made after a major disaster (Martin, 1993). For example, the 1971 San Fernando, 1985 Mexico City, and 1989 Loma Prieta earthquakes all led to major modifications to seismic codes, particularly in earthquake prone areas (Nordenson, 1993). Thus, building age is important because older structures were built under outdated codes. Moreover, the building inventory in the eastern and central United States is different from that in the western part of the country, with the former regions having many more URM structures (Jirsa, 1993).

7.2.5 Building Contents Protection

For most hazards, protecting buildings from damage also protects the contents from harm. One exception is flooding when the property owner has elected to protect a structure by wet floodproofing. Wet floodproofing allows water to enter a structure so the basement walls will not collapse from hydrostatic pressure. The building contents will be sacrificed unless they are protected in some other way. Similarly, earthquake shaking can cause damage to contents even if the structure is not compromised. For example, seismic shaking can cause light fixtures to tear loose from their mountings, heavy items with high centers of gravity to topple, and glass windows and doors to fracture. These unrestrained objects can cause damage to other building contents and injury to building occupants. Building contents strategies can be implemented in three different ways. *Internal movement* involves moving property to a safer location, usually a higher elevation, within a building. *Protection in place* involves shielding contents from hazards that have entered the structure. *Contents stabilization* is used to prevent property from moving.

Outdoor furniture, sheds, storage tanks, and air conditioners can also become hazards in high wind if they are located outdoors. Light materials can become deadly missiles when propelled by a strong wind. For example, a 2 × 4 stud can penetrate a concrete masonry wall when traveling at 100 mph (FEMA). Other items are too heavy to fly through the air but are significant rollover hazards. Their moderate speed is compensated by their heavy weight to breach windows, doors, and even walls. Light items can be moved indoors. Heavier items can be protected in place by anchoring them securely to the ground. Securing items also protects light fixtures, furniture, storage tanks, generators, air conditioners and

FOR EXAMPLE

Earthquakes in Los Angeles

California, a state prone to earthquakes, has a seismic safety commission that works on developing building codes that reduce earthquake hazards. The commission also provides information about how people can retrofit their homes for earthquakes. Los Angeles, for example, has many homes built prior to 1940 that were built without being bolted to their foundations. Some of the homes built after 1970 are vulnerable as well because weak bracing materials were used on the cripple wall. The cripple wall is the short wall that connects the foundation to the first floor of the house.

heaters from seismic hazard. Such measures can reduce human casualties and property damage (American Institute of Architects, 1992). For a more comprehensive list of methods for protection, see the American Red Cross at www.redcross.org/services/prepare/0,1082,0_77_,00.html.

SELF-CHECK

- Define **channelization.**
- Describe the difference between a dam and a levee.
- Define **industrial hazard controls.**
- Describe the difference between elevating on continuous foundation walls and elevating on open foundation walls.

7.3 Applying Hazard Mitigation Measures

Table 7-1 indicates that source control is suitable for eight hazards. Flood and landslides can be controlled at the source by reforestation and other measures that increase infiltration into the soil. Floods also can be controlled by other measures such as detention ponds. Accidental industrial fires, explosions, toxic chemical releases, and radiological release can be controlled by system designs and operating procedures that reduce the likelihood of extreme temperatures and pressures that will overwhelm containment systems. Deliberate radiological, chemical, or biological incidents caused by sabotage or terrorist attacks can be controlled by stopping the perpetrators from implementing their plans.

Table 7-1: Applicability of Mitigation Measures by Hazard

Hazard	*Source control*	*Protection works*	*Land-use practices*	*Building construction practices*	*Building contents protection*
Severe storms/cold				×	×
Extreme heat				×	
Tornado				×	
Hurricane			×	×	
Wildfire	×	×	×	×	
Flood	×	×	×	×	×
Storm surge		×	×	×	×
Tsunami		×	×	×	×
Volcanic eruption		×	×	×	×
Earthquake			×	×	×
Landslide	×	×	×	×	
Structural fire/conflagration	×		×	×	
Explosion	×		×	×	
Chemical spill or release	×		×	×	
Radiological release	×		×	×	
Biological incident	×			×	

Protection works are suitable for six of the hazards. In the case of wildfire, this takes the form of fire breaks that subdivide areas with large amounts of fuel. For inland floods, storm surges, tsunamis, mudflows and lava flows from volcanic eruptions, protection works consists of dams that impede the flow and dikes that confine it. In addition, inland floods can be mitigated by channelization. Landslides can be mitigated by installing drain fields that remove ground water from unstable slopes and retaining walls that stabilize the toe of the slope.

Land-use practices are suitable for twelve hazards. Hurricane wind and all of the hydrological hazards produce variations in hazard exposure within jurisdictions. Land-use practices reduce vulnerability to these hazards. Land-use practices also reduce vulnerability to the geophysical hazards. In the case of volcanic eruptions, vulnerability can be reduced by limiting development in areas subject to lava flows, mudflows, and pyroclastic flows. In the case of earthquakes, land-use practices can restrict the construction of buildings close to known faults, on unstable slopes, or on soils prone to failure. Land-use practices cannot avoid the vulnerability of infrastructure that must cross these faults. Land-use practices can control the type and density of housing development and the number of roads in landslide-prone areas. This reduces the load on unstable slopes.

Land-use practices are also quite relevant for the technological hazards. You can reduce population densities close to technological facilities and major transportation routes. Land-use practices can also avoid placing critical facilities such as school, hospitals, and nursing homes close to hazardous facilities. Land-use practices have no practical application if there is no significant variation in hazard exposure throughout a jurisdiction. This is especially true for severe storms, extreme heat, and tornadoes. There is no way land-use practices can control exposure to biological incidents.

Building construction practices are suitable for all of the hazards. People and property can be protected from severe storms and extreme heat and cold by requiring levels of insulation appropriate to the climate. Vulnerability to these hazards can also be reduced by decreasing air infiltration into buildings. This also decreases inhalation exposure to toxic chemical releases. Protection from high wind, the seismic shaking of earthquakes, and the blast effects of explosions can be reduced by better building designs, stronger materials, and better connections among building systems. Vulnerability to flood, storm surge, tsunami, and volcanic mudflow can be achieved by elevating structures or floodproofing them. Protection from volcanic ashfall focuses on strengthening support for the roof system. Building construction practices can reduce vulnerability to wildfire and structural fire/conflagration by using fire resistant materials. Landslide protection can be increased by anchoring foundations in deeper, stable soil layers.

Building contents protection is applicable to seven hazards—severe storms, severe cold, flood, storm surge, tsunami, volcanic eruption, and earthquake. People and property can be protected from severe storms and cold by providing additional insulation to plumbing and other systems containing water that might freeze. Physical vulnerability to flood, storm surge, tsunami, and mudflows following volcanic eruption can be reduced by elevating furniture, appliances, and HVAC systems, or by building flood walls within the structure. Contents can be protected against earthquakes by bolting heavy objects, installing latches on cabinets, and providing lips on shelves.

7.3.1 Reducing Technological Hazards

The federal approach to reducing toxic chemical hazards focuses on local emergency planning and the community right-to-know (RTK). **RTK provisions** require that a community is informed when a facility stores hazardous substances in amounts that are greater than EPA thresholds. This requirement prompted many facilities to review their chemical inventories. This led to facilities reducing the quantity stored on site or switching to a less hazardous chemical. Laws adopted since the 9/11 terrorist attacks have weakened the RTK provisions. This is because of the legitimate concern that information about toxic chemical inventories might be used by terrorists for attacks. Unfortunately, weaker RTK provisions also hamper residents' access to information and their ability to pressure industry to reduce chemical hazards.

Local governments are expected to develop standards on hazardous facility siting in their land-use plans. However, there is no federal guidance or incentives to do so. There aren't any sanctions for failure to do so. The federal government has also not promoted hazard resistant building construction practices. This is a missed opportunity. Chemical hazard mitigation can also be achieved by adopting more strict energy conservation practices (Lindell, 1995).

7.3.2 Reducing Natural Hazards

The federal government actively promoted mitigation approaches to natural hazards in the 1990s. FEMA funded programs that provided information about these hazards. The government supported the development of innovations such as the tornado safe room and methods of floodproofing. A program of hazard mitigation was created during the latter part of the 1990s. Project Impact (PI) was a way to encourage everyone to work together to reduce the potential for losses due to natural disasters. The idea was to fill in some of the holes left in hazard mitigation by the "patchwork" of federal, state and local regulations (Mileti, 1999, p. 7). PI was based on small incentives and large amounts of persuasion. It had no real sanctions for nonperformance. At its peak, there were more than 250 PI communities.

The three guiding principles for PI were as follows. Preventive actions must be decided at the local level. Private sector participation is vital. Long-term efforts and investments in prevention measures are essential. Activities for PI fell into four *phases:* Partnership, Assessment, Mitigation, and Success. Phase one involved private sector organizations as partners in the program. In phase two, communities performed a hazard assessment. Phase three was a crucial stage, when specific mitigation and preparedness projects were selected and implemented. In phase four, the success of PI projects was to be communicated to all citizens. Despite its promise, PI was never evaluated and a new administration discontinued the program in 2001.

Even though PI no longer exists, local governments can devise their own hazard mitigation programs by using the same basic principles. Incentives for hazard mitigation measures could include government grants, loans, tax deductions, and tax credits for money for mitigation projects. Of course, government cannot afford to support a mitigation program only through grants. Loans for repayment of principal would be less expensive. This would limit the subsidy to the amount of interest the government would have to pay for the use of the money. Tax deductions would allow taxpayers to deduct all or part of the cost of hazard mitigation projects from their income before taxes. Tax credits would allow deductions of all or part of the cost of the projects from the tax bill. These options have not been explored. Indeed, given the country's current military commitments and economic circumstances, such incentives are unlikely to be adopted in the foreseeable future.

The only market mechanism that seems to have any prospect of success in promoting hazard mitigation is hazard insurance. One of the interesting consequences of recent disasters is that the insurance industry has taken an active political role in encouraging individuals, businesses, and governments to undertake mitigation measures in order to reduce insurers' financial exposure.

Having insurance has sometimes been identified as a mitigation strategy, but this is not accurate. Instead, it is a mechanism for spreading the financial risk posed by hazards. Thus, it is a *recovery preparedness* measure. That is, insurance reimburses the policyholder for the monetary value of property that has been damaged or destroyed. This is in contrast to land-use practices and building construction practices that prevent damage from occurring in the first place. Even though hazard insurance is not a method of mitigation, it can provide an incentive for hazard mitigation if insurance premiums are structured to reflect the actual risk of a given building in a hazard prone area.

7.3.3 Creating Disaster Resilient Communities

Several challenging trends have prompted a reconsideration of approaches to hazard mitigation. One of these is increased hazard exposure. Global climate change is seen in weather patterns, an increase in the number of extreme weather events, and a rise in sea levels. There is also a global drive toward urbanization, increasing industrialization of agriculture, and decreasing fertility of land. As a result, small farms are being consolidated into large agribusinesses. The displaced farmers are absorbed into cities where they have little access to jobs, education, or health care. Many of the fastest growing cities are located in areas, such as coastal areas of the Pacific Rim, that are exposed to multiple hazards.

Another trend is increased social vulnerability. Some of these trends involve demographic shifts. The wealthiest countries are seeing a rapid increase in the median age of their populations, whereas the poorest are still experiencing population growth. In addition, many countries have experienced a general increase in income inequality, and the United States is no exception. These trends bode ill for the future because the elderly and the poor are among the most vulnerable groups in disasters.

In response to these problems, the *Brundtland Report* proposed the concept of **sustainable development** that meets the needs of the present without compromising the ability of future generations to meet their own needs (UNWCED, 1987). Sustainable development recognizes the limits of nature. Disasters are a sign that current development practices are not viable over the long term. At the core of the debate over whether to rebuild low-lying parts of New Orleans is the issue of sustainable development. Sustainable development emphasizes the creation of communities that are less likely to experience major disasters and can recover from them if they do occur. The United States has become more concerned with sustainability as the costs of disaster response and recovery have risen. There is a connection between disaster reduction and sustainable development. Reports by the President's Council on Sustainable Development (1996) called for ending subsidies to development in floodplains. Other reports have proposed that development should take into account the natural variability of our planet, including seasonal and longer-term cycles (National Science and Technology Council, 1996).

To change development styles, people need to reorient their understanding of development. The first change is to realize development is not the same as growth (Daly, 1995). A city can cease to increase in size, but this does not necessarily mean it stops changing, improving, and developing its capacities. The current economic system is built on the satisfaction of consumers' desires. However, human desires are infinite and the resources of the planet are not. What is needed is a system that preserves the advantages of free market economies yet places increased emphasis on the satisfaction of human *needs* (rather than wants) and fairness.

One measure of sustainability is the "ecological footprint" (Rees, 1992; Wachernagel and Rees, 1995). This is an estimate of the land and water needed to support a particular pattern of consumption and development. Societies differ in the amount of land used to support each individual. For example, each North American has an ecological footprint of 5 hectares (12.5 acres), but this amount of resources is only available to us because other countries' citizens have much smaller ecological footprints.

Although changing public policies to adopt sustainable development is a challenge, recent progress is encouraging. In many areas, a "culture of prevention"

FOR EXAMPLE

Sustainability and Hazard Mitigation

Principles of sustainability and hazard mitigation can sometimes conflict. Such conflicts will occur if poorly designed hazard mitigation programs reduce the standard of living of the poor. For example, public acquisition of properties in floodplains can lead to increases in rents that, in turn, reduce the supply of affordable housing. Such conflicts must be addressed through comprehensive approaches and open decision-making processes that seek answers to the root problems.

is arising as people become more aware of the how they have increased their vulnerability. Knowledge is shared across national boundaries. Governments at all levels are attempting to address these problems. The poor and vulnerable are the focus of many programs. These efforts are likely to reduce disaster vulnerability as local governmental and personal resources increase. Settlements are also increasingly located in less dangerous areas.

SELF-CHECK

- Define RTK and list its legal provisions.
- Describe how local governments can devise their own hazard mitigation programs.
- Explain how having insurance is a recovery preparedness measure.
- Explain the concept of **sustainable development.**

SUMMARY

This chapter looked at hazard mitigation strategies and mitigation measures. It explained how building construction practice, land-use practices, and other mitigation legalities affect mitigation in communities. It is your responsibility to also know how to manage the risks of natural and technological hazards by reducing them. Finally, you cannot just know hazard mitigation strategies. You need to know how they apply to different hazards. In other words, you need to know how to prepare for the bad things that can happen. Mitigation strategies are important because they can ultimately save lives and property.

KEY TERMS

Capital Improvements Program (CIP) A program used to plan community infrastructure and critical facilities.

Channelization The process of deepening and straightening stream channels.

Dams Elevated barriers sited *across* a streambed that increase surface storage of floodwater in reservoirs upstream from them.

Elevating on continuous foundation walls A method used to raise a house slightly higher than the base average projected flood height, increasing the height of the basement walls and providing secure storage.

Elevating on open foundations A method used in which a structure's foundation only supports the structure at critical points, allowing high velocity water flow and breaking waves to pass under the structure with minimal resistance.

Eminent domain Power held by the government that can force private owners to sell their property to the government at a fair market value if the property is to be used for a public purpose.

Floodwalls Water barriers that are built of strong materials such as concrete. They are more expensive than levees, but they are also stronger.

Industrial hazard controls Community protection works that are used to confine hazardous materials flows.

Land-use practices Alternative ways in which people use the land. Residential, commercial, and industrial development of urbanized areas are especially important in determining disaster impacts.

Landslide controls Methods for reducing shear stress, increasing shear resistance, or a combination of these two.

Levees Elevated barriers placed along a streambed that limit stream flow to the floodway.

Overtopping The flow of water over the top of a levee. Once this happens, the water begins to erode a path that allows increasing amounts of water to flow through the opening.

Piping A penetration through a dam or levee that occurs when an animal burrow, rotted tree root, or other disturbance creates a long circular tunnel through or nearly through the structure.

RTK provisions A legal requirement that requires handlers of dangerous chemicals to inform neighboring communities when they store hazardous substances in amounts that are greater than EPA thresholds.

Seepage erosion A form of erosion that occurs when the height of the water in the river puts pressure on water that has seeped into the riverbed, under the levee, and into the soil on the landward side of the levee. The resulting flow of water can eventually cause boils of muddy water that erode a path for the water to flow underneath and then behind the levee.

Sustainable development A concept stating that the needs of the present must be met without compromising the ability of future generations to meet their own needs.

Trusses Engineered systems that are internally braced to provide maximum strength at minimum weight.

Wave action A destructive condition that causes levee failure by attacking the face of the levee and scouring away the material from which it is constructed.

ASSESS YOUR UNDERSTANDING

Go to www.wiley.com/college/lindell to evaluate your knowledge of the basics of reducing risks.

Measure your learning by comparing pre-test and post-test results.

Summary Questions

1. The federal government cannot intervene directly in local land-use or building construction practices. True or False?
2. States must update hazard mitigation plans within six months of a presidential disaster declaration as a condition for receiving federal disaster assistance. True or False?
3. Which of the following does not influence land-use practices?
   (a) risk communication
   (b) incentives
   (c) environment
   (d) sanctions
4. What does a comprehensive mitigation plan assesses in the community?
   (a) geography and history
   (b) demographic trends and economy
   (c) historic preservation and environmental protection
   (d) all of the above
5. Laws adopted since the 9/11 terrorist attacks have strengthened the RTK provisions. True or False?
6. Which of the following were guiding principles for Project Impact?
   (a) Long-term efforts and investments in prevention measures are essential.
   (b) Preventive actions must be decided at the local level.
   (c) Private sector participation is vital.
   (d) all of the above
7. What practices are suitable for all types of hazards?
   (a) protection works
   (b) building construction practices
   (c) land-use practices
   (d) all of the above

Review Questions

1. What is eminent domain and why it is debated in federal court?
2. What is the role of the FHMO?
3. What are the five categories of mitigation strategies?
4. What are the four types of levee failure mechanisms?
5. What is a capital improvement program is used for?
6. What are the four types of flood control?
7. What is the purpose of zoning?
8. What are two hazards that can be mitigated by source control?
9. What types of mitigation strategies can people use to protect against storm surge and tsunamis?

Applying This Chapter

1. The federal government cannot intervene directly in local land-use or building construction practices. However, it wants to change these practices because it pays for much of the high cost of disaster recovery. Years ago, the federal government tried to reduce losses by reducing hazard exposure. In the case of floods, this led to a program of dams and levees. Unfortunately, flood losses continued to increase so the federal government has more recently tried to intervene indirectly. States must update hazard mitigation plans within six months of a presidential disaster declaration as a condition for receiving federal disaster assistance. How else might you suggest the federal government attempt to intervene in local land-use or building construction practices? Give two specific examples.
2. You are a building inspector in your local community. The emergency manager has asked you to create a report on building construction practices. What should you include in the report? Write a report that includes these elements.
3. A group of community leaders has asked you to describe the mitigation strategies that a local government could implement without any outside assistance. What do you say to your audience?

YOU TRY IT

Hazard Mitigation and Eminent Domain

In what circumstances do you think the government should take private land for hazard mitigation?

Buying a Home in Los Angeles

If you were going to buy a home in Los Angeles, what construction features would you look for to ensure you are well protected from earthquakes?

Affordable Housing

New Orleans officials have decided not to rebuild the 9th Ward due to the fact that homes were located below sea level behind the levees and they don't want another hurricane to claim the lives of residents. However, many people lived in the 9th Ward because they and their extended families have lived there for decades. Moreover, some of them have jobs in that area and they can't afford housing elsewhere. How would you balance the reduction of hazard vulnerability, the preservation of neighborhood integrity, and the provision of affordable housing?

8

MYTHS AND REALITIES OF DISASTER RESPONSE

How People and Communities Respond in an Emergency

Starting Point

Go to www.wiley.com/college/lindell to assess your knowledge of the myths and realities of disaster response.
Determine where you need to concentrate your effort.

What You'll Learn in This Chapter

- Myths about human behavior in a disaster
- The therapeutic community response to disasters
- Warning methods and the ways they differ
- Variables that affect evacuation rates
- Who rescues victims during disasters
- Household behavior in emergencies
- Stress effects and health consequences for people who live through disasters

After Studying This Chapter, You'll Be Able To

- Examine the myths of disasters and compare the myths to the facts about human response to disasters
- Distinguish among fear, panic, and shock
- Analyze people's beliefs about information sources
- Demonstrate an understanding of household evacuation
- Coordinate official and unofficial search and rescue teams
- Analyze how a community will respond in a disaster
- Prepare for the stress effects and health consequences of disasters

Goals and Outcomes

- Argue against myths on how people respond in disasters
- Design a plan for convergence
- Design a warning dissemination plan
- Create a plan to work with and organize volunteers and emergent organizations
- Analyze evacuation time estimates
- Predict how people will respond to disasters
- Analyze which population segments will experience psychological stress

INTRODUCTION

The way you handle an emergency depends on many factors. One of the biggest factors is how you believe people will behave during times of crisis. Even those who have never been in a disaster hold beliefs about how people respond to disaster. Many of these beliefs are myths; they are either false or are overly generalized because they only apply to a small minority of the population. Less prominent, but just as damaging, are some specific misconceptions and incorrect assumptions about households' emergency response. Misconceptions are *explicit* beliefs that are easy to identify and correct. By contrast, incorrect assumptions are *implicit* and are more difficult to correct because they are more difficult to detect. Myths, misconceptions, and incorrect assumptions all produce false images of the household response process. You must know how people truly react during a disaster so you can plan your response appropriately. This chapter discusses both the myths and the realities of household response to emergencies.

8.1 Myths of Household Response to Disasters

Many believe, perhaps due to media depictions, that disaster victims act irrationally. They flee in panic, wander aimlessly in shock, or comply obediently with the authorities. Following impact, they cannot protect themselves or others. They are incapable of protecting their property from further damage. Thus, they need assistance from governmental agencies or nongovernmental organizations (NGOs), such as the American Red Cross. The breakdown of social order leads to looting and increased crime rates. Consequently, martial law must be declared to restore order. Moreover, concerned citizens must travel to the area to donate blood, food, and clothing. Indeed, decades of movies, novels, and press coverage of disasters emphasize the general theme that a few "exceptional" people lead the masses of frightened and passive victims to safety (Wenger, 1980). Thus, conventional wisdom holds that victims respond to disasters with shock, passivity, or panic (Dynes and Quarantelli, 1985; Perry, 1983).

Studies have repeatedly shown that the majority of disaster victims do not react in this way (see Tierney, Lindell, and Perry, 2001, for a recent summary). Few develop shock reactions. Panic rarely occurs. People act in what they believe is their best interest, given their limited understanding of the situation. Most people respond constructively by seeking information and progressing through a logical sequence of steps. Behavior is generally prosocial as well as rational. After a disaster strikes, uninjured victims are usually the first to search for survivors, care for those who are injured, and assist others in protecting property from further damage. When they seek assistance, victims usually contact informal sources such as friends, relatives, and local groups. They are more likely to contact these peer groups than governmental agencies or quasi-official sources such as the Red Cross. Antisocial behaviors such as looting are rare. Indeed, crime rates even tend to *decline* following disaster impact. People rush to the disaster scene to help. Even those who live far away send money and supplies. The picture that emerges of disaster victims is one of responsible activism, self-reliance, community support, and adaptation the situation as best they understand it, using whatever resources are available. Victims are typically supported in their efforts by official organizations and resources. They are also supported by contributions from other people not directly affected by the event.

The myths are not just wrong, they are dangerous. Emergency managers make poor decisions if they believe these myths. For example, the belief that people will panic becomes a reason to withhold information about a threat. In fact, people are less likely to comply when they have vague or incomplete information. Therefore, emergency managers' belief in the panic myth can lead to fewer people taking protective action. This example illustrates why it is important to review studies that describe people's actual disaster response patterns. The behavioral record is very clear with respect to three commonly held beliefs about disaster victims' reactions—disaster shock, panic flight, and homogeneity of victim response.

8.1.1 Disaster Shock

There have been reports of victims going into a psychological shock that is characterized by docility, disoriented thinking, and insensitivity to cues in the immediate environment. Menninger (1952) reported that some flood victims experienced emotions such as apathy, confusion and disbelief. There have been many studies since then identifying cases where shock symptoms have appeared. Melick's (1985) review of studies conducted between 1943 and 1983 led to three important conclusions:

1. Shock appears most frequently in sudden events involving widespread destruction, traumatic injuries, or death (Fritz and Marks, 1954; Murphy, 1984; Melick, 1985).
2. When the symptoms do appear, few people are affected. Fritz and Marks (1954) reported that 14% of victims showed evidence of the early symptoms associated with shock. Most reported only mild symptoms. These symptoms might include uneasiness or trouble sleeping. For example, Moore (1958) reported that 17 to 30% of families exposed to a tornado had at least one member who had "emotional upset." Taylor's (1977) study of another tornado reported that 27% of the victims had "trouble sleeping."
3. Shock lasts for a maximum of a few hours or days. It is rare for shock to last longer. This is not to say that the symptoms vanish. Depending upon

> the circumstances, studies have concluded that situational anxiety, phobia, and depression can persist for years (Gleser, Green, and Winget, 1981). However, these disorders are psychological conditions that are distinct from disaster shock.

Disasters are not associated with increases in mental health problems in the affected population. For example, following the 1978 floods in Rochester, Minnesota, Ollendick and Hoffman (1982) reported one-third of their sample of victims claimed they functioned better after their disaster experience. You should expect disasters to produce several minor psychological consequences but very few major ones. Singer (1982, p. 248) summarizes findings with the following generalization.

> Reports of actual experiences reveal that most persons respond in an adaptive, responsible manner. Those who show manifestly inappropriate responses tend to be in a distinct minority. At the same time, most people do show some signs of emotional disturbance as an immediate response to a disaster, and these tend to appear in characteristic phases or stages.

Disasters are significant events for some victims. Reactions include sleep disruptions, anxiety, nausea, vomiting, bedwetting, and irritability (Houts, Cleary and Hu, 1988). In a very few cases, serious psychological consequences such as long-term grief, depression, and psychoses ensue (Erikson, 1976). However, such symptoms are most likely to follow a sudden disaster that is without a rational explanation. Moreover, these conditions are most commonly found in people who have chronic psychological problems. This makes it difficult to link them directly to any disaster. Thus, certain disaster experiences could aggravate preexisting conditions.

People who experience disaster shock either recover on their own or with short-term counseling. Such short-term stress reactions do not interfere with disaster victims' abilities to act responsibly on their own or to follow instructions. Isolated cases of immobilizing shock affect some people, but such reactions are very rare and are not typical of the population as a whole (Wert, 1979). In summary, disaster shock rarely occurs. Therefore, it will not hamper a community's emergency response or disaster recovery.

8.1.2 Panic

Perhaps the most stubborn myth about human response to disasters is that there is widespread **panic.** In general, panic can be defined as "an *acute fear reaction* marked by a loss of self-control which is followed by *nonsocial* and *nonrational* flight behavior" (Quarantelli, 1954, p. 272). Although such panic is a staple of horror movies, it is a rare response to disasters. Fear is not the same as panic. People's fear of disaster impacts motivates them to take actions that will avoid those impacts. Thus, it is only when fear reaches an overwhelming level—panic—that people take nonrational or nonsocial actions.

The myth of panic is reinforced by incorrect interpretations of people's reactions in disasters. First, people think panic is common because victims often label their immediate reaction to the situation as one of "panic" when interviewed by the news media. A statement by a victim such as, "When I saw the funnel cloud, I panicked" indicates they are referring *only* to the acute fear reaction. Subsequent statements from the victim usually describe rational responses. For example, a victim might go on to say, "I grabbed the baby out of the upstairs bedroom and ran down to the basement just before the house collapsed." This latter statement is often ignored.

Second, observers also misinterpret the state of mind of disaster victims. For example, a news story might assert that the victims of a motel fire found dead in a hall storage closet got there because they "panicked." Crawling through the zero visibility heavy smoke, the victims probably wrongly concluded that the first unlocked door they encountered in this unfamiliar hallway was the door to the stairwell. Once they realized their mistake, it might have seemed safer to remain in the closet. Or, advancing flames may have blocked their exit. In short, an error of judgment does not provide evidence of panic.

Third, even when victims are successful in avoiding death, observers often interpret any attempt to flee the hazard as evidence of panic. Yet it is difficult to see why anyone would assume that it is anything other than *rational* to want to put distance between oneself and a life-threatening event. In such cases, those affected are assessing a threat and taking an immediate protective action.

Of course, some examples of panic flight cannot be explained away as observer errors. While it is rare, panic flight does occur under certain circumstances. In research dating back to the early 1950s, analysis of situations in which panic flight took place identifies several conditions that must occur, probably simultaneously, in order to evoke mass panic flight (Fritz, 1957; Quarantelli, 1981; Drabek, 1986). These are:

- ▲ A perception of immediate and extreme danger.
- ▲ The existence of a limited number of escape routes.
- ▲ A perception that the escape routes are closing, necessitating immediate escape.
- ▲ A lack of communication.

It is important to recognize that these conditions are defined in terms of an individual's beliefs. The conditions are based on what those at risk believe to be true at the time, not what others know after the fact. It is also important to note the distinction between the occurrence of an event and the potential for dangerous consequences resulting from that event. In this connection, Quarantelli (1954, p. 274) has observed the following:

> Coal miners entombed by a collapsed tunnel who recognize they will have sufficient air until rescuers can dig through to them do not panic. [Panic occurs in reaction] to the immediate dangerous consequences of possible entrapment rather than to being trapped as such.

In summary, panic has sometimes been documented in response to both natural and technological disasters, but it is not a common reaction to any type of disaster. When panic flight is observed, it seems to involve a relatively small proportion of the people exposed to the threat and does not usually persist for any period of time. Panic does not always materialize even in cases where conditions support its emergence. For example, Johnson (1988) reported that evacuation was orderly and altruistic responses were common during the 1977 Beverly Hills Supper Club fire in Kentucky where 160 patrons died. Similar findings have been reported in other fires.

8.1.3 Homogeneity of Victim Response

A common misconception is that there is a single "public" or "population." In fact, there are many publics or population segments. Each differs in their hazard knowledge, family roles, and household resources. In particular, you need to distinguish among residents, transients, and special facility populations. These population segments differ in their willingness and ability to evacuate (Urbanik, 2000). Residents are those who live or work in a risk area. Transients consist primarily of those who stay in commercial lodging facilities such as motels. However, in some disasters you also have to consider day visitors. This population segment can be a significant concern in rapid onset disasters such as terrorist attacks and accidents at nuclear power plants. However, day visitors will not be a problem for hurricanes and other hazards with ample forewarning because they

FOR EXAMPLE

Reaction in the Twin Towers during 1993

Before the 9/11 terrorist attacks, the World Trade Center was bombed in 1993. The evacuation took six hours and was hampered by the loss of power in the stairwells. Despite these problems, Aguirre, Wenger, and Vigo (1998) reported the evacuation of the World Trade Center in 1993 was tense but orderly, with no reports of panic.

will leave early. People in special facilities such as schools, hospitals, nursing homes, and jails must also be analyzed separately. Their patterns of warning and evacuation are different from those of residents and transients (Urbanik, 2000).

SELF-CHECK

- Define panic.
- Describe three conclusions about shock in Melick's study?
- Explain why an emergency manager's belief in the panic myth can lead to fewer people taking protective action.
- Explain how population segments differ in their willingness and ability to evacuate.

8.2 Socially Integrative Responses

Disasters sometimes produce a shift in values that results in stronger communities. Wenger (1972) has documented the decrease in socializing, the curtailment of buying luxury goods, and a decline in social control problems following disasters. At the same time, there is an increase of mutual support among victims and others in stricken communities (Wilmer, 1958; Fritz, 1961; Boileau et al., 1979). These behaviors produce the therapeutic community response (Fritz, 1968; Midlarsky, 1968). People in the disaster impact area work hard and long to help others. Thus, at least in the immediate postimpact period, a disaster tends to bring a community closer together.

8.2.1 Convergence

The therapeutic community response is related to convergence. Often a disaster site is overwhelmed with volunteers and donations. The positive impact of convergence is the increased resource base. Convergence also improves the morale of victims. Victims see that others care. They believe that they can overcome the disaster.

Although convergence provides resources, it can also hamper response. For example, fire departments from distant communities created a strain when they appeared at a Louisiana crash site (Kartez and Lindell, 1989). The authorities not only had to handle the crash, they had to integrate the additional responders. Donations also arrived unannounced. Donations continued to arrive for days and weeks. These donations and volunteers, although potentially an asset, were a liability because they were unanticipated. It is essential to develop donations management procedures. One such plan is Supply Management (SUMA) developed by the Pan American Health Organization. Donations management procedures allow for integration of volunteers. They also create a routine for receiving, storing, and using material and equipment.

8.2.2 Resources

A second aspect of the positive social response is a sympathetic behavior from the public. This is related to but distinct from convergence. We are referring here to charity. Many volunteers direct help to victims in the form of needed clothing, food, and lodging. Perhaps the earliest documentation of this type of response is found in Prince's (1920, p. 137) study of an explosion in Halifax, Nova Scotia.

The idea was to take the refugees into private homes. It became the thing to do. And there was social pressure to take in the refugees. Social pressure worked effectively upon all who had an unused room.

A more recent example is the response of the public to Hurricane Katrina. Thousands of people across the nation opened their homes to displaced victims. Several studies have shown the extent to which people not directly impacted by a disaster support the victims. Such charity is seen as a normative response in societies worldwide. What is important is the positive social climate created by such altruism.

The result of these processes is a therapeutic social system. The outpouring of personal warmth and direct help provides support to many victims (Barton, 1969). Of course, this does not provide complete support for victims. Nor does it entirely diminish the negative psychological consequences of disaster. Disasters are crisis experiences for many victims. Terrorist attacks also elicit extreme outpourings of help to the victims. The amount of support given following the 9/11 terrorist attacks is an extreme example. It is essential to recognize disasters can cause positive as well as negative effects.

FOR EXAMPLE

Donations after 9/11

New York was flooded with donations after the 9/11 attacks. Food, medical supplies, toys, and clothes from well-intentioned people came from all over the world. Many of these donations were not needed. For example, many pounds of dog food were donated for the search-and-rescue dogs, but they have a very specialized diet. Warehouses in New York still house some of these donations that were not needed.

Early researchers saw the therapeutic community as "an outpouring of altruistic feelings and behavior beginning with mass rescue work and carrying on for days, weeks, possibly even months after the impact" (Barton, 1969, p. 206). Regrettably, not enough research has been completed to know if Barton is correct. The therapeutic community may not be a lasting condition (Quarantelli and Dynes, 1976, 1977). Consensus following disasters is a short-lived phenomenon. For example, six months after the 9/11 attacks, there was conflict on how to distribute compensation funds. There is agreement that a therapeutic community develops in the short-term aftermath. It promotes positive psychological outcomes for disaster victims. Also remember, however, that it will be short-lived.

SELF-CHECK

- **Explain the importance of developing donation management plans.**
- **Describe the advantages and disadvantages of socially integrative responses.**

8.3 Warnings

It is commonly assumed that authorities are the first to warn risk area residents. It is also assumed that authorities provide almost all the information about a disaster. However, this is not entirely true. People also rely substantially on the news media, peers (friends, relatives, neighbors, or coworkers), and environmental cues. People observe the behavior of others to assess the need for protective actions such as evacuation. They also seek confirmation of warnings, regardless of the initial warning source. They try to verify the information they have received and to work out the logistics of response. Indeed, in a major disaster, the volume of calls into and within the impact area can overload the available telephone circuits.

The process of warning dissemination generates a distribution of times at which households first receive a warning. Figure 8-1 shows the cumulative distribution of warning receipt over time of the type reported by Rogers and Sorensen (1988). The curve is nonlinear so the rate of warning receipt first increases and later decreases over time. For example, 39% of households receive a warning in the first hour. Another 47% of households receive a warning in the second hour. Only 12% of households receive a warning during the third hour, and 1% of households receive a warning in the fourth hour. The last few people take a long time to receive a warning.

Figure 8-1

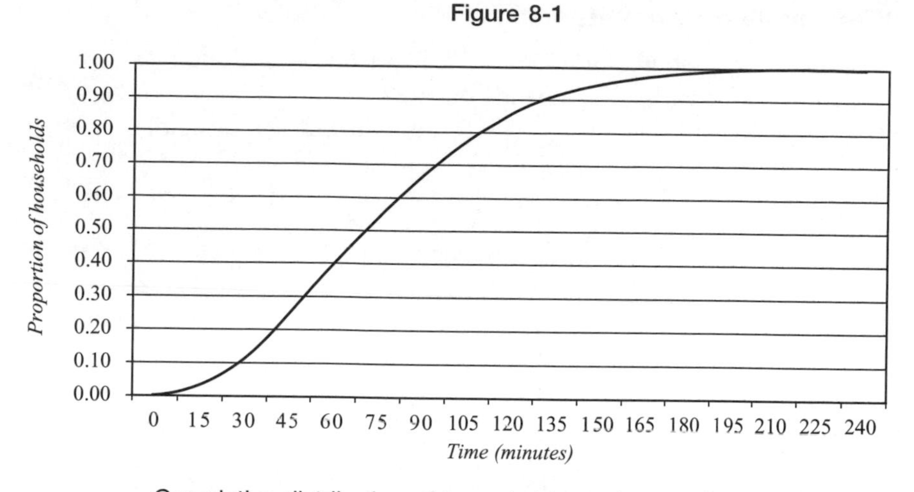

Cumulative distribution of household warning receipt over time.

Whether and when people receive a warning has been the subject of substantial research over the past five decades (Lindell and Perry, 2004). Warning sources differ in their accessibility. These differences can vary by community. For example, Lindell and Perry reported that the majority of those at risk from the eruption of Mt. St. Helens received their first warning from peers (see Table 8-1). In Toutle (a town very close to the volcano), 58% of the population received their first warning from peers. In Woodland (which was farther away), 47% received their first warning from peers as well. The greatest differences between communities were associated with the news media. The news media was responsible for providing the first warning for 39% of the Woodland residents but only 6% of the Toutle residents.

Table 8-1: Source of First Eruption Warning by Community

	Toutle		*Woodland*	
	Number	*Percent*	*Number*	*Percent*
Saw environmental cues	27	30	12	14
Authorities	6	6	0	0
News media	6	6	34	39
Peers	51	58	41	47

8.3.1 Different Warning Methods

The extent to which any source provides the first warning—and the rate at which the entire population is warned—depends on the communications channels that each source uses. Lindell and Perry (1992) summarized the available warning methods as:

- ▲ Face-to-face
- ▲ Route alert (loudspeaker broadcast from a moving vehicle)
- ▲ Siren
- ▲ Commercial radio and television
- ▲ Tone alert radio
- ▲ Telephones
- ▲ Newspapers

These warning methods differ in a variety of ways, including the:

- ▲ Ability to get a message to the group(s) that are most at risk.
- ▲ Ability to get people's attention as they go about their daily activities.
- ▲ Ability to provide specific information.
- ▲ Ease with which a message can become distorted.
- ▲ Number of people who receive the message over time.
- ▲ Requirements for people sending and receiving warnings.
- ▲ Feedback.

Sources using channels that can reach more people are able to warn more people. However, radio and television cannot provide warnings unless they are turned on. This limits their effectiveness when people are asleep or outdoors. Sirens are not effective when people are engaged in noisy activities or are asleep. Thus, the effectiveness of a given warning method varies with the types of activities in which people are engaged. Tone alert radio has been consistently effective in providing rapid warning dissemination. This is because it turns on to receive a warning message when a National Weather Service radio station broadcasts a special tone.

Many hazards have environmental cues such as smoke coming from a building or noise from an explosion. However, for those hazards without cues, people must rely on human sources of information. These information sources could be authorities, the news media, or peers. The time it takes each household to receive a warning from authorities depends upon the warning mechanisms the authorities use. For example, warnings disseminated by route alert depend on the number of vehicles dispatched and the speed at which they travel. By contrast, warnings disseminated by tone alert radio depend upon who owns these devices.

The time it takes each household to receive a warning from the news media depends on whether people are paying attention to this source. Access to the news media varies to some extent among households. Some people seem to have radio or television turned on all day long, whereas others rarely listen or watch. Access varies even more by time of day—with the highest levels being reached during the day and early evening (Lindell and Perry, 1992; Rogers and Sorensen, 1988).

The time it takes each household to receive a warning from peers depends on the extent to which people interact with others. This is correlated with a number of variables. Specifically, social integration with both relatives and friends and coworkers is negatively correlated with age (Perry, Lindell, and Greene, 1981), socio-economic status (Alvirez and Bean, 1976), and ethnicity (Bianchi and Farley, 1979). Ethnicity introduces some complexities because minority households are more likely to be extended families. Thus, ethnic minorities are more integrated into kinship networks, which speeds warning reception. On the other hand, there is a greater probability that some family members may not be present, which slows their response until they can all be accounted for.

There is substantial variation across communities in the relative importance of different warning sources to different ethnic groups. Perry and Mushkatel (1986) found that Hispanics in three communities were consistently more likely to receive their first warning from peers. However, the role of the news media as a warning source for this ethnic group varied among communities. Differences across communities were even more striking for African-Americans. In one community, African-Americans received their first warning from authorities. However, in another community, almost no African-Americans received their first warning from the same source. There were also differences across communities for Caucasians. In one community, the majority of Caucasians received their first warning from the news media. However, in another community, peers were the most common source of first warning.

8.3.2 Personal Risk Assessment and Response

People look to authorities for information about what protective actions to take and when to take them. However, they also rely on the news media, peers, and environmental cues. They also use their own pre-existing beliefs about appropriate protective actions. Perry and Greene (1983) reported the four most important reasons for evacuating during the Mt. St. Helens eruption were

1. Environmental cues (29.1%).
2. Authorities' evacuation recommendations (26.6%).
3. Relatives' evacuation recommendations (20.3%).
4. Observations of neighbors leaving (12.7%).

FOR EXAMPLE

Warnings

When issuing warnings, be sure to know what language is spoken and understood by residents of the risk area. Do not assume everyone speaks English. For example, in Brooklyn, New York, emergency officials have mapped the languages spoken in different neighborhoods. There are ten different languages spoken in Brooklyn alone.

Personal experience has no consistent effect on evacuation (Baker, 1991). There is no evidence that false alarms reduce the likelihood of future evacuation (Dow and Cutter, 1998). However, false alarms do cause decreased confidence in official warning sources. To avoid this problem, authorities should explain that forecasts and warnings cannot be made with complete certainty. Indeed, Sorensen concluded "[t]he likelihood of people responding to a warning is not diminished by what has come to be labeled the *cry wolf* [italics added] syndrome if the basis for the false alarm is understood" (2000, p. 121).

Demographic characteristics of a population are not very useful in explaining people's warning responses. It is the source and content of the warning message that largely determine response to warnings.

SELF-CHECK

- Name three ways that warning methods differ.
- Identify the major sources of information that people rely on in disasters.

8.4 Evacuation

Evacuation trip generation refers to the number and location of vehicles evacuating from a risk area. People often assume that all vehicles in a risk area will evacuate, but this is not the case. The number of evacuating vehicles can be estimated using procedures developed by traffic engineers and disaster researchers (Lindell and Prater, 2005). There are three factors affecting trip generation that you can estimate from U.S. Census data:

- ▲ Size and distribution of the resident population.
- ▲ Number of persons per residential household.
- ▲ Size and distribution of the population dependent on public transportation.

In addition, data from two variables can be collected from your local convention and visitors' bureau:

- ▲ Size and distribution of the transient population.
- ▲ Number of evacuating vehicles per transient household.

Finally, there are four variables that must be estimated from behavioral research:

- ▲ Number of evacuating vehicles per resident household.
- ▲ Number of evacuating trailers per resident household
- ▲ Percentage of residents' protective action recommendation (PAR) compliance/spontaneous evacuation.
- ▲ Percentage of transients' PAR compliance/spontaneous evacuation.

You can estimate population size and distribution with geographical information systems. For example, you can overlay risk area boundaries onto census block group boundaries to compute each risk area's residential population (Lindell et al., 2002b). In addition, risk area residents are also at home during different times of the day (Alam and Goulias, 1999). That is, homes are completely occupied at some times, partially occupied at other times, and sometimes completely empty. Over 90% of the population is indoors at home from 10:00 p.m. to 6:00 a.m. but only about one-third is there from 10:00 a.m. to 3:00 p.m. (Klepeis et al., 2001). Evacuations initiated during daytime hours should include time for travel from work or school to home. Such variation is especially important for the evacuation of areas around nuclear and chemical facilities. These areas can have incidents with short forewarning (Hobeika, Kim, and Beckwith, 1994; Urbanik, 2000). However, the forewarning of a hurricane leads families to stay home in anticipation of evacuation (Lindell, Lu, and Prater, 2005). For hazards with a long forewarning, you probably do not need to adjust for variation in household activity.

People often assume that all households have their own private vehicles. However, this is not accurate. As we saw for Hurricane Katrina, as much as one-third or more of the households in some cities are dependent on public transportation. This can be as much as 15% or more of the population in some coastal counties exposed to hurricanes. This level of dependence on public transportation can be found in both rural and urban areas. Many of these people will evacuate with peers (Lindell et al., 2005). Thus, the number of buses needed to evacuate the transit dependent is likely to be smaller than expected. However, the time required to mobilize buses, pick up evacuees, and travel out of the risk area might be greater than the time required to evacuate households in cars. This is especially true if transit dependent households must share a limited number of buses with inhabitants of special facilities.

You should also estimate the size and distribution of the transient population. You can do this from local convention bureau data on the number of hotel rooms (Hobeika et al., 1994; Lindell et al., 2002b). In addition, Hobeika et al. (1994) considered the number of campsites and their occupancy rate. You must account for seasonal variation in the transient population. For example, coastal areas have high rates of tourist occupancy on holiday weekends during the summer. They have much lower rates during the week after Labor Day. Occupancy rates are often reported by month rather than by week or day. Analysts must make assumptions about variation within a month.

Some analysts have assumed 70% to 80% of the registered vehicles would be used in a daytime evacuation and 90% would be used in a nighttime evacuation (Southworth and Chin, 1987). These analysts also contended that only 75% of vehicles would be used in rapidly developing incidents. Unfortunately, there is not any data to support such assumptions. Instead, it is more logical to base the estimated number of evacuating vehicles on the number of evacuating households. This is because fifty years of disaster research has identified the household as the basic unit of evacuation. Indeed, households that are separated when a warning is given almost always attempt to reunite before leaving the risk area (Drabek, 1986; Lindell and Perry, 1992; Tierney et al., 2001). Data on the number of evacuating vehicles per residential household have been collected. The number varies substantially from county to county, with a low of 1.10 and a high of 2.15 (Lindell et al., 2005) but the most probable range is about 1.2 to 1.5. Data on the number of evacuating trailers per residential household is an important consideration. Many households in coastal areas load boats onto trailers to take when they evacuate. These trailers take space on the highway and should be included in estimating traffic demand.

Warning compliance refers to the percentage of those warned to evacuate who actually do so. Spontaneous evacuation refers to the percentage of those who were *not* warned to evacuate but do so anyway. Many evacuation analysts have assumed that there will be 100% evacuation compliance and no spontaneous evacuation. By contrast, Lindell et al. (2002) used data on expected evacuation rates based on the severity of the storm and of residents' risk areas. The evacuation percentages in the original data were subject to sampling error. The data can be smoothed statistically to yield the percentages in Table 8-2 (Lindell and Prater, 2005).

Table 8-2 is consistent with the findings of previous research. It shows that risk area residents' rate of evacuation is not ideal. The ideal pattern is 100% evacuation in areas advised to evacuate and 0% spontaneous evacuation.

Table 8-2: Smoothed Percentages of Households Expecting to Evacuate for Hurricanes in Category 1 through Category 5 (by Risk Area)

Risk Area	*Category One*	*Category Two*	*Category Three*	*Category Four*	*Category Five*
1	45.9	63.7	87.8	98.2	100.0
2	35.9	53.7	77.8	88.2	91.4
3	31.1	48.9	73.0	83.4	86.6
4	28.2	46.0	70.1	80.5	83.7
5	26.5	44.3	68.4	78.8	82.0

Dow and Cutter (2002) reported 65% compliance during Hurricane Floyd in South Carolina. Prater et al. (2000) reported 34% compliance in Hurricane Bret. Lindell et al. (2005) reported that evacuation rates in Hurricane Lili ranged from 11.7% to 86.8% across five jurisdictions. These rates decayed as a function of distance from the point of landfall. These findings are consistent with Baker's (1991) report that evacuation compliance rates in 15 studies varied significantly from one storm to another at a given location.

Data on spontaneous evacuation is sparse. Baker (1991) reported spontaneous evacuation to range from 20% to 50% of residents in areas of "low risk." There appears to be no available data on the percentage of transient households complying with an evacuation warning or spontaneously evacuating. It is reasonable to assume tourists will produce 100% compliance with a hurricane warning, regardless of their risk area.

8.4.1 Departure Timing

Departure timing refers to the rate at which evacuating vehicles enter the evacuation route system over time. It is common for people to assume that people will leave immediately after receiving a warning, but this is not the case. Instead, there are two distinct population groups, residents and transients, whose departure timing must be estimated from behavioral research. For each of these groups, it is important to estimate the percentage of early (before an official warning) evacuating households and the distribution of departure times for households leaving after an official warning is issued.

Data from Lindell et al. (2005) show the distribution of times when households decided to evacuate from Hurricane Lili. The storm had a Category 4 intensity and was on a steady track. Because of these two factors, authorities expected

Figure 8-2

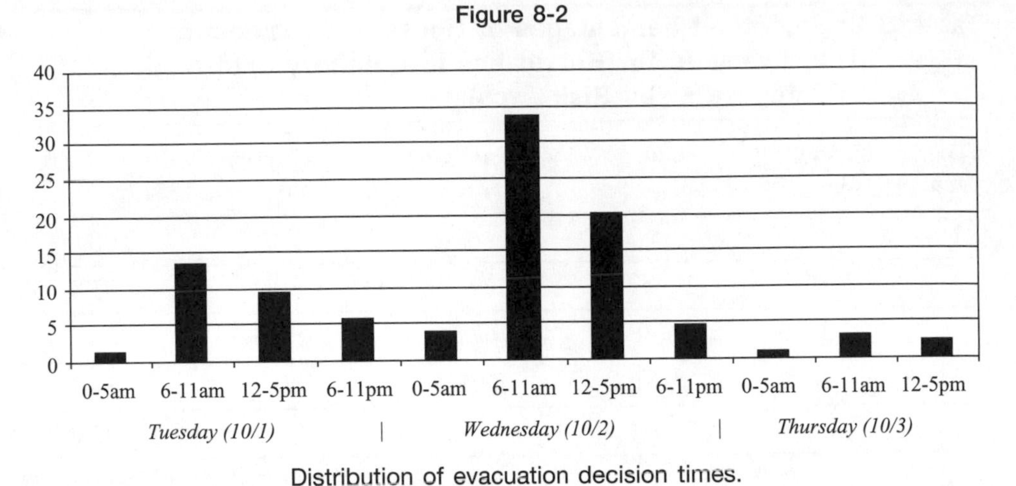

Distribution of evacuation decision times.

the National Hurricane Center to announce a hurricane warning on Wednesday morning. Accordingly, they announced on Tuesday evening that they would initiate an evacuation on Wednesday morning. Figure 8-2 shows that almost two-thirds of the households decided to evacuate before the official warning. There was a tendency for people to make their decisions early in the morning. This is so they would have the maximum number of hours of daylight in which to evacuate.

We do not have a great deal of data on household departure time distributions. There are warning and preparation times from four floods and the Mt. St. Helens eruption (Lindell and Perry, 1992). In addition, Sorensen and Rogers (1989) reported warning and preparation times from two hazardous materials spills. Researchers have used such data to construct departure time distributions for hurricane evacuations by combining warning and preparation time distributions from two different situations (Lindell, et.al, 2002). To approximate a rapid warning, they used data from the 1980 eruption of Mt. St. Helens (Lindell and Perry, 1992). The use of this data is supported by a survey conducted the month before the eruption. The survey showed that 34% of all residents checked the news for information about the volcano 2 to 3 times per day. Another 56% checked more than four times per day (Greene, Perry, and Lindell, 1981). This level of hazard monitoring mirrors that in hurricane risk areas during the days before landfall.

To estimate hurricane preparation times, surveys were conducted that asked coastal residents the time it took to:

- Prepare to leave work.
- Travel from work to home.
- Gather household members.
- Pack travel items.

▲ Install storm shutters.
▲ Secure their home before evacuating from a hurricane.

These expected preparation times were later compared with data on actual preparation times during Hurricane Lili. They were closely correlated (Kang, et. al., 2004). The researchers also estimated the departure times for transients. Based on Drabek's (1996) work, the researchers assumed transients would be warned at a faster rate than residents and would also prepare to evacuate more rapidly than residents.

8.4.2 Destination/Route Choice

Finally, there are four factors defining evacuees' destination/route choice:

▲ Ultimate evacuation destination, the place where evacuees want to stay until they can return home.
▲ Proximate destination, the point at which the evacuees leave the risk area.
▲ Route choice, the roads evacuees take to get out of the risk area.
▲ Primary evacuation route utilization, the percentage of evacuating vehicles using official evacuation routes.

Some analysts have assumed that evacuees choose their routes dynamically when they encounter traffic queues. Specifically, Sheffi, Mahmassani, and Powell (1981) contended drivers have a "myopic view" of the alternative routes that could take them to their proximate destination. They theorized that drivers select the least congested road they encounter at each intersection. However, reports from recent hurricane evacuations (Dow and Cutter, 2002; Prater et al., 2000) show that evacuees tend to take the most familiar routes inland (especially interstate highways). They overload those routes and ignore unused capacity on

FOR EXAMPLE

Evacuation of Houston

Often in evacuations there are traffic jams and long waits to leave the risk area. To evacuate traffic for Hurricane Rita, local officials opened up all interstate lanes to outgoing traffic. The evacuation time out of Houston to other cities such as San Antonio, Austin, and Dallas was still many hours more than it would have taken under normal circumstances. Trips that would normally take 3 to 5 hours were taking 15 to 25 hours and sometimes even more.

alternate routes. It seems likely that some drivers persist in following predetermined routes whereas other drivers modify their routes when they encounter heavy traffic. There is insufficient data to determine what the relative proportion of each type of driver is and what conditions affect these proportions.

SELF-CHECK

- Define **evacuation trip generation.**
- Describe how drivers choose the routes they take when evacuating a city.

8.5 Search and Rescue

The societal response to disasters can be understood in reference to Figure 8-3. The innermost circle is the *total impact* zone where casualties and damage are the greatest. Immediately adjacent to the total impact area is the *fringe impact* zone. In this zone, casualties and damage are significant but not overwhelming. The

Figure 8-3

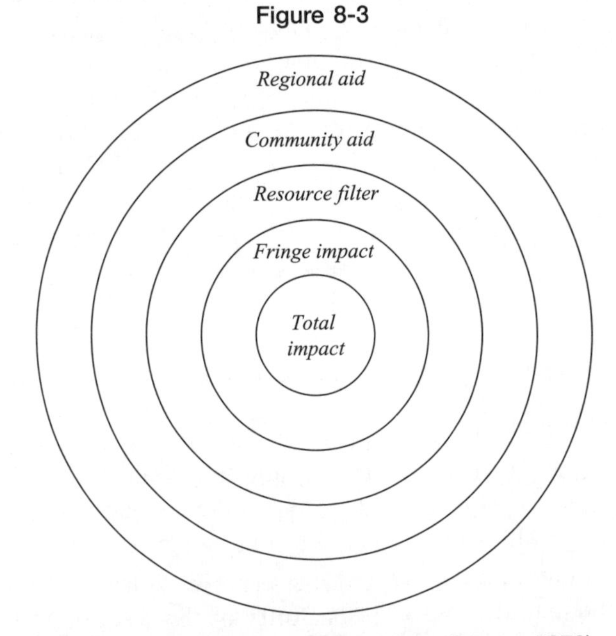

Disaster impact zones (adapted from Dynes, 1970).

next ring is the *resource filter* zone. Through this zone, information passes from the inner (total impact and fringe) zones to the outer (community aid and regional aid) zones. Material resources pass from the outer zones to the inner zones. The *community aid* zone is the area from which assistance is drawn for minor disasters whereas the *regional aid* zone is needed to support response and recovery from major disasters.

Survivors in the impact area are usually the first to respond to disaster impact. They search for those who are trapped in building debris and rescue the victims if possible. They provide preliminary treatment. They transport the injured to hospitals. As time passes, there is increasing involvement by those in the fringe impact area and other zones (Dynes, 1970; Form and Nosow, 1958). These volunteers' search and rescue efforts can be highly effective when the local building stock comprises wood frame and unreinforced masonry (URM) structures. However, collapsed buildings made of steel reinforced concrete require trained teams with specialized equipment (see Figure 8-4).

8.5.1 Medical Transport

It is commonly assumed that victims are sent to hospitals in ambulances. However, 46% of casualties reach hospitals in their own vehicles or those of peers or bystanders (Quarantelli, 1983). The vast majority (75%) of the victims are transported to the nearest hospital. Consequently, this hospital is usually overloaded even though other hospitals are sent few or no patients. A study of fourteen disasters found an average of 67% of casualties was treated in a single hospital despite the fact that the affected communities had from 3 to 105 hospitals (Golec and Gurney, 1977). You must work with your local EMS agency to determine how to allocate the injured among the available hospitals.

8.5.2 Shelter Use

People also assume that most evacuees stay in mass care facilities (shelters). In fact, only a minority do so. Mileti, Sorensen and O'Brien (1992) examined 23 evacuations. In doing so, they learned that 14.7% of the evacuees went to shelters across all disasters. The smallest percentage of evacuees using shelters was 5% when evacuations were early in the morning with good weather and effective traffic routes. Conversely, the demand for shelters is likely to be 20% or higher when the evacuation zone has a high proportion of low income or minority households and the evacuation takes place in darkness, bad weather, and traffic congestion.

In general, evacuees prefer to avoid mass care facilities. For example, a 2001 hurricane planning analysis found the majority of the respondents to the survey expect to stay with friends and relatives (46.3%) and 32.9% expected to be in

Figure 8-4

The Coast Guard rescued many stranded hurricane victims from their rooftops in the aftermath of Hurricane Katrina.

commercial hotels or motels (Lindell et al., 2001). Another 4.3% of the respondents expected to stay in campers or trailers, 3.2% expected to stay in second homes, and 9.8% indicated they don't know where they would stay or did not respond. Only 3.4% expect to stay in public shelters. These findings are consistent with the findings of previous evacuation research.

8.5.3 Emergency Responder Role Abandonment

Another common concern is that emergency responders will abandon their professional duties to protect their families. When given the choice of protecting the public or their families, it is assumed they will protect their families. This generally is not accurate. Quarantelli's (1982) examination of the Disaster Research Center's studies of hundreds of emergencies showed no evidence of role abandonment. This is because emergency responders usually develop family emergency plans. In addition, many emergency response agencies develop family protection plans to prevent conflict.

The performance of the New Orleans police in Hurricane Katrina seems to be a well documented exception to the general rule that emergency response personnel do not abandon their roles. Future research will undoubtedly be conducted to explain why police performance in this situation was so different from all the previously studied disasters. In addition, Lindell et al. (1982) cautioned that conclusions about emergency personnel role abandonment will not necessarily apply to emergency response auxiliaries. These are people who are asked to perform their normal duties but aren't trained to perform them under emergency conditions. For example, you might plan for school bus drivers to evacuate people who do not have cars. You should not just *assume* that these drivers will perform their duties in an emergency. Instead, you should have any emergency response auxiliaries make an explicit commitment to perform their jobs in an emergency. You should encourage them to make the same preparations as other responders.

8.5.4 Volunteers and Emergent Organizations

Volunteers can be important assets during emergency response and disaster recovery, but you need to organize them. To be effective, volunteers must be placed into groups that perform tasks that are within the scope of their abilities. They must also be able to interact effectively with the rest of the emergency response organization. Dynes's (1970) typology of disaster organizations provides a useful way to think about volunteers. As Table 8-3 indicates, tasks can be characterized as normal or novel. Normal tasks are routine. Novel tasks are ones that members

Table 8-3: Types of Organized Behavior in Disasters

		Tasks	
		Normal	*Novel*
Organizational structures	Normal	Established	Extending
	Novel	Expanding	Emergent

perform only in disasters. Similarly, organizational structures can also be characterized as normal or novel. Normal organizations are ones in which members conform to their normal roles and authority relationships. Novel organizations are ones in which or members develop new roles and authority relationships.

These contingencies produce four different types of disaster response organizations.

- ▲ **Established organizations** perform their normal tasks within normal organizations. For example, police directing evacuation traffic.
- ▲ **Extending organizations** perform novel tasks within normal organizations. For example, crews from public works agencies digging through rubble to extricate trapped victims.
- ▲ **Expanding organizations** perform their normal tasks within novel organizations. For example, Red Cross volunteers working under the supervision of permanent staff to operate mass care centers.
- ▲ Finally, **emergent organizations** perform novel tasks within novel organizations. For example, neighbors forming a team to search for and extricate trapped casualties after an earthquake.

Drabek et al. (1980) extended this typology by noting that organizations of all four types must often interact with each other in novel ways. They termed such structures **emergent multi-organizational networks** (EMONs). EMONs typically comprise professional and volunteer personnel from government agencies within local government. EMONs also include representatives from state and federal agencies and representatives from nongovernmental organizations (NGOs) and the private sector. These groups have different organizational titles, organizational structures, training, experience, and legal authority. Because of this, EMONs have frequently experienced severe difficulties in communicating with each other and coordinating their responses to disasters. When all emergency responders are trained according to the Incident Command System or Incident Management System, they can focus their efforts on saving lives and protecting property rather than on debating who is in charge of the emergency response.

SELF-CHECK

- Name three disaster impact zones.
- Explain whether emergency responders abandon their roles in a disaster.
- Name the four types of disaster response organizations.
- Explain **EMON**.

8.6 Basic Principles of Household Behavior in Emergencies

People in a disaster respond in ways that seem logical to them given their limited information and their options. People generally act adaptively and sometimes even heroically. For example, consider the case of passengers in United Airlines Flight 93 on September 11, 2001. Unlike the passengers on the flights hijacked earlier that day, they understood the hijackers' goal. They organized and attacked their hijackers. They chose to crash the flight in a Pennsylvania field rather than allow it to be flown into a building in Washington DC. Interpreting the results of decades of disaster research permits the identification of three distinct patterns of expected citizen response to such events.

8.6.1 Information

The first principle is that people threatened by disaster have multiple sources of information. None of these sources is considered completely credible. Nor is any single source expected to have all the information a household needs to protect itself. This can produce confusing and conflicting information unless different sources coordinate their risk communication. For example, Figure 8-5 depicts ratings that residents of Longview, Washington, and Kelso, Washington, made in 1985 about the degree of hazard knowledge held by themselves, their peers, the news media, and the government. They were asked to make these judgments about hazard knowledge separately for three hazard agents. The first was a volcanic eruption of Mt. St. Helens, which was about 40 miles east of their communities. The second was a chlorine release from a truck or train on nearby

Figure 8-5

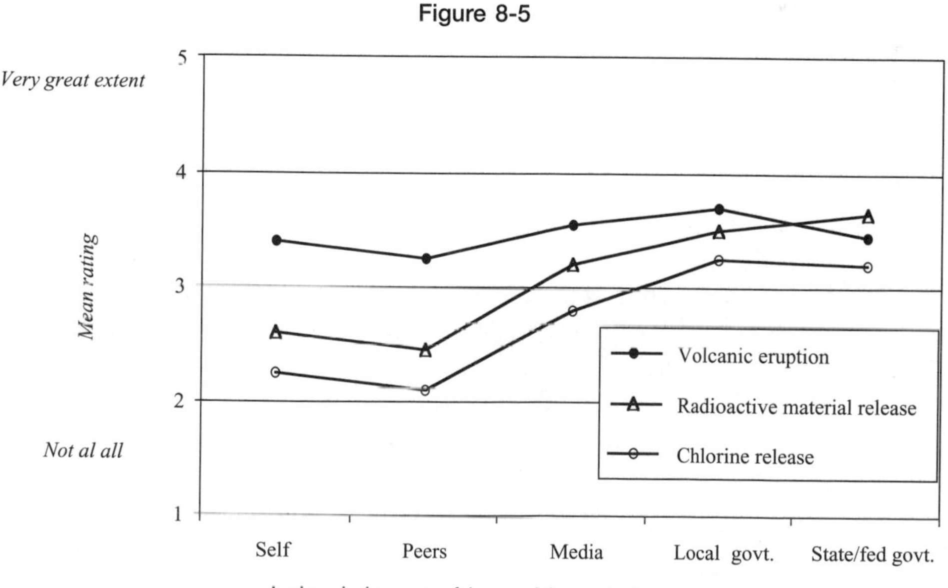

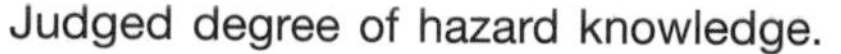
Judged degree of hazard knowledge.

transportation routes. The third was a release of radioactive materials from the Trojan nuclear power plant, which was less than 10 miles away. The figure indicates most people thought they knew more about hazards than their peers. They also thought they knew less than the news media and government. However, the news media were judged less knowledgeable than government. And, the differences in knowledge ratings are smaller for the more familiar volcanic hazard. Interestingly, the knowledge ratings were consistently higher for radiation hazard than for chlorine hazard. This could be because federal licensing regulations required the nuclear power plant to distribute emergency information brochures annually. It is important to note that none of the sources was rated as extremely knowledgeable about any of the hazards (i.e., ratings of 4 or above). Nor were any of the sources rated as severely lacking in knowledge (i.e., ratings of 2 or below). There were significant differences among sources, but each was credited with some degree of knowledge. Thus, conflicts in the information provided by different sources would be difficult for people to resolve.

8.6.2 Fear

The second basic principle is that people generally experience *fear.* They do not experience debilitating shock or panic. Fear is a normal human reaction to conditions that threaten them or their loved ones. Fear rarely is so overwhelming that it prevents people from responding. However, it does impair people's ability to reason through complex, unfamiliar problems. Fear is especially high when people lack information about the consequences of hazard exposure. Technological hazards and terrorist events involving chemical, biological, or radiological agents inherently involve unknown consequences. Many of these agents are undetectable by normal human senses. People cannot tell if they are being exposed. In addition, some of these agents have long latencies. This means it takes many years for symptoms to develop. They can result in dreaded conditions such as cancer and birth defects (Slovic, Fischhoff, and Lichtenstein, 1980).

It is important for you to address people's concerns directly. This is done most effectively by providing information. You should not try to give people a university education on the topic. Instead, provide clear, direct, relevant information about the hazard agent. Inform people of its potential personal consequences. In addition, people should also be told what the authorities are doing to protect them from the threat. They should be told how they can receive additional information. Contrary to popular fiction, the path to fear reduction is to provide—not withhold—information (Quarantelli and Dynes, 1985).

8.6.3 Protective Action

The third basic principle is that people take action when they think they are at risk. The initial response to a threatening situation might be to seek information.

Those who conclude they are at risk take protective action. It is important that official warning messages include recommended protective actions. If authorities do not provide recommended actions, people take action anyway. They take the most appropriate actions they already know, or are told about by peers or the media. They use what resources are available to them. Figure 8-6 illustrates some of the factors that people might consider if told they are threatened by a toxic chemical release (Lindell and Perry, 1992). First of all, it is important to recognize that many people would not even think of sheltering in-place and expedient respiratory protection unless these were specifically mentioned in a warning message. Second, many people would not know how to implement sheltering in-place. Nor would they know how to implement expedient respiratory protection, even if told to do so. Third, evacuation seems to be the best option because it is believed to be more effective than either of the other two protective actions. It is also attractive because it takes little more skill than sheltering in-place and expedient respiratory protection. However, evacuation is thought to require more time, effort, and money and produce more obstacles than the other protective actions. The presence of advantages and disadvantages is likely to make people hesitate to evacuate. In addition, research has found that people have concerns about evacuation in physically destructive hazards such as hurricanes. For example, some evacuees from Hurricane Lili were concerned that evacuation would expose their homes to looters and to storm damage. They believed they could prevent these outcomes if they remained at home (Lindell et al., 2005).

Figure 8-6

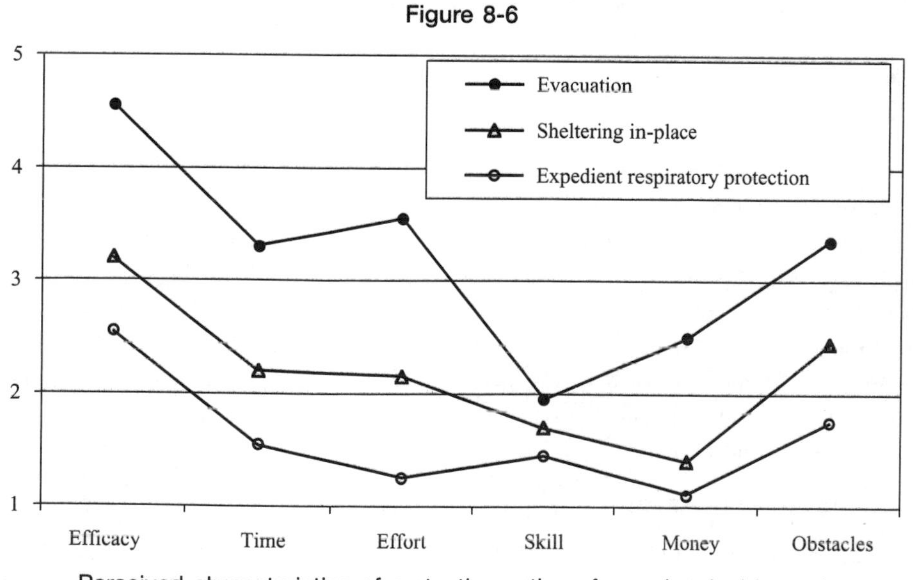

Perceived characteristics of protective actions for a chemical hazard.

You might consider some of these ratings to reflect misperceptions. For example, people might believe sheltering in-place and expedient respiratory protection are much more similar to evacuation in their efficacy and different in terms of their time requirements. Alternatively, you might want to know why sheltering in-place received such high ratings on obstacles. If one obstacle is not knowing how to shelter in-place, you could provide information about how to do this in a risk communication campaign. This would make sheltering in-place more attractive.

A message not accompanied by guidance for protective action fails to provide opportunities to reduce fear. In providing protective action recommendations, you might need to explain why the action will be effective. For example, tell people *why* quarantine at home will reduce their exposure to a biological hazard such as smallpox. This information accomplishes two things. First, it gives people a rationale for complying with official instructions. Second, it discourages people from inventing other apparently reasonable alternative actions. These actions might not be protective.

8.6.4 Compliance

The fourth basic principle is that some of those at risk will comply with authorities' recommendations. However, their compliance is rarely automatic. The level of compliance is contingent upon a variety of other factors:

- Information source.
- Message content.
- Receiver characteristics.
- Situational characteristics such as threat familiarity and urgency for response.

In cases such as seasonal floods, risk area residents are likely to be familiar with the threat and its environmental cues. In these cases, compliance with protective action recommendations from authorities is likely to be lower than with less familiar threats such as toxic chemicals. When those at risk are familiar with threat agents, they understand, or think they understand, the danger. They know when and where it will materialize. They know what should be done about it. Their own assessment of the situation might cause them to reject official recommendations. They will, at the least, thoroughly examine the basis for recommendations. When threat familiarity is low, people are more likely to accept the assessment of authorities. This is because they have little or no personal experience with the threat. Also, when the warning claims that the impact will come soon, people comply more readily. This is because there is no time for reflection.

This is a research-based conclusion from disaster literature (Lindell and Perry, 1992). In times of extreme stress, people look to government for guidance. When the agent of destruction is unfamiliar or intangible, or when the consequences appear overwhelming, people's expectations of protection and help are especially pronounced. Thus, compliance tends to be higher with technological and terrorist threats. For example, national opinion polling following the 9/11 attacks indicated substantial increases in levels of trust in government. The combination of citizen concern and a tendency to feel that taking action is important sets the stage for attention to messages from emergency authorities and enhances compliance. As has been the case in other types of disasters, people return to their normal skeptical attitudes toward government. Nonetheless, there is a window of opportunity in the height of crisis and for some time thereafter.

During the response phase, people appear to carefully comply with directions from police and fire personnel. For example, in Phoenix, Arizona, during March, 1999, women were believed to have been exposed to anthrax. They underwent nude decontamination by male hazardous materials technicians. This happened in a decontamination shelter without roof covering while news helicopters hovered above. One person mentioned concern with modesty, but none of the victims hesitated to follow instructions. Since that time, the Phoenix Metropolitan Medical Response System (MMRS) acquired enhanced decontamination shelters. They now have the ability to deploy "all female" decontamination teams. The incident stands as an example of compliance with emergency instructions when the threat is unfamiliar and time for action is limited.

The expectation of compliance also places a special responsibility upon local authorities. Namely, authorities must manage responsibly. They must have current, ongoing vulnerability assessment. They must have detection and prediction systems for threats when technologically possible. Response plans must be in place and they must be capable of executing those plans. In the absence of such plans, people will hold authorities responsible through the political process and possibly through the courts.

8.6.5 Spontaneous Evacuation

The fifth basic principle is that some of those who are not at risk will also comply with protective action recommendations. Spontaneous evacuation has been reported in response to hazard agents as varied as nuclear power plant accidents, and hurricanes. Spontaneous evacuation occurred during Hurricane Rita in 2005. Approximately 1.8 million people evacuated from the Houston/Galveston area. This is despite the fact that only about 700,000 people were in the risk area. The evacuation was likely greatly affected by the devastating impact of Hurricane Katrina on New Orleans three weeks earlier. Moreover, people were aware of the plight of the New Orleans evacuees because Houston was the site of the largest mass care operation for the Katrina evacuees. Problems with the response to Katrina were being reported every day in the news media. The Hurricane Rita evacuation makes it clear that large scale spontaneous evacuation is not limited to unfamiliar hazards such as radiation.

FOR EXAMPLE

Three Mile Island Evacuation

Ten times as many people evacuated during the nuclear power plant accident at Three Mile Island as were designated in evacuation advisory (Lindell and Perry, 1992). When only 15,000 people were issued an advisory, 150,000 actually evacuated.

SELF-CHECK

- Describe how people seek information about impending disasters.
- Explain how fear affects people.
- Describe the basic principle about people's action during a disaster.
- Explain the difference between warning compliance and spontaneous evacuation.

8.7 Expectations Regarding Stress Effects and Health Consequences

Psychological consequences rarely prevent people from responding in the short-term. However, the experience of any disaster can have long-term consequences for a few of the victims (Perry, 1985). You need to work with mental health professionals to anticipate disaster shock and traumatic responses. There may be cases of post-traumatic stress disorder (PTSD) among some population segments. Other difficulties can be depression and "survivor syndrome." The research literature shows that such long-term consequences are more likely to arise among

- ▲ People who have witnessed death or handled the dead.
- ▲ People who have been exposed to large scale property destruction.
- ▲ People whose relatives, neighbors or friends have been seriously injured or lost their lives.

However, people become depressed even if less severe conditions occur. As authorities move from response recovery and reconstruction, they should anticipate the need for referrals for crisis counseling. Some people also need other

FOR EXAMPLE

The Holocaust

Search the internet for information about the Holocaust to learn about survivors of one of the most heinous crimes of the twentieth century. Many survivors, including concentration camp survivors, experienced posttraumatic stress disorder with symptoms of anxiety and depression as a result of extreme guilt for surviving.

short-term therapy to reduce long-term negative consequences. Attention can also be given to victims' needs for economic support. They need a sense of closure and a way to fit the disaster experience into a worldview that allows a transition to a stable life (Perry and Lindell, 1997).

8.7.1 Expectations for Health Consequences

One of the least studied phenomena following disasters is the tendency of victims to develop physical health symptoms (Bourque, Russell, and Goltz, 1993). Perry (1979) observed that studies dating back to Prince's (1920) research on the Halifax, Nova Scotia, explosion indicated victims developed both psychological responses and physical health responses. Even in the apparent absence of psychological symptoms, victims and nonvictims have developed physical health problems following disasters. Some of these health problems are unrelated to the disaster agent. Tichner (1988) reported that disaster survivors one year after the event reported more health problems compared to nonvictims. Taylor (1977) found that tornado victims showed higher levels of headache, nausea, and emergency room visits. Logue and his colleagues (1979) found higher levels of emergency room visits following hurricane exposure. They also found higher incidents of gastritis, constipation, bladder problems, and headache. Smith, Handmer, and Martin (1980) reported higher levels of heart disease symptoms among flood victims. Janerich (1981) found (also among flood victims) higher levels of spontaneous abortions, leukemia, and lymphoma. There is no direct link between natural disasters and the types of nonimpact related physical health problems cited above. However, there is a time linkage between the disaster event and the onset of symptoms. This condition leaves open the possibility of direct—but unknown—causality of physical health symptoms. Or there is an indirect—and unstudied—link through psychological processes (Logue, Hansen, and Streuning, 1981). As Melick (1985, p. 196) concluded, "Uniformly victims have indicated poorer post disaster health." However, further studies are needed using better research designs to better explain this phenomenon.

8.7.2 Expectations About Adaptive Behavior

According to Dynes's (1983) *emergent human resources* model, you should assume that many risk area residents have disaster-relevant competencies. In addition, you should rely on households' existing patterns of social and task behavior rather than expect them to learn new ones. Finally, you should use existing authority structures and communications channels. Know what percentage of the population can perform different emergency response actions and what resources households need to respond effectively. Avoid stereotyping all households as being identical. Recognize the differences among households and, especially, avoid the error of assuming other households are just like yours. Recognize that many households are not self-reliant nuclear families who have reliable cars, can (and will) evacuate, and have credit cards to pay for their evacuation expenses. Finally, you should recognize the problems that can arise from volunteer "mass assault." You can avoid the confusion arising from convergence of massive amounts of resources. Devise special units to organize volunteers, and develop special locations and procedures for donations management. Ask people to donate money to organizations such as the Red Cross rather than send supplies.

SELF-CHECK

- **Describe the differences between the conditions experienced by Holocaust survivors and the conditions experienced by survivors of the 9/11 attacks on the World Trade Center and the survivors of Hurricane Katrina.**
- **Identify the groups most likely to experience psychological consequences of disasters and describe the treatment they need.**
- **Identify the link between disasters and health problems.**

SUMMARY

You must know how people truly react during a disaster so you can plan your response appropriately. This chapter shows you the differences between the myths and the realities of household response to emergencies. People may experience fear when disaster hits, but as this chapter shows, they also seek information and respond by taking protective action, regardless of their circumstances. In addition, those who don't experience the disaster, often respond by volunteering to help. Because you can predict some behaviors, it's important you develop plans to respond to those behaviors. As emergency managers, you have to disseminate information to those who seek it, and you have to ensure appropriate warnings, solutions for protection, and even plans for organizing the people who show up to help after the disaster are given in a timely manner.

KEY TERMS

Emergent multiorganizational networks (EMONs)	A group of organizations whose interactions develop in response to the demands of a disaster rather than being planned beforehand.
Emergent organizations	A disaster response organization that performs novel tasks within novel organizations.
Established organizations	A disaster response organization that performs normal tasks within normal organizations.
Evacuation trip generation	The number and location of vehicles evacuating from a risk area.
Expanding organizations	A disaster response organization that performs normal tasks within novel organizations.
Extending organizations	A disaster response organization that performs novel tasks within normal organizations.
Panic	An acute fear reaction marked by a loss of self-control which is followed by nonsocial and nonrational flight behavior.

ASSESS YOUR UNDERSTANDING

Go to www.wiley.com/college/lindell to evaluate your knowledge of hazards, vulnerability, and risk analysis.
Measure your learning by comparing pre-test and post-test results.

Summary Questions

1. Antisocial behaviors such as looting are common during and after disasters. True or False?
2. Disasters are not associated with increases in mental health problems in the affected population. True or False?
3. The positive impact of convergence is the increased resource base and increased morale. True or False?
4. It is the source and content of the warning message that largely determine response to warnings. True or False?
5. What percent of casualties reach a hospital on their own or through the help of a friend or bystander?
   (a) over 75%
   (b) 27%
   (c) 46%
   (d) less than 10%
6. Information passes from the inner zone through the outer zone through which of the following impact zones?
   (a) total impact
   (b) fringe impact
   (c) community aid
   (d) resource filter
7. If authorities do not provide recommended actions, people will not take action. True or False?
8. What can you expect from people when they respond to disaster?
   (a) People threatened by disaster have multiple sources of information and will seek to find information.
   (b) People generally experience *fear.*
   (c) People will take action when they think they are at risk.
   (d) all of the above
9. Emergent organizations
   (a) perform novel tasks within novel organizations.
   (b) perform novel tasks within normal organizations.

(c) perform normal tasks within novel organizations.
(d) perform normal tasks within normal organizations.

10. Red Cross volunteers working under the supervision of permanent staff to operate mass care centers is an example of an
    (a) extending organization.
    (b) expanding organization.
    (c) emergent organization.
    (d) established organization.

Review Questions

1. Why can myths about disasters be dangerous?
2. What evokes panic flight?
3. What is the therapeutic community?
4. What are five warning methods?
5. What four factors define evacuees' destination/route choice?
6. What are the different types of disaster response organizations?
7. What contingencies affect the level of warning compliance?
8. What are three distinct patterns of expected citizen response to disasters?

Applying This Chapter

1. As part of an initiative to educate the public about the dangers of myths that are related to disasters, you are writing an article for a Web site. Outline the key points you would make in the article.
2. You live in the Midwest and are watching national news coverage of a hurricane disaster off the coast of Florida. The media predicts major damage and casualties. What kind of response can you expect from the surrounding Florida communities? What kind of response can you expect from those living in your part of the country?
3. You need to warn people of an approaching hurricane? What types of warning systems do you use and why?
4. You are evacuating a city; what factors do you consider when estimating evacuation times?
5. Is the rate of compliance likely to be higher with a threat of flooding or with a threat of a nuclear power plant release of radioactive materials? How about the rate of spontaneous evacuation? Why?
6. You are tasked with organizing volunteers who have shown up to help after a tornado has destroyed a community. How would you organize the volunteers?

7. As an emergency manager, you need to understand the effects of stress factors on people who survive disasters in which they lost loved ones or in which they experienced guilt from survival. Research how stress affects those who experienced a traumatic experience, and then describe how the survivors of the tragedy cope with the effects.
8. As an emergency manager, you have been asked to describe adaptive behavior of people responding to disaster. Your audience assumes that the average person will evacuate from the disaster area before the disaster hits. You know there are many potential problems because of individual circumstances. What problems do you suggest discussing with your audience? Why is it important to explain these problems of adaptive behavior and what cannot be assumed about individual circumstances?

YOU TRY IT

Handling Volunteers and Donations

What steps can you take before a disaster to ensure you and your staff are not overwhelmed with volunteers and donations?

Community Response

After a major disaster, what type of response can you expect from the immediate community? From distant communities?

Medical Transport

What steps do you take to ensure injured victims are transported by ambulance to the appropriate hospital?

9

PREPAREDNESS FOR EMERGENCY RESPONSE

Organizing a Response

Starting Point

Go to www.wiley.com/college/lindell to assess your knowledge of hazards, vulnerability, and risk analysis.
Determine where to concentrate your effort.

What You'll Learn in This Chapter

- ▲ Principles of emergency planning
- ▲ Basic emergency response functions
- ▲ Common structures for emergency response
- ▲ Common structures for emergency preparedness
- ▲ Federal guidance for emergency operations plan (EOP) development

After Studying This Chapter, You'll Be Able To

- ▲ Examine the guiding principles of emergency planning
- ▲ Compare and contrast the four basic emergency response functions: emergency assessment, hazard operations, population protection, and incident management
- ▲ Compare and contrast different organizational structures for emergency response and preparedness
- ▲ Examine the basic principles of the Incident Management System (IMS)
- ▲ Examine the federal guidance for EOP development

Goals and Outcomes

- ▲ Manage resistance to the planning process
- ▲ Write an EOP
- ▲ Evaluate emergency response functions
- ▲ Select organizations for emergency response
- ▲ Select organizations for emergency preparedness
- ▲ Organize an emergency operations center (EOC)

INTRODUCTION

Your goal in a disaster is to prevent casualties and damage. How is this done effectively and efficiently? How do you plan for a disaster? In most cases, local governments have the resources to meet disaster demands without any outside assistance. Even when catastrophes such as Hurricane Katrina strike, local governments must be able to be self-sustaining for a significant period of time before state and federal assistance arrive. In this chapter, we look at ways to build community emergency preparedness by taking preimpact actions that establish a state of readiness to respond to extreme events threatening the community. In this chapter, you will evaluate the guiding principles of emergency planning. You will examine emergency response functions. You will then assess organizational structures for both emergency response and emergency preparedness. Finally, you will assess the purpose of an EOP and the federal guidance for EOP development.

9.1 Guiding Principles of Emergency Planning

Planning is a *process*. You have to develop a plan, make sure that individuals and teams have the skills they need, critique the team's performance, and make any adjustments as necessary (Dynes, Quarantelli, and Kreps, 1972; Kartez and Lindell, 1987, 1990). Emergency response planning varies among communities. Some communities have a formal process that is assigned to a specific department and has a specific budget. The EOP and standard operating procedures (SOPs) are written. Other communities have an informal process. In these communities, responsibility for emergency planning is poorly defined and there may be a limited budget that is dispersed among many agencies. There might not be a written plan or SOPs. To some extent, the type of emergency planning in place depends on the size of the community. Larger communities with many resources and personnel, and perhaps higher levels of staff turnover, tend to have formalized processes. They rely more heavily upon written documentation and agreements. In smaller communities, the planning process might generate few written products and rely principally on informal, personal relationships. Formalization of the planning process also varies with the frequency of hazard impact. In communities subject to frequent threats, response to hazards may be routine. A community that is frequently flooded, for example, may not feel the need to write a plan. Officials may believe that everyone knows the fire department evacuates residents of low-lying areas when the floodwater reaches a certain street.

Despite the differences in the planning process, there are some consistencies in emergency planning. The following are fundamental principles of community emergency planning that are related to high levels of community preparedness (Quarantelli, 1982):

- ▲ Emergency planners should anticipate both active and passive resistance to the planning process and develop strategies to manage these obstacles.
- ▲ Preimpact planning should address all hazards to which the community is exposed.
- ▲ Preimpact planning should elicit participation, commitment, and clearly defined agreement among all response organizations.
- ▲ Preimpact planning should be based upon accurate assumptions about the threat, typical human behavior in disasters, and likely support from external sources such as state and federal agencies.
- ▲ EOPs should identify the types of emergency response actions that are most likely to be appropriate but encourage improvisation based on continuing emergency assessment.
- ▲ Emergency planning should address the linkage of emergency response to disaster recovery and hazard mitigation.
- ▲ Preimpact planning should provide for training and evaluating the emergency response organization at all levels—individual, team, department, and community.
- ▲ Emergency planning should be recognized as a continuing process.

9.1.1 Managing Resistance to the Planning Process

The first principle listed above states that some people will not want to participate in emergency planning. Others will resist the process (Auf der Heide, 1989; McEntire, 2003; Quarantelli, 1982). Many people do not like to think about disasters occurring. A common concern is that it takes away resources that are needed for more pressing needs, such as road repairs and school expansion. Planning mandates help but are not enough. Planning activities require strong support from one of three sources. The first is the chief administrative officer (CAO), who has legitimate, reward, and coercive power. Another source of support is an issue champion, also known as a **policy entrepreneur**, who has the expertise and legitimacy to promote emergency planning. The third source of support is a disaster planning committee that can mobilize the support of many different community organizations (Lindell et al., 1996; Prater and Lindell, 2000). However, acceptance of the need for planning does not eliminate conflict. Organizations seek to preserve their autonomy, security, and prestige. They resist collaborative activities that can threaten these objectives (Haas and Drabek, 1973). Emergency planning needs power and resources (especially personnel and budget). Every unit within an organization wants its "proper role" recognized and a budget that reflects its role.

9.1.2 Adopt an All-Hazards Approach

If there are separate plans for each hazard, you need to incorporate them into one plan. Identify the types of threats your community might face. Then examine the extent to which different hazard agents make similar demands. When two hazard agents are similar, they will likely require the same type of response. This provides multiple use opportunities for personnel, procedures, facilities, and equipment. This simplifies the EOP by reducing the number of functional annexes. In addition, it simplifies training and enhances the reliability of organizational performance during emergencies. When hazards have different characteristics and require different responses, hazard-specific appendixes are required for the functional annexes.

9.1.3 Promote Multiorganizational Participation

Ask other groups to participate in the planning process and commit to the planning process. This obviously should include public safety agencies such as emergency management, fire, police, and emergency medical services. It should also include groups that are potential hazard sources. For example, it should include hazardous materials facilities and hazardous materials transporters. It should also include groups that must protect sensitive populations such as schools, hospitals, and nursing homes. All of these organizations must coordinate because they have different capabilities and vulnerabilities. To perform their functions effectively, emergency response personnel must be aware of each emergency organization's:

- ▲ Mission.
- ▲ Organizational structure.
- ▲ Style of operation.
- ▲ Communication systems.
- ▲ Procedures for allocating scarce resources.

9.1.4 Rely on Accurate Assumptions

Emergency planning must be based on accurate knowledge of threats and the likely human response. You already know that you must identify the hazards your community faces. Know the characteristics of these hazard agents—their speed of onset and scope and duration of impact. Also know these hazard agents' potential for producing casualties and property damage. Then determine which geographical areas are exposed to those hazards. Identify the facilities and population segments in those risk areas.

When it comes to identifying hazards, planners and officials frequently recognize the limits of their expertise in doing so. They recognize that they lack accurate knowledge about the behavior of geophysical, meteorological, or technological hazards. They know they need help from experts. Unfortunately, the same cannot always be said about accurate knowledge regarding likely human behavior in disasters. The problem is not so much that people don't know what is true, but that what they do "know" is false. Belief in disaster myths can hamper the effectiveness of emergency planning and response. These myths misdirect the allocation of resources and information. For example, officials sometimes withhold information because of concern that people will panic. This approach is counterproductive. Research has shown that people are more reluctant to comply with recommended protective actions when they are provided with vague or incomplete warning messages. Thus, believing the myth of panic can lead to exactly the opposite result that officials are trying to achieve. For these reasons, the planning process must include knowledge about people's behavior during emergencies.

Finally, plans must be based on accurate assumptions about aid from outside the community. In major disasters, hospitals might be overloaded. Destruction of telecommunication and transportation systems could prevent outside assistance from arriving for days. Restoration of disrupted water, sewer, electric power, and natural gas pipeline systems could take much longer. Consequently, households, businesses, and government agencies must be prepared to be self-reliant for as much as a week.

9.1.5 Identify Appropriate Actions While Encouraging Improvisation

Even though an EOP should identify the actions that are most likely to be appropriate, it also should emphasize flexibility. Those involved must be encouraged to improvise based on their assessment of disaster demands (Kreps, 1991). Much emphasis has been given to the idea that careful planning promotes quicker response. Rapid response is important. However, it is not the only objective of emergency planning. The appropriateness of response is as important as the speed of response (Quarantelli, 1977). In turn, the best response is based on continuous and accurate assessment. In the high pressure situation accompanying an imminent threat, it is difficult to appear to be "doing nothing." However, it is important to recognize that the best action might be to mobilize emergency personnel and actively monitor the situation. Gathering further information is better than taking unnecessary actions.

The planning process should address principles of response in addition to providing detailed operating procedures. An overemphasis on detail causes four problems:

1. The anticipation of all contingencies is simply impossible (Lindell and Perry, 1980).
2. Specific details tend to become out of date quickly. You then have to constantly update the plan (Dynes et al., 1972).

3. Very specific plans often contain so many details that all emergency functions appear to be of equal importance. This causes response priorities to be unclear or confusing (Tierney, 1980).
4. More detailed plans are more complex plans. This makes it more difficult for people to understand how everyone's roles fit into the overall response. If they don't understand everyone's roles, they can't implement the plan effectively.

9.1.6 Link Emergency Response to Disaster Recovery and Hazard Mitigation

Response and recovery are not two distinct and different phases (Schwab, et al., 1998). Some residents will still be engaged in emergency response tasks while other residents have moved on to disaster recovery tasks. Moreover, officials will be working on emergency response tasks at the same time they must begin working on disaster recovery. Therefore, preimpact emergency response planning should be linked to preimpact disaster recovery planning. Linking these two planning processes speeds the recovery. The coordination between emergency response planning and disaster recovery planning can be achieved by having the committees responsible for these two activities work together.

9.1.7 Conduct Thorough Training and Evaluation

Training and evaluation must be part of the EOP. First, explain the EOP to the administrators and personnel who will be involved in the emergency response. Second, train all those who have emergency response roles. This includes fire, police, and emergency medical services personnel. There also should be training for personnel in hospitals, schools, nursing homes, and other facilities that might need to take protective action. Finally, you should involve the population at risk. They must be aware that planning for community threats is underway and know what is expected of them. They need to know what is likely to happen in a disaster. They need to know what organizations can and cannot do for them. It is also essential that training include tests of the proposed response operations. Drills and exercises allow people to better understand each other's professional capabilities and personal characteristics. Furthermore, exercises test plans and procedures, staffing levels, personnel training, facilities, equipment, and materials all at the same time. Finally, exercises produce publicity. This publicity reassures the public that planning for disasters is underway and that preparedness is being enhanced.

9.1.8 Adopt a Continuous Planning Process

Planning is a continuous process. Threats, staff, facilities, and equipment change over time. Therefore, the planning process must detect and respond to these

FOR EXAMPLE

Hazard Modeling

One of the principles of emergency preparedness is to rely on accurate assumptions. You may need to work with others and simulate the disaster to know what the effect of the hazard will be. For example, before Hurricane Katrina, emergency officials worked with hazard modeling software and simulated the effects of a Category 5 storm on the New Orleans levees. They learned from this simulation that the levees would fail during a hurricane that was greater than a Category 3.

changes. Unfortunately, this is not always recognized. Wenger, Faupel, and James (1980, p. 134) found that "there is a tendency on the part of officials to see disaster planning as a product, not a process." Planning does require written documentation. However, planning is also made up of factors that are difficult to document. These include the knowledge about resources available from governmental and private organizations. They also include learning about emergency demands and other agencies' capabilities. They include establishing collaborative relationships with other organizations. Tangible documents and hardware simply do not provide a sufficient representation of what the emergency planning process has produced. By treating written plans as final products, you might believe you are prepared for an emergency when you are not (Quarantelli, 1977). As time passes, the plan sitting in a red three-ring binder on the bookshelf looks just as impressive as it did the day it was published. Yet, many changes have taken place. For example, new hazardous facilities might have been built and others decommissioned. New neighborhoods might have been built. Agencies might have been reorganized. The potential for change dictates that plans and procedures be reviewed periodically, at least on an annual basis.

SELF-CHECK

- Explain the similarities and differences between creating an EOP in a small city and a large city
- Define **policy entrepreneur.**
- Explain why two similar hazard agents will likely require the same type of response.
- Name the five characteristics that emergency response personnel must understand to promote effective multiorganizational participation.

9.2 Emergency Response Functions

You must analyze your organization's capability to perform basic response functions. There are four basic emergency response functions:

- ▲ **Emergency assessment:** Detecting a threat, predicting its potential impact, and determining how to respond.
- ▲ **Hazard operations:** Taking actions to limit the magnitude of the disaster impact.
- ▲ **Population protection:** Taking protective actions to minimize the number of casualties.
- ▲ **Incident management:** Mobilizing and directing resources to respond to an emergency.

This chapter describes the emergency preparedness actions you must take *before* a disaster so you can implement them *during* a disaster. These emergency preparedness actions involve identifying the personnel, procedures, facilities, equipment, materials, and supplies the emergency response organization needs. The operational aspects of implementing these functions will be discussed later.

9.2.1 Emergency Assessment

You need to decide how your community will *detect and classify* each threat to which your community is exposed. Some natural hazards, such as earthquakes, are detected and classified by local agencies. Other hazards, such as hurricanes, are detected and classified by specialized federal agencies. Incidents at fixed site facilities are usually detected and classified by plant personnel. Transportation incidents are detected by carrier personnel, local first responders, and passers-by.

You should review your hazard vulnerability analysis (HVA) to determine how detection is likely to be achieved and how authorities will be informed. Locally detected hazards require you to ensure that the necessary detection systems are established and maintained. For hazards detected by others, ensure that a hazard alert can be called in to a warning point. This center must be staffed around the clock, so you will probably use your community's dispatch center.

Another part of emergency assessment is *hazard monitoring*. You must continually monitor the hazard and project its future status. The technology for monitoring varies by hazard. In many cases, continuing information is provided by the detection source. For example, the National Hurricane Center provides hurricane updates every six hours. Similarly, plant personnel should provide continuing information about a plant emergency and the likelihood of a hazardous materials release.

Environmental monitoring is also needed when hazardous materials can be spread by wind or water. Toxic chemicals, radiological materials, and volcanic ash are carried downwind. This means you must establish procedures to obtain current weather reports and forecasts of future weather conditions. This information allows you to monitor changes in wind direction, wind speed, and atmospheric stability. You can then identify the risk areas and determine if they are likely to change over time. Environmental monitoring is also needed for hazmat spills into bodies of water. The speed and direction of river, lake, or ocean currents determine which sections of the shoreline will be affected. Identify the sources of this information before an emergency occurs.

▲ **Damage assessment** begins by identifying the boundaries of the impact area. It then proceeds to estimate the total amount of damage to buildings and infrastructure in the impact area. Damage assessment is needed to begin the process of requesting a presidential disaster declaration. You need to start before an emergency to identify the staff, equipment, and procedures you will use for damage assessment.

▲ **Population monitoring and assessment** identifies the size of the population at risk. This is extremely important if the number of people in the risk area varies over time.

Many communities have athletic contests and festivals in areas of high hazard exposure. You should maintain a calendar of major events so you know if special warning and evacuation procedures are necessary for these occasions. You also need to work with schools, hospitals, and nursing home administrators to establish procedures for monitoring special facility evacuations. Finally, you need to work with police and transportation officials to establish procedures for monitoring the flow of evacuation traffic and rerouting it if traffic jams develop.

9.2.2 Hazard Operations

Preparing for hazard operations varies significantly between hazards. In some cases, hazard operations require equipment that is normally available within the community. For example, structural fires require routine equipment but wildfires often require special brush trucks for hazard source control. Area protection works are another type of hazard operations best illustrated by elevating levees during floods. You need to plan in advance to identify the sources from which you can get the sandbags you need. Some hazard agents require preimpact preparation for emergency response operations to implement building construction practices. For example, communities exposed to earthquakes should identify sources of shoring materials that will keep severely damaged buildings from collapsing. Finally, you need to plan in advance to develop procedures for contents protection practices. For example, you might identify sources of plywood for protecting windows from wind damage.

9.2.3 Population Protection

You need to prepare for emergencies by developing procedures for *protective action selection.* For some hazard agents, there is only one recommended protective action. For example, people threatened by tornadoes should shelter in-place. Those faced with inland floods and tsunamis should evacuate. However, in toxic chemical and radiological releases, the appropriate protective action depends on the situation (Lindell and Perry, 1992; Sorensen, Shumpert, and Vogt, 2004).

You should also devise procedures and acquire equipment to warn the risk area population. During slow onset incidents, such as hurricanes, there is likely to be adequate time for different types of warning mechanisms such as route alerting, where emergency vehicles announce warnings over loudspeakers as they drive through neighborhoods. You may even have enough time for emergency personnel to deliver warnings door-to-door. However, rapid onset incidents such as toxic chemical releases might require using siren systems to alert people to turn on their radio or television sets. You should also prepare for *search and rescue* by providing special training and equipment to organized teams. Other population protection tasks such as impact zone access control/security, and medical care also require special equipment and procedures.

9.2.4 Incident Management

Managing hazards involves the same tasks regardless of the hazard agent. Agency notification and mobilization require equipment, such as pagers, and personnel, such as duty officers who are available around the clock. You must also develop procedures to ensure that key personnel are notified quickly. Establish a space that will be used as the emergency operations center (EOC). Mobilization of emergency facilities and equipment is facilitated by actions such as storing critical documents in the EOC. Communication and documentation are supported by the acquisition of radios, telephone systems, and personal computers. These tasks also are supported by procedures to record and route messages. You can also prepare for many of your emergency response organization's specific activities such as analysis/planning, internal direction and control, logistics, finance/

FOR EXAMPLE

Preparedness for Population Protection Actions

Some hazard agents such as earthquakes require special preparation. After an earthquake, you will probably need to implement urban search and rescue. Heavy construction equipment will be needed to stabilize buildings, extricate victims, and protect building contents from further damage.

administration, and external coordination. Identify the ways in which personnel will perform tasks or have reporting relationships that differ from the ones they encounter in normal conditions. You can devise organization charts, task checklists, telephone lists, and other job performance aids that will assist them in their duties. You can prepare for providing public information to the media. Identify a joint information center (JIC). Provide extra phone lines for media personnel. Develop background information about the community. Develop information on its hazards. Provide information on your organization.

SELF-CHECK

- **Define damage assessment.**
- **Describe hazard source control.**
- **Explain how population protection preparedness is different for a toxic chemical release and an earthquake.**
- **Describe the purpose of a joint information center (JIC).**

9.3 Organizational Structures for Emergency Response

Organizational structures for emergency response should be based on two basic principles. First, the structure used to respond to everyday emergencies should form the basis of a larger structure to deal with disasters. Second, the local response structure must be flexible. It must be able to expand as additional external resources are added to match the demands of the disaster. The most common structures for emergency response organizations are called the Incident Command System (ICS) and the Incident Management System (IMS).

9.3.1 Incident Command System and Incident Management System

For many years, the federal government provided state and local governments with criteria for evaluating their EOPs. However, it did not require or even recommend a specific structure to meet those criteria. In part, reluctance to do so was based on the principle that state and local governments should be allowed to meet the planning criteria in any way that they deemed appropriate. Moreover, state governments differ from each other in their structures and resources, as do local governments. Forcing a single structure on all emergency response organizations seemed doomed to fail. The federal government's performance-oriented approach produced a very logical outcome. Jurisdictions differed in their emergency response organizations' structures, positions, resource names, and procedures. This even hampered cooperation among identical agencies (e.g., fire departments) from neighboring communities within a single state.

Following a series of wildfires in Southern California in 1970, fire departments joined to address this problem. Their concerns were as follows:

▲ Lack of a common organizational structure.
▲ Inadequate emergency assessments.
▲ Poorly coordinated planning.
▲ Uncoordinated resource allocation.
▲ Inadequate interagency communications at incident scenes.

This led to the development of the Incident Command System (ICS), which can be summarized in terms of seven basic principles (Irwin, 1989).

1. **Standardization:** All communities must use a common emergency response organization structure with standardized names and functions for subunits.
2. **Functional specificity:** There is a division of labor so each of these units is assigned a specific function to perform.
3. **Manageable span of control:** Subunits are established to limit the number of personnel directly supervised by each unit manager. This is usually five subordinates but can range from three to seven.
4. **Unit integrity:** People from a given professional discipline (e.g., police or fire) are assigned to the same unit.
5. **Unified command:** A single incident commander (IC) manages most incidents. A unified command team manages response when multiple agencies have responsibility for a given incident.
6. **Management by objectives:** Senior incident managers develop action plans that include specific, measurable objectives. They evaluate their effectiveness by monitoring the achievement of these objectives.
7. **Comprehensive resource management:** The IC or unified command team directs the allocation of all resources to response tasks. This includes personnel, facilities, vehicles, and equipment.

Over the next decade, ICS received increasing support. Unfortunately, ICSs tended to be region-specific. By the 1980s, the fire services in particular became concerned that responding departments needed a common ICS to increase the effectiveness of response to larger incidents. FIRESCOPE (Firefighting Resources of Southern California Organized for Potential Emergencies) developed a version of ICS that was funded, adopted, and promoted by the FEMA (FEMA, 1987).

FIRESCOPE ICS is a planning-based response system that combines planning functions with the functions of an EOC. The planning and coordination is achieved by a multiagency coordination system (MAC). It is operated by a team of agency directors and divided into two subsystems. The first is a software-based fire information management system that stores fire-relevant data. The second is a coordination system that implements policy devised by the MAC. The EOC component of FIRESCOPE comprises sections that address field operations, logistics, planning and finance.

This version of ICS was designed specifically for large-scale incidents, especially fire services in Southern California. It was a major improvement over previous systems (Coleman and Granito, 1988; Lesak, 1989). Later, Alan Brunacini (1985, 2002) enhanced the FIRESCOPE system with support from the National Fire Protection Association. He changed it so ICS could be used as readily in small events as large ones. Brunacini changed the command function to include specialized advisors. He expanded the operations function to include routine departmental response demands. He also included connections to a municipal EOC and police incident commanders. The revised structure was called the Incident Management System (IMS). Its advantage was that daily use in all incidents—minor and major—would enhance its effectiveness for major incidents. IMS is now widely used in the American fire services. It is also used in Canadian, British, and Australian fire services.

The major advantage of IMS is to make all resources available for every incident. This is true whether the incident is a routine emergency or a large-scale disaster. The resources are provided automatically as the IC escalates the response to meet the emerging incident demands. The IMS itself is a field structure that can manage resources at multiple impact scenes from an incident command post. In such cases, the IMS might not be supported by activation of an EOC, especially in minor incidents. In disasters that have a widespread impact, a jurisdiction's EOC can assume the role of the on-scene incident command post. This would be likely in response to a biological attack which impacts might not be detected until long after the attack. The strength of using IMS as the basis for emergency and disaster response lies in its ability to quickly and effectively initiate emergency operations. Every incident is initially addressed by trained and equipped emergency responders guided by an IC. These personnel are always on duty and respond to all calls. Whether the incident is known to be a disaster or initially appears to be a routine incident that becomes a disaster, IMS is an organizational structure for response that can be expanded to fit situational demands. In summary, the IMS is a flexible structure for assembling resources and directing emergency response efforts that is adaptable to any type of hazard. It can address large, complex, incidents as effectively as small, routine incidents. Unfortunately, there has been no research on the effectiveness of IMS or ICS. Ultimately, the use of IMS relies on the belief that using the seven basic principles will be effective.

9.3.2 Basic IMS Principles

IMS is a system based on responsibilities assigned to specific standardized positions. This means agencies must select and train their staff to perform all the duties associated with these positions. Any responder may assume the role of IC. In practice, however, the IC is usually the first arriving company officer or Battalion Chief. The fundamental principle of IMS is that there must always be one and only one IC at every incident scene. The most senior officer who is first to arrive at the incident assumes command. Once established, command may be transferred to other more senior officers as they arrive. Figure 9-1 shows a fully implemented IMS structure that would be appropriate for a major disaster. IMS size and composition expand as the IC seeks to meet incident demands. The structure begins with the assumption of command and the designation of specialized units to address the hazard at the scene. This includes handling agent-generated demands that address the threat itself. It also includes response-generated demands that support the responders and coordinate with other agencies.

IMS uses the terms sections, branches, and sectors to describe different size groupings of personnel, equipment, and apparatus. In Figure 9-1, Incident Command is shown with five sections directly attached to it. The five sections are planning, operations, safety, administration, and logistics. They are staffed based on incident

Figure 9-1

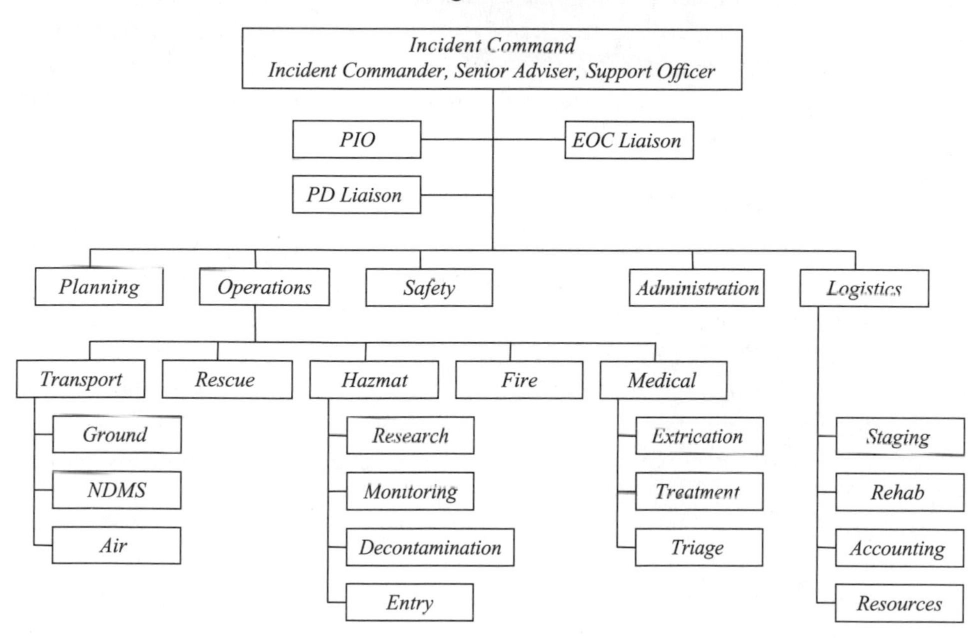

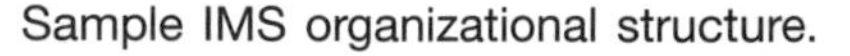
Sample IMS organizational structure.

size and conditions. Section chiefs in the incident command post work with the command staff to formulate an overall emergency response strategy. The section chiefs then direct and monitor operations. Branch and sector officers implement tactical operations. In a fully implemented IMS, branches are established under sections and are functional tactical areas relevant to each section. For example, Figure 9-1 shows five branches under the operations section. These are transport, rescue, hazardous materials, fire, and medical.

The naming of branches follows the specific activity they perform; the number of branches depends on the intensity of the demand for each of the functions needed in the incident response. Thus, IMS for an urban earthquake would include a heavy rescue branch. Sectors are defined beneath branches and execute specific tasks. Typically, sectors contain fire companies or special teams. Branches and sectors are activated in response to (or, better still, in anticipation of) the incident demands. Hence, in small hazardous materials incidents where few victims are present, the medical branch would be only a single unit and is called a medical sector. In events where there is no fire, the fire branch would not be activated. Although basic principles of IMS are easy to grasp, more advanced concepts provide a sophisticated method of allocating responsibility for response strategy, tactics, and tasks (Brunacini, 2002; Carlson, 1983).

As indicated earlier, there are some differences between IMS and ICS structures. Under ICS, there is neither a senior advisor nor a support officer. Instead, there is a scientific officer in the command section, and safety is staffed by a single officer in the command section rather than by a separate section. Moreover, ICS has only a single liaison officer rather than separate police and EOC liaisons. Finally, ICS defines finance and administration as two separate sections rather than combining them as in IMS.

9.3.3 IMS Implementation

In larger incidents, the IC may be supported by a support officer and a senior advisor. Senior officers fill these two additional roles within the IMS command section as they arrive at the scene. After assuming command, the IC establishes a command post. Throughout the incident, the IC performs seven activities listed in Table 9-1. Through these duties, the IC develops and maintains the strategy and resources that are needed to terminate the incident. The IC assigns duties to the senior advisor and support officer. These include reviewing, evaluating, and recommending changes to the incident action plan. In particular, the senior advisor focuses on overall management or "big picture" issues. This officer monitors the overall incident. He or she evaluates possible responses to current and future demands. The senior advisor then determines the need for activating additional branches or sections. The senior advisor also evaluates the need for working with other departments and groups. The support officer provides additional

Table 9-1: Incident Command Activities

1. Conduct initial situation evaluation and continual reassessments.
2. Initiate, maintain, and control communications.
3. Identify the incident management strategy, develop an action plan, and assign resources.
4. Call for supplemental resources, including EOC activation.
5. Develop an organizational command structure.
6. Continually review, evaluate, and revise the incident action plan.
7. Provide for continuing, transferring, and terminating command.

assistance. He or she assists with determining priorities and also provides direction with regard to safety. This officer assists with written plans for control and accountability. He or she evaluates the viability of the response organization and span of control. The support officer also evaluates the need for additional resources and assigns logistics responsibilities.

When there is a major emergency, extra steps must be taken. Most jurisdictions provide for the command staff to be supported by an on-scene public information officer (PIO) and a police liaison. In addition, there is an EOC liaison who is responsible for coordination between the incident scene and the EOC. The goal of an articulated command is to spread the functions to specialists. This permits effective communication with responders on scene and emergency authorities off scene. It also allows the IC to focus on the incident demands.

The command staff establishes a public information sector to handle the news media. This sector provides the information the news media needs to accurately report the status of the incident and the response to it. The staff PIO directs the sector and establishes a media area that does not impede operations. The PIO gathers information about the incident. In a major incident, the on scene PIO coordinates with the EOC PIO and PIOs of other agencies. This ensures consistent, accurate information. It also avoids the release of sensitive information.

Some incidents, particularly terrorist attacks, may require law enforcement actions. In these cases, the command staff assigns a police liaison sector. The IC may establish a communications link with the police command post or request that a police supervisor be assigned to the fire command post. The police liaison sector deals with all activities requiring coordination between the two departments. This includes traffic control, crowd control, incident scene security, evacuations, and crime scene management.

The command staff delegates responsibility for implementing their response strategies to the five section chiefs. The planning section is charged primarily with forecasting incident demands and other planning functions. The planning section serves as the incident commander's "clearinghouse" for information. In chemical, biological, or radiological (CBR) incidents, this function is particularly critical. This is because information from a variety of specialists will flow to the scene. The planning section relays information from these sources to the command staff.

The operations section deals directly with all hazard source control activities at the incident site. In addition, it is responsible for the safety of personnel working within the section. A critical administrative duty of the operations section is to establish branches, creating and overseeing as many branches as needed, depending on the demands of the incident. Branches typically include the primary operational functions of transport, rescue, hazmat, fire, and medical. The transport branch is responsible for transporting injured persons from the incident scene to hospitals for definitive care. The rescue branch is charged with search and rescue and extrication of firefighters who become lost, trapped, or endangered. This branch may oversee a large number of units serving as rapid intervention crews (RIC units) depending on the size of the incident. RIC units have the exclusive responsibility of rescuing emergency responders. In addition, an evacuation branch can be created to deal with endangered citizens.

The hazardous materials branch typically houses four sectors representing four principal functions. These are research, monitoring, decontamination, and site entry. In a hazmat incident, the hazardous materials branch addresses critical response priorities. These are identifying the hazard agent; designating hot, warm, and cold zones; and coordinating with law enforcement resources for site access control and special services. To assist in agent identification, this branch is supported by the planning section, on-scene toxicology specialists, and other specialized personnel. An entry team sector is responsible for hot zone entry and is supported by a backup team sector. The latter is present for relief or rescue of the entry team. Decontamination of victims can begin with the first units on scene. The hazardous materials branch assembles decontamination lines and performs decontamination.

The fire branch is charged with the management and suppression of fires. It also, when appropriate, operates sectors. Sometimes fires occur in context of other hazard agents such as explosives or hazardous materials. In these cases, the fire branch works with the IC to identify priorities. The fire branch operates in a defensive posture until other hazards have been addressed. It then shifts to offensive operations to extinguish fire. In the operational phase, the fire branch operates a safety sector that includes one company in reserve for rapid rescue of trapped firefighters. Building-related sectors are used in high-rise incidents to control access and conduct inside firefighting. Directional sectors are established for both defensive and offensive attacks. The fire is fought until the IC declares that it has been controlled and flames have been knocked down. Then an overhaul sector is established to extinguish any remaining active fire. The overhaul sector remains at the site as long as needed to extinguish spontaneous combustions.

The medical branch addresses removal and treatment of patients. The extrication sector is responsible for locating and removing patients to treatment areas. A triage sector performs the initial assessment of patient conditions and treatment needs. In a hazmat incident, this function may be performed before, during, or after decontamination. The toxicity of the agent determines victim assessment. In the case of nerve agents, toxicity determines when patients should be given antidotes. Triage and initial treatment may also be performed within the extrication sector. Similarly, depending on the agent, antidote administration may be appropriate at the earliest moment. In such cases, treatment begins before or during mass decontamination. When time is not critical to survival, antidote administration may take place at treatment areas. Triage tags are used to categorize patient injuries and record treatments administered. Triage tag numbers also become tracking numbers for patients.

Behavioral health operates as a sector within the medical branch. These personnel and units may be assigned in a variety of activities at the scene. The on-scene behavioral health coordinator works through the medical branch officer while maintaining liaison with the planning section and the EOC. Behavioral health units may oversee and assist patients. This can occur while patients are awaiting decontamination, being decontaminated, in treatment, and during transportation.

The transportation branch can expand to as many as four sectors as incident demands escalate. Transport north and south represent different directional movement points for ground transportation to local hospitals or mass care facilities. This movement may involve different vehicles, as appropriate to patient needs. If the fire department does not operate its own ambulance system, formal agreements should be established for transport vehicles from local EMS providers and ambulance services. The air sector moves patients by rotary wing aircraft if this is safe, given the hazard agent involved and the requirements of the patients' conditions. Finally, the National Disaster Medical System (NDMS) sector prepares patients in accordance with the local NDMS plan.

The safety section is staffed by a safety officer. This officer mobilizes the unit and maintains safe operations at the incident scene. The safety officer's primary task is to develop and implement plans for rescue and incident scene safety. The safety officer also must implement environmental cleanup after operations have ended. In large incidents, additional personnel support the safety officer. They monitor reports from all incident scenes. They report progress to the command section. If safety section personnel discover a pattern of unsafe practices, the safety officer is authorized to stop operations at an incident scene.

The administration section focuses on procurement, cost recovery, liability, and risk management. These activities involve contracting with vendors to deliver services that cannot be provided by the responding agencies. They also establish resource-sharing agreements among responding agencies. They document casualties and property damage to settle later claims.

The logistics section oversees many functions. There are four principal branches under logistics. They are staging, accountability, rehabilitation, and resources. The staging branch oversees the initial arrivals of unassigned units. The accountability branch tracks the responding units and individual crews to insure their safety. The rehabilitation branch monitors deployed personnel. They address both physical and psychological needs. This sector uses specialized equipment. They provide food, fluids, and debriefing for personnel. Finally, the resource sector oversees all equipment. They provide any needed communications equipment. They handle repairs and resupply. In a hazmat incident, this sector moves antidotes, medical supplies, and equipment to the scene.

In summary, the IMS is a flexible structure. Its value lies in the close linkage between emergency plans and operations. To respond to a threat, it is imperative that the emergency response organizations adapt to the specific demands of each incident. The IMS both reflects and directs the capabilities of the organizations that respond to the incident. Planning processes that account for the local IMS have greater flexibility. They have a greater likelihood of being successfully implemented. The advantage of IMS over the earlier ICS is that it provides for a better accounting of the activities that must be performed away from the incident scene. For example, IMS addresses activities such as warning and evacuation that are not addressed within ICS. Unfortunately, these activities must all be addressed by the operations section. For example, an evacuation branch would coordinate the movement of people from risk areas adjacent to the scene. It would also coordinate information releases. This requires the operations chief at an incident scene to be responsible for branches or sectors that he or she cannot supervise directly. Assignment of these activities to the operations chief could violate the principle of manageable span of control. One person cannot do everything. The operations chief could be put into a position of supervising warning, evacuation, and mass care branches. This is in addition to supervising the transport, rescue, hazmat, fire, and medical branches.

You can easily see from the complexity of IMS that it can only be executed effectively if emergency responders are thoroughly trained before incidents occur. This obviously means fire, police, and emergency medical services must be trained. In addition, senior officials and personnel from other organizations also need some ICS/IMS training if they perform emergency response duties. Depending on the nature of your community's hazards, this might include public works, the American Red Cross, the Salvation Army, and other support service organizations.

9.3.4 Emergency Operations Centers

An EOC is a facility located in a safe area that provides support to responders at the scene of an incident. This support might be requests for information, personnel, equipment, or supplies. An EOC is important because resources are often widely dispersed throughout a jurisdiction. Specialized resources might only be located in other jurisdictions or higher levels of government. The specific resources needed to respond to a particular type of incident at a given location cannot be predicted in advance. Moreover, many organizations participate in the incident response. Each organization must have a capability for receiving and processing information about the incident. This means you should allocate enough people to an EOC so no one is overwhelmed. However, you should not assign too many people to an EOC or response operations will become bogged down by noise and congestion. Decision-making authority should be given to organizations close to the incident site. This is because they have better knowledge of local conditions. However, greater technical knowledge and resources are usually available at higher levels. Thus, the EOC is used to provide close coordination among organizations at all levels.

An EOC should be at a location that provides ready access to those who are essential to a timely and effective response. This includes both those who have technical knowledge as well as those who have policymaking responsibilities. An EOC must have enough space to support the response functions that take place within it. Moreover, it must provide a layout that places its staff close to the equipment, information, and materials they need. Previous guidance and practice indicates EOC designers must perform the tasks listed in Table 9-2.

During task 1, a design team should be established. This team should contain expertise from emergency preparedness, architecture, ergonomics, and information technology. During task 2, the design team should examine the EOP and SOPs to identify the functional teams. The team should examine the positions to be staffed within each team, and how the positions are related. In addition, the design team should assess the flow of resources. It should especially assess the flow of information. Static information such as plans, plant layouts,

Table 9-2: EOC Design Tasks
Task 1: Establish the EOC design team
Task 2: Analyze the organization of the EOC
Task 3: Assess the flow of resources associated with each position
Task 4: Determine the workstation requirements for each position
Task 5: Assess the environmental conditions needed to support each position
Task 6: Determine the space needs for each position
Task 7: Develop a conceptual design for the EOC
Task 8: Document the design basis for the EOC

and evacuation routes can be gathered ahead of time and stored for easy retrieval. Dynamic information about hazards must be collected, routed to those who need it, and processed quickly and accurately. Both static and dynamic information can be conveyed in three different formats. These formats are verbal, numeric, and graphic. The difficulty in sharing some types of information can combine with the volume of information to strain the capacity of EOC staff to perform their functions. Advanced telecommunication technologies such as electronic mail and computer-based information displays are needed to manage the flow.

The design team must recognize that flows of personnel are intense during the EOC's initial activation and shift changes. Moreover, some positions require a considerable amount of movement. Many organizations have analysis teams whose leaders link their teams with an executive team or emergency director (e.g., mayor or city manager). The team leaders need to move back and forth between groups. Because of this frequent movement, EOCs must be designed to ensure the team leaders remain informed about events that take place in one group when they are with another group. These leaders must not disrupt others as they move back and forth.

As Table 9-2 indicates, the design team should determine the workstation requirements for each position during task 4. They should provide seating and surfaces that can be adjusted for different workers. Similarly, keyboard heights and computer viewing angles also should be adjustable.

During task 5, the design team should assess the environmental conditions needed. All positions within the EOC are likely to have similar needs for heating,

FOR EXAMPLE

Designing an EOC

Like many other organizations, the U.S. Nuclear Regulatory Commission (NRC) first established an EOC by starting with one room. They assigned a duty officer to a communications console where emergency notification would arrive from any nuclear power plant having an emergency. As the agency gained experience during incidents and exercises, it added adjacent offices to the EOC. Some of this space was dedicated to the EOC. For example, an adjacent conference room became the work space for the executive team. Other rooms were normally used as office space but were taken over by analysis teams during incident response. Almost everyone complained about the EOC's noise and congestion. The division director in charge of the EOC hired a team to design a new EOC. By following the steps in Table 9-2, this team developed a design that worked much better. In fact, NRC regional offices and other federal agencies used the same team (and the same procedures) to redesign their EOCs.

ventilation, and air conditioning. However, there can be differences in the need for lighting and noise suppression. Variation in lighting needs can be accommodated by providing locally controllable task lighting. Noise suppression can be achieved with absorbent material.

During task 6, the design team should determine the space needs for each position. The space needed for each position depends on the amount of horizontal workspace and also required by the requirement for circulation space for people to move around easily. Variation in the staffing needs for different types of incidents requires a design that provides flexibility in space allocation. In most cases, this flexibility can be provided by open space designs with moveable partitions between team areas.

During task 7, the design team's architect can use the information flow to construct an adjacency matrix. This describes the degree to which each of the EOC teams needs to be located close to each of the other teams. The adjacency matrix, together with the information from the space analysis, can be used to develop an idealized layout. In most cases, this idealized layout must be adapted to the physical constraints of an existing building in which the EOC will be constructed. During task 8, the design team should document the design basis for the EOC. The design team should prepare a design basis document. This summarizes the results of their analyses and the resulting design. This document should be reviewed by managers and by a committee representing each team that will staff the EOC. The design review will provide an opportunity for users to verify the accuracy of the design basis and provide a benchmark against which subsequent proposals for EOC renovations can be assessed.

SELF-CHECK

- List four of the seven principles of the Incident Command System (ICS).
- Explain the differences between the IMS and ICS structures.
- Name the seven activities that the IC performs during an incident.
- Explain why an EOC is important.

9.4 Organizational Structures for Emergency Preparedness

There are many organizational structures that can help you in emergency response. Three of the most important are the Metropolitan Medical Response System (MMRS), Urban Areas Security Initiative (UASI), and National Incident Management System (NIMS).

9.4.1 Metropolitan Medical Response System

The Department of Health and Human Services (DHHS) started the MMRS in 1997. By March 2004, 124 city and regional MMRS programs had been established. These tend to be concentrated in high population density areas and other areas that are terrorist targets. Forty-three states have at least one MMRS program. This program covers a large percentage of the American population.

The initial purpose of this program was to enhance local efforts to manage large mass casualty incidents arising from terrorist use of weapons of mass destruction (Perry, 2003). The mission was driven by the fact that, for local governments, federal help for terrorist attacks is 48 to 72 hours away. The goal is to ensure that cities can operate by themselves until support arrives. Another goal is for the city to develop a strong local incident management system that can effectively integrate federal resources. The program objectives have evolved over time in constructive ways. The focus has come to include CBR agents as well as any other agent that could produce large numbers of casualties. It has become firmly established as an all-hazards program. The two most distinctive features of the program are funding and organization. DHHS provided funding directly to the cities. This eliminated concerns about funding losses to intermediate government levels and increased purchasing flexibility for the cities. The constraint is that cities must create programs that include broad participation. Municipal departments (not just fire and police), county and state agencies, and the private sector (e.g., hospitals) must all work together. The funding conditions are a benefit. The organizational issues, however, are significant challenges.

MMRS links multiple response systems. Horizontal linkages involve working with other departments in the same jurisdiction. For example, first responders, public health, law enforcement, must work together along with behavioral health services. There also are vertical linkages. For example, public health participation involves city, county, and state agencies. Private sector organizations are included in the planning process to establish contact with groups that provide critical services in mass casualty incidents. Examples of these groups include hospitals, funeral director associations, and environmental clean up companies. Cities must have a plan on how they will receive and use federal assets. This includes being able to receive and distribute pharmaceuticals from the national stockpile.

Whether by design or not, the MMRS program has imposed a comprehensive emergency management process on recipient cities. Cities are required to:

- Operate an incident management system.
- Link the IMS to a jurisdictional EOC.
- Enhance mutual aid agreements with surrounding communities.
- Integrate county and state agencies.

▲ Conduct joint planning, training, and exercising on a continuing basis.
▲ Conduct a full-scale exercise with federal evaluation.

In March 2003, responsibility for the MMRS program passed to the Department of Homeland Security (DHS). The challenge for the program is to sustain its funding and policy priorities. Through the years of DHHS oversight, the agency continued to fund established MMRS cities. However, the amount of funding to cities varied each year. An indicator of the strength of the program and proof of serious commitment is that cities kept their programs alive, even during years of small federal allocations. They did this by making difficult choices about the distribution of local resources. Funds to sustain the existing MMRS programs have continued under DHS, but future federal support is not guaranteed. No new MMRSs have been established since 2003, and the funding for this program was reduced in 2005.

9.4.2 Urban Areas Security Initiative

In July of 2002, President Bush approved the National Strategy for Homeland *Security*. Part of this strategy is the Urban Areas Security Initiative (UASI). In late 2003, President Bush increased UASI funding to over $4 billion. Seven new urban areas were approved for funding in 2003 and another 50 in 2004. DHS added seven more in 2005 but discontinued funding for seven urban areas that were funded in 2004. The financial awards are substantial. They range from a high of more than $207 million to New York City to a low of $5 million given to Louisville, Kentucky. In addition, 25 mass transit systems were funded in 2004.

The purpose of UASI is to prevent, respond to, and recover from acts of terrorism. In launching UASI, high-threat, high population density areas were identified. The level of funding for each area has been based in part upon vulnerability assessments and needs assessments. UASI does not impose one generic model on participating urban areas. Instead, it requires local governments located around a designated core city to cooperate in developing a strategic plan for terrorist attacks anywhere in the urban area. UASI then authorizes program expenses across five areas:

1. Planning.
2. Equipment acquisition.
3. Training exercises.
4. Management.
5. Administration.

UASI has the advantage of providing substantial funding for local needs. Moreover, it allows local choice in planning, administration, and funding. Another positive point is that 47 core cities of the 50 UASI urban areas already had existing MMRS programs. This means that they had already engaged in substantial emergency planning. They had an existing structure on which to build further capability. UASI has also been the target of complaints, including concern that federal authorities tightly define authorized expenditures within a budget category. It is time consuming and difficult for local governments to account for all of the expenses. There is also concern that the pass-through mechanism from federal to state and then to local agencies is complex and administratively demanding. Finally, if UASI is to succeed, there must be high levels of continuing cooperation among federal, state, county, and municipal governments. There must also be cooperation among the municipal governments within each urban area.

9.4.3 The National Incident Management System (NIMS)

NIMS was created in response to the terrorist attacks of September 11, 2001. NIMS is a standardized system for managing emergency preparedness and emergency response that builds on the ICS/IMS framework. One of the reasons for establishing a national system is that different jurisdictions and different professions had developed different versions of IMS or ICS. Quite obviously, having multiple versions defeats the basic purpose of having a standardized system. The U.S. Department of Homeland Security (2004) issued the documentation for NIMS on March 1, 2004. DHS required all federal agencies to adopt NIMS immediately. It required all state and local organizations to adopt NIMS by 2005 if they wished to qualify for federal preparedness funding.

The similarity between the names NIMS and IMS makes it sound like NIMS would be very much like the IMS described in the previous section. In fact, NIMS is broader than IMS because it addresses emergency preparedness as well as emergency response. There are six components to NIMS.

The first component, labeled *command and management,* includes the basic features of IMS plus a definition of "multiagency coordination systems" and "public information systems." This makes NIMS more similar to California's Standardized Emergency Management System than to a conventional fire service IMS. What DHS identifies separately as multiagency coordination systems and public information systems overlaps the structure of traditional fire services incident management. The conventional fire service IMS accommodates the need to link with incident management systems operated by different classes of agencies and governments (e.g., public works, EMS, law enforcement, hospitals) and includes joint information systems to disseminate incident information to the public (Brunacini, 2002). These same features also characterize all municipal MMRS programs.

The second component of NIMS is labeled *preparedness.* It "involves an integrated combination of planning, training, exercises, personnel qualification and certification standards, equipment acquisition and certification standards, and publication management processes and activities" (U.S. Department of Homeland Security, 2004, p. 4). Just as with any other version of IMS, you need to plan, train, and exercise, as well as develop mutual aid pacts. NIMS is more restrictive than other versions of IMS because DHS plans to issue standards and test personnel to certify their ability to perform "NIMS-related functions." NIMS will also have a certification process for equipment. Finally, the preparedness component specifies that all forms used during incident response must comply with federal standards. This includes the forms for the incident action plan, organization assignment list, and many others.

The *resource management* component of NIMS is complex and extensive. NIMS requires local jurisdictions to inventory resources according to a standardized system. The NIMS resource typing system establishes specific definitions for each type of resource. It also lists rules for determining what resources are needed for an incident, as well as how they are to be ordered, mobilized, tracked, reported, and recovered. Finally, there is a section requiring certification and credentials for personnel.

The final three components of NIMS are less well defined than the first three. The communication and information management component establishes standards for communications at an incident and specifies processes for managing incident information. The supporting technologies component exhorts locals to acquire and continually review the availability of new technology for incident management. The ongoing management and maintenance component "establishes an activity to provide strategic direction for and oversight of the NIMS, supporting both routine review and the continuous refinement of the system and its components over the long term" (U.S. Department of Homeland Security, 2004, p. 6).

It is difficult to evaluate NIMS at this stage of implementation. It appears that disaster research was considered minimally, if at all, in the process of generating NIMS. It is unclear what other guidance was solicited by DHS, from whom, or how it was incorporated. What appears to be an even greater concern

FOR EXAMPLE

Financial Investments

The Phoenix Fire Department has made a tremendous investment in emergency preparedness. It operates a command training center for certifying its own command officers (and others from the surrounding region), but the simulation models, props, computers, and software require a substantial financial investment. Smaller jurisdictions will find it much more difficult to meet the requirements of the NIMS resource management component.

to many local emergency managers and responders is the detail with which processes and protocols are specified within NIMS. In what could be regarded as a significant understatement, Christen (2004, p. 96) states that in the fire service, "not everyone is happy with national standards and protocols that supersede local preferences." More important is the question of whether such detailed specification promotes or retards the effective and efficient management of emergencies and disasters.

On a practical level, the likelihood that NIMS can be implemented successfully is difficult to estimate. There is no doubt DHS can impose its requirements on agencies accepting federal disaster preparedness funding. However, effective implementation is typically much more difficult than official adoption. The ICS command and management component is similar to the IMS that is used by most large fire services agencies in the United States. Consequently, it is likely to be implemented effectively by those organizations. However, implementation by public works departments, hospitals, and law enforcement agencies is more difficult to estimate. To implement NIMS effectively, these organizations must make significant commitments to training, drills, and exercises.

NIMS implementation also presents other important challenges for both DHS and local jurisdictions. One important problem is that DHS will need immense resources to produce standards and annually test and certify every command officer in the United States. If equipment must also be certified, the task will be even more daunting. Even if the certification and testing were passed to local jurisdictions, many would be severely burdened and would likely see the process as another "unfunded federal mandate." For local agencies that do not routinely use ICS or IMS, the additional resources required to comply with NIMS will be substantial. Perhaps in anticipation of a challenging transition, DHS has created a "NIMS Integration Center." There is a Web site at www.fema.gov/nims to answer questions regarding system adoption. In addition, the DHS offers multiple online classes that address both NIMS and basic ICS.

SELF-CHECK

- **Briefly describe the three organizational structures that can help you in emergency response.**
- **Explain the goals of the MMRS program.**
- **What is the purpose of UASI?**
- **Compare NIMS and IMS.**

9.5 Federal Guidance for EOP Development

For many years, the federal government provided state and local governments with criteria for evaluating their EOPs. Some of this guidance was developed for specific hazards, such as nuclear power plant incidents, and toxic chemical incidents whereas other guidance had an all-hazards approach. The guidance for chemical hazards (National Response Team, 1987, 1988) appears to have been derived from the earlier guidance for radiological hazards (Nuclear Regulatory Commission/FEMA, 1980). However, there are marked differences between the guidance for these two hazards and the all-hazards guidance (FEMA, 1988, 1990, 1996). Of course, no emergency manager wants to develop one EOP for chemical/radiological incidents and another EOP for all other hazards. That would violate the all-hazards approach. Consequently, the presentation below attempts to explain the relationships between these two different sources of guidance for EOP development.

9.5.1 EOP Components

EOPs should have four basic components: basic plan, functional annexes, hazard-specific appendices, and standard operating procedures and checklists.

Basic Plan

The first page of every plan should contain the date of the original plan and the dates of all plan revisions arranged chronologically. Typically, copies of EOPs are provided to multiple offices and organizations. You must take utmost care to ensure that all people and organizations on the plan distribution list always have the most current version of the document. As a rule, the provisions of the general plan are captured under seven separate headings:

1. **Purpose:** This briefly states the mission of the plan. It summarizes the basic plan.
2. **Situation and assumptions:** This reviews the community's vulnerability analysis. It also describes any policies that limit the authority of the emergency response organization.
3. **Concept of operations:** This describes the sequence of emergency response activities.
4. **Organization and assignment of responsibilities:** This describes the structure of the emergency response organization.
5. **Administration and logistics:** This describes the policies for expanding the emergency response organization through mutual aid and volunteers. It also addresses policies for identifying resource needs, acquisition of additional resources, tracking resources allocation, and compensation.

6. **Plan development and maintenance:** This defines how the plan will be reviewed and updated.
7. **Authorities and references:** This addresses the legal and administrative basis for the EOP. It refers the reader to other documents, such as the hazard vulnerability analysis, for further details.

Functional Annexes

The definition of the *functional annexes* is perhaps the most problematic aspect of writing an EOP. Each **functional annex** should describe how the emergency response organization will perform a function needed to respond to disaster demands. Thus, the annexes of an all-hazards EOP should collectively list all of the emergency response functions needed to respond to all hazards. Unfortunately, no single federal guidance document provides a comprehensive list. In fact, these guidance documents aren't even consistent with each other in their lists of emergency response functions. Table 9-3 contains four columns, corresponding to four sources of federal guidance. The first column lists the four organizational functions Lindell and Perry (1992) derived from guidance for nuclear power plant emergency preparedness (NUREG-0654). The second column lists the organizational subfunctions associated with each of the four basic functions. The third column lists the guidance for chemical emergency preparedness (NRT-1).

The fourth column contains FEMA's (1996) *SLG-101* list of typical annexes. Each item in the SLG-101 list has been assigned to the basic emergency response function that seems most appropriate. The SLG-101 functions appear to be relatively similar to those from NUREG-0654 and NRT-1 but provide less emphasis on emergency assessment. Presumably, this reflects a greater concern about accurate emergency assessment during nuclear and chemical plant accidents.

The fifth column maps ICS sections onto the basic emergency response functions. ICS does not address the basic emergency response functions at the section level, although some of these are addressed at the level of branches and sectors. The most notable difference between column 4 from the previous three columns are ICS's much greater emphasis on hazard operations and incident management than on emergency assessment and population protection. The differences in emphasis should come as no surprise, given the origins of the different systems. NUREG-0654 and NRT-1 were written by federal regulatory agencies to guide preparedness for large scale hazmat incidents. ICS and IMS were developed by firefighters to guide response to fires.

As an emergency manager, you should be aware of the differences in the emphasis these systems place on different emergency response functions. As you work on your community's EOP, you should not be surprised to find these differences in emphasis among the people you work with. Those whose jobs

Table 9-3: Typologies of Emergency Response Functions

Organizational functions	*Organizational subfunctions*	*NRT-1 functions*	*Local plan annexes (SLG-101)*	*ICS functions*
Emergency assessment				
	Threat detection/ emergency classification	Ongoing incident assessment		
	Hazard/ environmental monitoring	Ongoing incident assessment		Planning
	Population monitoring and assessment			
	Damage assessment		Recovery	
Hazard operations				
	Hazard source control	Containment and cleanup	Firefighting or fire/ rescue; hazmat/oil spill	Operations
	Protection works	Public works	Public works/ engineering	Operations
	Building construction			Operations
	Contents protection			Operations
			Utilities	Operations
Population protection				
	Protective action selection			
	Population warning	Warning systems and emergency public notification	Warning	

Table 9-3: ***(continued)***

Organizational functions	*Organizational subfunctions*	*NRT-1 functions*	*Local plan annexes (SLG-101)*	*ICS functions*
	Protective action implementation	Personal protection of citizens	Evacuation/ transportation; radiological protection	
	Impact zone access control/ security	Law enforcement	Law enforcement	
	Reception/ care of victims	Human services	Shelter/ mass care; human services	
	Search and rescue	Fire and rescue	Search and rescue	
	Emergency medical care	Health and medical	Health/ medical services	
	Hazard exposure control	Response personnel safety		
Incident management				
	Agency notification/ mobilization	Initial notification of response agencies	Warning	
	Mobilization of emergency facilities/ equipment			Planning
	Communication/ documentation	Responder communications	Direction and control	
	Analysis/ planning			Planning
	Internal direction and control	Direction and control	Communication	Command

	Public information	Public information/ community relations	Emergency public information	Command
	Finance/ administration	Resource management	Resource management	Planning; Logistics; Finance/ administration
	Logistics		Donations management	Logistics
	External coordination	Direction and control		Command
			Legal	

focus on structural fires and wildfires will probably tend to emphasize hazard operations more than any other function. Those who are preoccupied with nuclear plant accidents or hurricanes are more likely to emphasize population protection.

You should not consider the lack of consistency in federal guidance to be a problem for your EOP. Local jurisdictions have the authority to decide how they will define their emergency response functions. Thus, your jurisdiction can organize its EOP annexes in the way that is most compatible with the hazards it faces and with its normal organizational structure.

Hazard-Specific Appendices

Hazard-specific appendices provide information about the ways in which the response to a particular hazard agent differs from the standard response. It is

FOR EXAMPLE

Organizations and Emergency Preparedness

A basic problem is that only a few organizations are specifically evaluated on their levels of emergency preparedness. For the rest, disaster is only a vague threat that "ought to be addressed someday" when more resources are available. Consistent with this characterization, organizations with explicit emergency management missions tend to have the highest degree of emergency preparedness. Moreover, larger organizations that have more human, material, and financial resources also tend to have higher levels of emergency preparedness.

important to avoid confusing specific types of threats (such as terrorist attacks) with general emergency response functions. Terrorist attacks can involve any one of four types of hazard agents—chemicals, biological agents, radiological or nuclear materials, or explosives. Each of these is a specific hazard that requires adjustments to some emergency response procedures and smaller adjustments to others. Thus, terrorist attacks should be addressed in hazard-specific appendices, not functional annexes.

Standard Operating Procedures and Checklists

Standard operating procedures and checklists list the steps that individuals and organizations will take to perform specific emergency response tasks. Some of these may be included in the EOP, whereas others may simply be referenced. The EOC should have a complete set of departmental plans and procedures, as well as checklists, maps, and other job performance aids.

SELF-CHECK

- Define **functional annex.**
- Name the four types of hazard agents that terrorists can use.

SUMMARY

As an emergency responder, you strive to prevent casualties and damage in a disaster situation. As shown, this can be done effectively and efficiently with the proper planning. Planning can be difficult, but identifying the way the basic response functions will be implemented creates an organized framework for tackling any type of emergency. Local governments often have the resources to meet disaster demands without any outside assistance. However, multiple groups should be able to work together in a disciplined way with the proper planning. As evidenced from our recent natural disasters such as Hurricanes Katrina and Rita and terrorist attacks such as 9/11, emergency managers must develop a state of readiness to respond to extreme events threatening their communities.

KEY TERMS

Damage assessment An evaluation that begins by identifying the boundaries of the impact area and proceeds to estimating the total amount of damage to buildings and infrastructure in the impact area. This information is used to support a request for a presidential disaster declaration.

Functional annex The part of an EOP that describes how the emergency response organization will perform a function needed to respond to disaster demands. The annexes of an all-hazards EOP should collectively list all of the emergency response functions needed to respond to all hazards.

Policy entrepreneur An issue champion who has the expertise and legitimacy to promote emergency planning.

Population monitoring and assessment The process of identifiying the population at risk.

ASSESS YOUR UNDERSTANDING

Go to www.wiley.com/college/lindell to evaluate your knowledge of hazards, vulnerability, and risk analysis.
Measure your learning by comparing pre-test and post-test results.

Summary Questions

1. Emergency planning should be recognized as a continuing process. True or False?
2. Hazard source control is a part of incident management. True or False?
3. IMS uses all the following terms to describe different size groupings of personnel, equipment, and apparatus *except* _______.
   (a) sections
   (b) branches
   (c) commands
   (d) sectors
4. Under the MMRS program, cities are required to integrate county and state agencies. True or False?
5. Which of the following is a component of NIMS?
   (a) preparedness
   (b) detection and classification
   (c) hazard monitoring
   (d) hazard source control
6. NIMS is broader than IMS. True or False?
7. The EOP's concept of operations
   (a) states the mission of the plan.
   (b) describes the sequence of emergency operations activities.
   (c) describes the structure of the emergency response organization.
   (d) defines how the plan will be reviewed and updated.
8. The EOC should have
   (a) departmental plans and procedures.
   (b) checklists.
   (c) maps.
   (d) all of the above
9. The most problematic aspect of writing an EOP is
   (a) getting assistance in writing it.
   (b) writing the mission.

(c) writing the functional annexes.
(d) writing the concept of operations.

Review Questions

1. What are the four problems caused by an overemphasis on detail in the planning process?
2. What are the four basic emergency response functions? Briefly explain each function.
3. What are the eight tasks that should be performed when designing an EOC?
4. How does MMRS link multiple response systems?
5. What are the seven headings under which the provisions of an EOP's basic plan are captured? Briefly explain each heading.
6. Why should EOPs be reviewed? How often?

Applying This Chapter

1. Weather forecasters are predicting a Category 5 hurricane will strike the Atlantic coast between Myrtle Beach, South Carolina, and Kitty Hawk, North Carolina. As an emergency manager, what steps should you take to prepare for this disaster?
2. Several explosions have occurred in your community. It is a suspected terrorist attack. Are there any extra steps you need to take around the incident scene in case it is later classified as a crime scene?
3. Two trucks collided on a busy highway. One truck was carrying livestock, while the other truck was carrying gasoline. What steps would you take to deal with this event?
4. You are working with other agencies to discuss community emergency preparedness. What should be the relative priority for planning, training, and exercising?
5. You are the local emergency manager for a small coastal town that is vulnerable to hurricanes. You need to design and organize an EOC. Where should the EOC be located? Why? Who will be in charge of it?
6. Your community is joining with others to fight a large wildfire that is burning near subdivisions on the outskirts of town. You are the senior advisor stationed at the incident command post. What are your duties?

YOU TRY IT

Public Information Officer

Your community was devastated by a hurricane. You brief the public information officer. What type of information does the public information officer (PIO) need to interact with the media?

EOC

You are the project manager of the design team for a new EOC. What factors should you consider when your team is designing the EOC?

EOP

You are writing the emergency operations plan for your community. Before you write the first section, what are the factors to consider when writing and organizing the plan? Are there times when it would be better to write a separate plan for each hazard?

10

ORGANIZATIONAL EMERGENCY RESPONSE

Handling an Emergency

Starting Point

Go to www.wiley.com/college/lindell to assess your knowledge of organizational emergency response.
Determine where you need to concentrate your effort.

What You'll Learn in This Chapter

▲ Emergency assessment activities in the response phase
▲ Hazard operations tasks for specific hazards
▲ Population protection measures and how to implement them
▲ Incident management tasks and their importance

After Studying This Chapter, You'll Be Able To

▲ Choose the appropriate location for performing each emergency response function
▲ Interpret an emergency classification system
▲ Analyze which actions are appropriate to protect the risk area population in different hazards
▲ Examine the seven specific functions that are the core of incident management

Goals and Outcomes

▲ Perform emergency assessment activities in the response phase
▲ Design a plan to protect the population and structures during hazards to which the community is vulnerable
▲ Manage the information flow within the emergency operations center (EOC)
▲ Manage and organize the work of the public sector, private sector, and nongovernmental organizations to successfully respond to a community-wide disaster

INTRODUCTION

The actions you take during emergency response draw upon the EOP and procedures developed during emergency preparedness. Of course, the demands of any specific incident can never be predicted with perfect accuracy. Consequently, your emergency response organization must always improvise. Too little improvisation may mean people are being too rigid; too much improvisation may mean you didn't prepare enough. As you respond to an emergency, you should focus on the four basic functions you must perform. These functions are emergency assessment, hazard operations, population protection, and incident management. This chapter looks at each function and shows how they are crucial to carrying out an effective response that limits casualties and damage.

10.1 Emergency Assessment

Emergency assessment activities in the response phase are directed toward intelligence—understanding the behavior of the hazard agent and the people and property at risk. This function involves the use of classification systems, protocols, and equipment that are developed or acquired prior to the threatened impact. Specific threats, their probabilities of impact, and policies for managing them are derived from the jurisdictional hazard vulnerability analysis. See Table 10-1 for description of the emergency assessment activities.

10.1.1 Threat Detection and Emergency Classification

Threat detection includes

- ▲ Recognizing that a threat exists.
- ▲ Assessing its magnitude, location, and timing of impact.
- ▲ Determining how to respond.

The first step is recognizing that a threat exists. You may detect it yourself or you may be notified of a threat. The source of notification often depends on the type of hazard agent. Examples include:

- ▲ The National Weather Service (NWS) notifying you that a severe storm is generating tornadoes.
- ▲ The plant operators informing you of a nuclear power plant accident in progress.
- ▲ A 911 operator letting you know a train derailment has caused a toxic chemical release.

Table 10-1: Emergency Response Functions and Specific Actions

Function	*Incident scene/ command post*	*EOC*	*Other locations*
Emergency assessment			
	Local threat detection and emergency classification	Regional threat detection and emergency classification	
	Local hazard monitoring	Regional hazard monitoring	
	Damage assessment	Environmental monitoring	
		Population monitoring and assessment	
Hazard operations			
	Hazard source control		
	Protection works		
	Building construction practices		
	Contents protection practices		
Population protection			
	Protective action selection	Protective action selection	Population warning
	Population warning	Population warning	Protective action implementation
	Search and rescue		Reception and care of victims
	Impact zone access control and security		Emergency medical care
	Hazard exposure control		
	Emergency medical care		
	Environmental surety		
Incident management			
	Agency notification and mobilization	Agency notification and mobilization	Public information

Table 10-1: *(Continued)*

Function	*Incident scene/ command post*	*EOC*	*Other locations*
Incident management (cont.)			
	Mobilization of emergency facilities/equipment	Mobilization of emergency facilities/equipment	Mobilization of emergency facilities/equipment
	Communication/ documentation	Communication/ documentation	
	Analysis/planning	Analysis/planning	
	Internal direction and control	Internal direction and control	
		Logistics	
		Finance/administration	
		External coordination	
		Public information	

After a threat is detected, you will often find it helpful to use an **emergency classification system.** An emergency classification system organizes a large number of potential incidents into a small set of categories. These categories link the threat assessment to the level of activation of the responding organization. Emergency classification systems are specific to each type of hazard. Their implementation depends on the state of technology regarding that hazard. The NWS uses *hurricane watch* as a category. A hurricane watch indicates the possibility of hurricane conditions (sustained winds of at least 74 mph) within a designated section of coast within 36 hours. A *hurricane warning* indicates possibility of hurricane conditions within 24 hours or less. These hurricane watches and warnings provide an indication of how soon to expect hurricane landfall. They are supplemented by storm's Saffir-Simpson category, which predicts the likely level of damage. Knowing that there is a hurricane warning for a Category 5 storm makes it very clear that the threatened jurisdictions should be making their final preparations before the storm strikes.

The same idea has also been applied to technological hazards at fixed site facilities. Such facilities routinely monitor their systems for changes in plant conditions. Such monitoring can detect hazardous conditions long before they threaten people and property (McKenna, 2000). In turn, this enables onsite and offsite personnel to respond rapidly (Lindell and Perry, 2006). However, monitoring alone is not sufficient. Onsite and offsite personnel should establish an emergency classification system to be sure plant operators will respond effectively when a situation becomes dangerous. Plant operators' ineffective response can usually be explained in one of four ways:

1. Monitoring devices often present confusing information during plant accidents. Plant personnel focus on trying to understand what is happening rather than notifying others.
2. Plant personnel fail to understand the implications of meter readings. They believe the situation is less serious than it actually is.
3. Plant personnel believe that they can control the situation before it worsens. It may take them a long time to realize conditions are beyond their control.
4. Plant personnel underestimate the amount of time local populations need to implement offsite protective actions. They don't realize that notification, warning, and evacuation take hours—not minutes.

An emergency classification system combats these problems. It provides specific, objective criteria for determining the severity of a threat. In turn, the system prescribes appropriate actions for:

- Emergency assessment actions to clarify the situation.
- Hazard operations to prevent or correct malfunctions.
- Population protection actions to minimize or avoid exposures.
- Incident management actions to activate and coordinate the emergency response organization.

Three factors are used in defining emergency classes. Each of these factors should be addressed in an emergency classification system:

1. **Hazard generation:** The emergency class should be higher when the probability of an extreme event is greater, its magnitude is greater, its point of impact is closer, and the time it will strike is sooner.
2. **Hazard transmission:** The emergency class should be higher when impacts are more readily transmitted into the surrounding environment and that environment has limited ability to absorb the impact. In most cases, explosive energy is dissipated and hazardous chemicals are dispersed with increasing distance from the source.
3. **Community vulnerability:** The emergency class should be higher when the community is more vulnerable to the threat. A given wind speed is more dangerous for wood-frame buildings with unprotected windows than for steel reinforced concrete buildings with window shutters. Toxic chemical releases are more dangerous for buildings that have high infiltration rates than for buildings that are tightly sealed.

In general, an emergency classification system is reliably implemented only if each emergency class is clearly defined by emergency action levels (EALs). An EAL is a specific event or condition that is easily recognized as an indicator of the severity of the emergency. For example, The U.S. Nuclear Regulatory Commission (NRC) and FEMA (1980) defined four classes of nuclear power plant emergencies:

1. An **unusual event** is defined as involving potential degradation of plant safety. No releases are expected unless other events occur.
2. An **alert** involves substantial degradation of plant safety. Releases are expected to be well below EPA exposure limits.
3. A **site area emergency** involves major failures of plant safety functions. Releases might exceed EPA limits onsite, but not offsite.
4. A **general emergency** involves substantial core degradation and the radioactive material might escape from the containment building. Releases might exceed EPA limits offsite. See Table 10-2 for a full description.

Each emergency class has a specific definition that is measured by predetermined EALs. These EALs have been developed by the nuclear utility and approved by the NRC inspectors. In turn, each emergency class defines the actions to be taken when it is declared. The system replaces subjective judgments made by one person during a threat with consensual objective judgments made by many before a threat occurs. Establishing a set of emergency classes does not itself provide a useful emergency classification system. For example, the first version of the Homeland Security Advisory System has five colors for different threat levels. It has not been very useful because no one but Homeland Security knows what conditions are used to determine the threat level. It is understandable that this information is withheld to prevent terrorists from using it to their advantage. Nonetheless, the population at risk has no way to determine what any given change in the threat level really means. For example, people have no way to know what the increase in threat is when the classification changes from *yellow* to *orange*. Even worse, the initial system failed to indicate what actions households, businesses, and communities should take in response to each threat level. The system became somewhat more useful after the different levels were linked to specific response actions. Interestingly, it was the Red Cross that identified what should be done by state and local government, airports and other critical facilities, and ordinary households and businesses (see www.redcross.org).

Emergency classification systems are important if you have chemical facilities located in your community. The NRC worked with power plants to devise their emergency classification system. You must work directly with local chemical facilities to develop their systems.

Another important assessment activity arises from the need to determine when an incident has been stabilized. This lets you know when you can declare that the

Table 10-2: Definition of a Nuclear Power Plant *General Emergency*

Class description
Events are in process or have occurred which involve actual or imminent substantial core degradation or melting with potential for loss of containment integrity. Releases can be reasonably expected to exceed EPA Protective Action Guideline exposure levels offsite for more than the immediate site area.
Purpose
Purpose of the general emergency declaration is to: (1) initiate predetermined protective actions for the public, (2) provide continuous assessment of information from licensee and offsite organization measurements, (3) initiate additional measures as indicated by actual or potential releases, (4) provide consultation with offsite authorities, and (5) provide updates of the public through offsite authorities.
Licensee actions (partial list)
1. Promptly inform state and local offsite authorities of general emergency status and reason for emergency as soon as discovered (parallel notification of state/local).
2. Augment resources by activating onsite technical support center (TSC), onsite operational support center (OSC), and near-site emergency operations facility (EOF).
3. Assess and respond.
4. Dispatch onsite and offsite monitoring teams and associated communications.
5. Dedicate an individual for plant status updates to offsite authorities and periodic press briefings (perhaps joint with offsite authorities).
6. Make senior technical and management staff available onsite for consultation with NRC and state authorities on periodic basis.
State and/or local offsite authority actions (partial list)
1. Provide any assistance requested.
2. Activate immediate public notification of emergency status and provide public periodic updates.
3. Recommend sheltering for a two mile radius and 5 miles downwind and assess need to extend distances. Consider advisability of evacuation (projected time available vs. estimated evacuation times).
4. Augment resources by activating primary response centers.
5. Dispatch key emergency personnel including monitoring teams and associated communications.
6. Dispatch other emergency personnel to duty stations within a five mile radius and alert all others to standby status.

emergency has ended. The apparent simplicity of such a declaration is misleading because there are many criteria for determining this. One criterion is when it is safe for the public to end protective actions such as evacuation. Another criterion is when response personnel can reduce levels of personal protective equipment use. Still another criterion is when emergency response personnel can be released.

You need to be able to explain why different termination points are used for different response functions. For example, the on-scene incident commander might discontinue the use of personal protective equipment by response personnel when an acid spill has been neutralized and is no longer a threat. However, you might want to wait to reopen the impact zone to evacuees until after police are ready to resume route patrols some hours later. The nature of the impact is part of all such decisions. In some incidents, the end of the danger may come much later than the end of response operations. For example, in some hazmat incidents, environmental remediation is needed. This is so contamination can be reduced to a level that is safe for population reentry. Assessment procedures should include criteria for incident termination.

10.1.2 Hazard and Environmental Monitoring

You must track the hazard agent over time to determine whether the threat is changing in its likelihood, magnitude, immediacy, or location of impact. You also need to monitor environmental conditions that might alter the disaster impact or the success of response actions. Hazard and environmental monitoring are closely linked. This is because some hazards *originate* in the atmosphere. Other hazards are *transmitted through* it. In the case of hurricanes, you can forecast the impact by monitoring data on the storm's current location, projected track, strike probability, intensity, size, and speed. In the case of toxic chemical releases, you can forecast the impact by modeling the release rate and duration together with the wind speed and direction, and atmospheric stability. In addition, you should anticipate the possibility of fires that could produce additional releases. You should also anticipate the possibility of precipitation and wind shifts. Rain can cause water soluble toxic chemicals to "wash out" of the atmosphere. Wind shifts will change the areas at risk.

You should address the technical and organizational provisions for monitoring and projecting the actual magnitude of the hazard. You should be able to do this at any point during the emergency. The capability for performing this task varies from one environmental hazard to another. Monitoring and projection is not possible for earthquakes due to the current lack of predictability of the hazard onset. However, this capability has long been present in riverine flood hazards. For hazards of regional scope, monitoring is provided by federal agencies. Examples include the National Hurricane Center, Pacific Tsunami Warning Center, and Centers for Disease Control and Prevention. For these hazards, you need equipment to receive hazard information. You need to maintain staff expertise to interpret and act upon that information. In the case of radiological and other hazmat, potential release sources are located within communities. Thus, the scope of impact is quite localized. Here too, you need equipment to receive hazard information provided by plant operators or hazmat carriers. You also need to maintain staff expertise to interpret and act on that information. In addition, you need to be able to acquire data about current and future weather conditions. Finally, you also should have the personnel and equipment to track hazmat plumes and measure the level of the hazard agents. Detection and identification equipment is becoming more sophisticated, compact, and mobile. The Nunn-Lugar-Domenici Act of 1996 made such equipment more affordable for local governments. Subsequent grants from the U.S. Department of Justice have continued this practice.

An active program of plume monitoring allows you to assess the magnitude of any human or animal exposures during a release. Plume monitoring also makes it possible to more precisely determine when it is safe to release people from protective measures such as evacuation and sheltering in-place. A key benefit of monitoring is identification of the communities at risk from plume exposure. This information makes it easier to notify people and to mobilize a response. It is very important that you record downwind concentrations of a hazmat release from fixed site facilities. Specific locations for monitoring from such facilities can be designated in advance. A standardized recording form can be established. Some states have done this to detect biological hazards.

In addition, some communities have established teams to document the location and movement of hazmat plumes. A plume monitoring team should be guided by a specific protocol. This protocol should indicate how personnel should identify and monitor concentrations of agents released. In communities with established hazmat response teams, the problem is substantially reduced. Most large fire departments take sophisticated equipment "on board" to the scene. This equipment includes biological agent detection kits. In some incidents, information about the nature of the material being transported, together with the visual inspection of the container, confirms that no threat exists. In other situations, only the use of the appropriate sensing instruments can be expected to clearly identify the location and magnitude of the hazard. Until the appropriate monitoring equipment arrives, responders should:

- Establish a perimeter around the scene.
- Prevent entry into the "hot zone."
- Decontaminate anyone who has been exposed.
- Maintain medical observation and isolation until released by a competent medical authority.

10.1.3 Population Monitoring and Assessment

You need to know how many people are in the impact area at any given time. If you have forewarning before a hurricane strikes, knowing how many people are usually there will help you manage an evacuation. Knowing if people have already left before you issue an evacuation warning will help even more. After a disaster strikes, you need to know how many casualties have occurred.

Census data is readily available to document the size and composition of the permanent population. However, you need to break down this data by age and ethnic composition. You also need to add estimates of the number of workers, tourists, and other transients (Lindell and Perry, 2003; Perry and Lindell, 1997). Departments of commerce routinely keep such information. Each of these segments respond differently during an emergency. Some are distinctive in terms of their *motivation* to comply with protective action recommendations. Others differ in their *ability* to comply. For example, permanent residents are less likely to evacuate than vacationers. However, school children, hospital patients, and occupants of assisted living units have a limited ability to evacuate.

Two other aspects of population monitoring are casualty assessment and responder accountability. They become important in the postimpact stage of emergency response. *Casualty assessment* is needed to determine how many people are missing and where they might be. This information is, of course, essential for search and rescue (SAR) teams. Accountability information allows SAR teams to "triage" structures and focus their operations (Olson and Olson, 1985). Clearly, SAR teams that know which buildings are empty can bypass these locations. Team members can then concentrate their efforts on buildings where rescue is possible. *Responder accountability* is needed for the same reasons. The emergency response organization, and especially the safety officer, should have a personnel roster and duty assignment chart that helps with this.

10.1.4 Damage Assessment

Damage assessment focuses on measuring the disaster impacts on public and private property. This function is most often thought of in terms of recovery. However, damage assessment is a continuing process that begins during emergency response. Emergency responders should perform **rapid damage assessment** to define the boundaries of the physical impact area and assess the intensity of damage within that impact area. This first stage of the damage assessment provides you with immediate information about the magnitude of the impact. After you know this, you can determine if you have enough resources for the incident demands. If not, you might need to recall off-duty personnel. If that won't give you enough emergency responders, you might request assistance from neighboring jurisdictions or the private sector. Another option is to request assistance from higher levels of government.

FOR EXAMPLE

Continuing Assessment

After your first damage assessment, you need to continue assessing the impact of the incident. For example, you might do an initial assessment of an earthquake. As an incident continues, a broader physical impact assessment should examine the potential for secondary threats. For example, you should explore the possibility that earthquake aftershocks would cause dam failures, hazmat releases, and landslides, or collapse buildings that were damaged in the initial shock.

In addition, you can use rapid damage assessment to decide how to redeploy resources as conditions change. For example, during response to a riverine flood, you can use damage assessment teams to identify weak spots in a levee that need to be reinforced.

SELF-CHECK

- Identify emergency response functions and their specific actions.
- Define **emergency classification system.**
- Explain the difference between hazard monitoring and environmental monitoring.
- Define **damage assessment** and **rapid damage assessment.**

10.2 Hazard Operations

Hazard mitigation is a strategy for providing passive protection at the time of disaster impact. However, some of these actions can be implemented during a response. That is, hazard operations actions have the same purpose as preimpact hazard mitigation actions. However, hazard operations actions are implemented only when the need arises. As is the case for preimpact hazard mitigation actions, the applicability of hazard operations actions varies considerably from one hazard to another. Moreover, hazard operations actions can be grouped into the same categories as the permanent hazard mitigation measures. The principal difference is that hazard operations measures must be able to be implemented rapidly. This eliminates all land-use practices (see Table 10-3). In addition, hazard operations measures are not feasible for hazards, such as tornadoes and earthquakes, which have insufficient forewarning.

As Table 10-3 indicates, there are a number of *hazard source control* measures that can be used to intervene at the stage of hazard generation. For example, wildfires can be suppressed by extinguishing them with water. Hazmat releases can be terminated by patching or plugging leaking storage tanks. In addition, *protection works* can be used to alter the hazard transmission process. For example, floods can be controlled by sandbagging and other methods of levee reinforcement. Similarly, earthmoving equipment can construct dikes to capture runoff from a hazmat spill. *Building construction practices* include the last-minute installation of plywood shutters over windows to protect them from wind and debris. Expedient *building contents protection* actions include the last minute

Table 10-3: Applicability of Hazard Operations Measures (by Hazard)

Hazard	*Source control*	*Protection works/ activities*	*Land-use practices*	*Building construction practices*	*Building contents protection*
Severe storms/cold					X
Extreme heat					
Tornado					
Hurricane				X	
Wildfire	X	X		X	X
Flood	X	X		X	X
Storm surge		X		X	X
Tsunami		X		X	X
Volcanic eruption		X		X	X
Earthquake					
Landslide	X	X			
Structural fire/conflagration	X			X	
Explosion	X			X	
Chemical spill or release	X			X	
Radiological release	X			X	
Biological incident	X			X	

FOR EXAMPLE

Preventive and Corrective Actions

There are often many preventive and corrective actions plant managers can take to prevent hazmat releases or limit their size. Nuclear power plants and chemical production facilities can inject cooling water, isolate leaking tanks and pipes, repair leaking pumps and valves, and patch damaged storage tanks. Some of these actions can prevent a release from occurring. Other actions can correct an equipment problem that is causing a release. In some cases, facilities can use water curtains to absorb vapors that are water soluble or flare towers to burn vapors that are flammable (Prugh and Johnson, 1988).

wrapping of water pipes when cold weather is forecast. Other measures of this type include moving furniture, equipment, and clothing to higher floors when flooding is forecast.

You should establish guidelines for choosing hazard operations actions. Your standard operations procedures (SOPs) should contain rules that define the conditions under which each hazard operations action should be used or avoided. They should also cross-reference any checklists required in implementing those actions. As a specific example, one method of mitigating a release of a flammable toxic gas is deliberate ignition. You should be clear about who can authorize such action. You should describe the conditions under which ignition should be attempted. You should describe when discretion is permitted, and when it should not be attempted. The authority for such decisions is different for different situations. For example, incidents at fixed-site facilities that are on private property are under the control of plant personnel. Transportation incidents that are generally on public property are under control of the incident commander.

SELF-CHECK

- Explain how hazard source control measures can intervene at the hazard generation stage.
- Explain why the applicability of hazard operations actions varies considerably from one hazard to another.

10.3 Population Protection

You must oversee the technical and organizational mechanisms by which the emergency response organization protects its own personnel and the public. Specific tasks include:

- ▲ Selecting protective action.
- ▲ Warning the affected population.
- ▲ Implementing protective action.
- ▲ Controlling hazard exposure.
- ▲ Controlling and securing impact zone access.
- ▲ Receiving and caring for victims.
- ▲ Conducting search and rescue operations.
- ▲ Providing emergency medical services.

Information collected through the emergency assessment function forms the basis for the population protection function. Much of your focus will be on determining which actions are appropriate in a specific situation. In addition, you must ensure that you have the resources needed to implement those protective actions.

10.3.1 Protective Action Selection

You must determine which population protection measures are likely to be effective. You must determine the time at which households and businesses should be advised to undertake them. The appropriate protective strategy varies with the:

- ▲ Type of environmental threat.
- ▲ Certainty of occurrence.
- ▲ Severity of impact.
- ▲ Immediacy of impact.
- ▲ Duration of impact.

In general, the threat decreases as distance increases from the point of impact. Thus, no protective action is required beyond a certain distance. For example, people should evacuate from areas close to a river that is expected to flood. No action would be required outside the expected flood zone. In many cases, you will find it difficult to define the boundary of the evacuation zone. In the case of floods, you will not know precisely which locations will be inundated. Similarly, you will not be able to predict the exact locations in which toxic gas levels will be life threatening. Scientific prediction tools can help but they cannot eliminate all uncertainty.

Moreover, there are situations in which the recommended protective action differs from one location to another. In the case of volcanic eruptions, people in areas threatened by heavy ash fall or mudflows should be advised to evacuate. However, people in areas threatened by light ash fall should be advised to shelter in-place and wear protective masks. It is only outside the ashfall zone that no action is required. In such situations, those receiving a warning that recommends different protective actions in different areas must know where they are located. However, people's perceptions of the risk are not accurate even when they are given maps to identify their risk areas (Arlikatti, et al., in press; Zhang, Lindell and Prater, 2004). You might need to use warning mechanisms that precisely target the areas in which each protective action recommendation is advised.

10.3.2 Population Warning

Warning is usually the first emergency response task to directly involve the public. You should ensure that someone who has the legal authority to issue public warnings is always available. Other warning tasks requiring explicit assignment include

- ▲ Deciding how the person or committee responsible for constructing warning messages announces them.
- ▲ Including a contact list for the organizations involved in warning dissemination.
- ▲ Identifying the person or committee responsible for constructing and ordering the all-clear signal.

Some jurisdictions construct "fill-in-the-blanks" messages for each hazard to which the community is exposed. These messages can be completed at the time a warning is issued.

During an emergency, specific warnings should be disseminated by expert and credible sources. They should describe the threat in terms of its location, severity, and expected time of impact. They should also recommend appropriate protective actions and indicate how to obtain additional information. These warnings must be conveyed in a timely and effective manner to all who are likely to be in the disaster impact area. Responders should disseminate these warnings using the mechanisms that the planning, training, and exercising committee analyzed during the preparedness phase. However, the choice of warning mechanism and message content of the warning must reflect the current emergency assessment. Consequently, personnel need to have been trained so they can exercise effective technical judgment. They should also know how to quickly contact technical experts.

10.3.3 Protective Action Implementation

Protective action implementation focuses on the performance of tasks that ensure those who want to comply with the authorities' protective action recommendations can, in fact, do so. In general, people might need the following to implement protective actions (Lindell and Prater, 2002):

- ▲ Money.
- ▲ Knowledge and physical skill.
- ▲ Facilities and vehicles.
- ▲ Tools and equipment.
- ▲ Time and energy.
- ▲ Social cooperation.

These resources are not distributed uniformly in any jurisdiction's population. You need to have some idea of which households lack the resources needed to implement protective action. It is helpful to know which demographic groups have very few emergency response resources. You should also be able to identify the neighborhoods in which they live.

When seeking to shelter in-place from toxic chemicals, the principal resource needed is an airtight structure. Thus, households living in newer homes find adequate protection because these structures generally provide better protection against air infiltration. Building occupants only need to know to shut doors and windows, turn off the HVAC system, and close the chimney flue. However, older homes in mild climates are less airtight. They do not provide as much protection. Consequently, households in these buildings can obtain additional protection by using duct tape and plastic sheets to seal an interior room, but these materials must be purchased and stored in advance. Similarly, households seeking to shelter in-place against hurricane or tornado wind need structures that will resist extreme wind pressure. They will be adequately protected if they have basements or safe rooms. By contrast, residents of mobile homes might need community shelters in their neighborhoods. They need enough forewarning to reach these shelters before a dangerous wind speed arrives.

Evacuation is a more complex response because it requires more resources. First, people need a mode of transportation so you must provide *evacuation transportation support* for those who lack personal vehicles. You must also provide mobility support for those who cannot walk to evacuation bus pickup points. Second, people need a route of travel so you must be able to provide *evacuation traffic management* to direct people to the appropriate evacuation routes. In addition, you must facilitate the orderly movement of vehicles along these routes. Third, people need an evacuation destination. The emergency response organization must be able to provide *mass care* for those who do not have people to stay with or who lack money for hotels. Finally, households need information. For example, households taking evacuation buses must know where the pickup points are located. Those who are using their own cars should be told which routes to take while evacuating—preferably before an incident occurs.

An effective evacuation protocol establishes the lead agency for the relocation effort. It lays out evacuation traffic management procedures that coordinate the timing and direction of evacuee movement. You should work with transportation officials and law enforcement personnel to select evacuation routes and establish procedures for maintaining a steady flow of vehicles. Procedures should designate the personnel and resources needed for traffic management. They should designate the locations where such resources are stored. Procedures should include contact information for the person who can release and deploy resources.

You should also make provisions for those who need transportation support. This routinely involves three population segments. The first segment consists of *transit dependent households* that do not have their own vehicles. They might not have access to another private vehicle, so you need to provide buses. You need to provide information about what personal items may be taken on these buses. In addition, you must broadcast information about where the buses can be boarded. It is logical to stage evacuation buses from local elementary schools. They are typically within easy walking distance of most households and their locations are well known. The second segment consists of *household members who have mobility limitations.* They require physical assistance or need medical support. Many of these people are distributed throughout the community. In many cases, authorities do not know their locations in advance. Home nursing providers probably will be willing to provide counts of the number of their homebound patients. This is true even if their patient confidentiality policies does not permit them to disclose patients' home addresses. The third population segment consists of *institutionalized populations.* In most cases, the staffs of jails, hospitals, and nursing homes manage the evacuation of their own clients. This should be coordinated by the lead evacuation agency. This ensures that multiple institutions are not relying on the same buses for evacuation or the same host facilities for reception of their clients. One subset of the institutionalized population, school children, requires advance planning to determine if students will be picked up by parents from school, returned to their homes, or evacuated as a group to a reception center where they will be reunited with their parents. You, school officials, and parents should discuss the alternatives in advance. You should remind parents and students of the procedure at the beginning of each semester of the school year.

The other population protection option, sheltering in-place, is somewhat simpler than evacuation but does have some complications. During a tornado, it is best to go to shelter in a basement. If that is not available, a closet or bathtub can provide some protection. Mobile homes provide little protection so people should seek a community shelter. For toxic chemical, radiological, or biological hazards, those in the risk area should seek safety in an existing structure. They might need to enhance an existing structure. In some cases, people can implement additional personal protection actions such as expedient respiratory protection. In the case of airborne chemical or radiological threats, for example, sheltering in a home may require

- ▲ Turning off HVAC systems.
- ▲ Sealing doors and windows with plastic sheets and duct tape.
- ▲ Breathing through a wet towel to filter out airborne particles or water soluble gases.

Successful implementation of in-place protection depends on effective communication. You can produce a timely reaction by the risk area population by explaining what structures provide the greatest protection. Let people know how long they should shelter in-place. If people are to shelter in-place for an extended period of time, broadcast continued information about the need to remain indoors. This is particularly important in public health threats. This is because periods of quarantine might be long (Perry and Lindell, 2003).

10.3.4 Impact Zone Access Control and Security

This element of population protection can be challenging because it depends on the scope and duration of impact of the hazard agent. The first step is to establish traffic control points on major routes at the perimeter of the impact area to prevent people from entering without authorization. The second step is to provide security patrols within the impact area. This can only be done to a certain extent. You don't want to unnecessarily expose law enforcement personnel to any hazards.

There are four reasons for implementing access control and security. The first is to prevent looting. People commonly overestimate the incidence of looting after disasters in the U.S. However, visible control points are a helpful deterrent. They also reassure evacuated property owners. Second, security measures ensure that people are not exposed to the hazard agent by inadvertently entering the impact area. Third, security measures allow responders to implement hazard operations and perform population protection tasks without being impeded. Finally, security measures limit the number of responders and risk area residents that might be affected by secondary devices planted as part of a delayed terrorist attack. This makes access control an essential part of exposure control for secondary hazards.

Impact zone access control and security should be referenced in the EOP. However, it does not need a lengthy discussion. This is because these tasks are addressed in police SOPs. In most cases, law enforcement is designated as the lead agency for this element, while other agencies coordinate with this lead agency before dispatching their personnel into the impact area. It is particularly important that the lead agency's authority be respected. Clear lines of communication must be maintained among all organizations involved in security. Another complication of impact zone security is that this location is considered a crime scene in a terrorist incident. In such cases, the FBI is the lead agency. Effective coordination among local, state, and federal agencies is critical. Effective coordination maintains security *and* preserves evidence that is retrieved from the impact area.

You should follow a security protocol that contains four elements. First, those in charge of security should be respected. Rules for relinquishing control to other agencies should be observed. Second, the type of controlled access described in the EOP should be implemented as it has been planned and exercised unless conditions require it to be modified. In general, there is a conflict among four goals:

1. Protecting responders and risk area residents from exposure.
2. Protecting risk area residents' property from theft.
3. Protecting risk area residents' property from further damage.
4. Allowing risk area residents to resume their normal activities.

The first two goals can best be achieved by minimizing the number of people in the impact area, whereas the last two goals are best achieved by giving risk area residents unrestricted access to the impact area. The access control policy that is implemented during emergency response should be based on preimpact plans and procedures but must be adapted to postimpact conditions. For example, changes might be made to the extent of control, the length of time the control will be in place, and the sizes and types of areas that are controlled. A volcanic eruptive sequence might require controlled entrance and exit to residences for weeks or months. This level of access control requires planning not just for perimeter security, but also for the escort and safety of residents given temporary access.

Third, the type of patrol or security surveillance system necessary to provide access control must be guided by the consequences of uncontrolled access. If radiological or chemical exposure might create severe negative health consequences, the level of access control should be substantial. Obviously, hazard exposure control must also be maintained to protect patrol personnel. Finally, authorities must specify procedures for allowing residents of the restricted area to return temporarily to their homes. Access control personnel should publicize the policy for allowing entry for such reasons as retrieval of medicines and care of farm animals.

Police should also establish a secure perimeter at the jurisdictional EOC. In addition, they provide security for other agencies within their jurisdiction. Hospitals generally have their own security personnel, although this is sometimes supplemented by local police officers. Sometimes hospitals establish surge facilities in auditoriums, gymnasiums, or convention centers. In such cases, local law enforcement should also establish a secure perimeter at these areas.

10.3.5 Search and Rescue

Search and rescue (SAR) activities often take place in loosely structured situations with uncertain exercise of authority (Quarantelli, 1980a). The time pressures are substantial and consequences of error are severe. All of these conditions pose special challenges.

In most disasters, victims and bystanders immediately initiate improvised SAR activities. In some cases, especially in developing countries, these volunteers might be the only people who can save victims. If the impact area is small, few are trapped in rubble, and the debris is easy to remove, bystanders can rescue many victims. For most urban disasters in developed countries, professional responders reach the scene and quickly contribute to the SAR effort. If the impact area is large, many victims are trapped, and the collapsed buildings are steel reinforced concrete, only trained emergency responders equipped for heavy rescue are successful. *Heavy rescue* uses specialized equipment to detect trapped victims and jack up the debris from steel reinforced concrete structures. It is an important capability after earthquakes and explosions. Unfortunately, the number of U.S. fire departments currently deploying heavy rescue units is small. For this reason, you should directly address the need for heavy rescue. This includes

- ▲ Assigning a lead agency.
- ▲ Maintaining a list of available heavy rescue equipment.
- ▲ Establishing decision criteria for prioritizing buildings in the event of multiple collapses.
- ▲ Creating a protocol for quickly obtaining services of victim location specialists.

In these efforts, the incident commander directs the emergency response. Meanwhile, the dispatch center or the EOC manages the resources. In major incidents, the incident commander should designate a SAR coordinator who uses a call list to activate SAR teams. It is usually desirable to clear victims and bystanders as part of providing access control and security at the incident scene. However, time pressures and a shortage of professional personnel might require using local volunteers. If volunteers are to be effectively involved, you need a procedure for registering them and organizing them into work groups. These volunteer SAR teams should be assigned to tasks that are appropriate to their skill level. In particular, there is considerable risk to rescuers working in collapsed high-rise buildings (see Figure 10-1). Consequently, this task should be limited to professional responders. Similarly, volunteers might become contaminated during hazmat incidents. At best, untrained volunteers should be considered only for lower risk tasks. Otherwise, they risk becoming additional victims who need attention. In any event, the SAR coordinator should maintain communications with mass care facilities, the emergency medical care coordinator, and the morgue.

Unplanned contacts with the mass media during SAR are an important problem (Drabek, Tamminga, Kilijanek, and Adams, 1981). These contacts can result

Figure 10-1

Search and rescue teams often include both professionals and volunteers.

in the release of victims' names, harassment of victims, and the release of incomplete or otherwise misleading information about disaster operations. To minimize these problems, SAR personnel should strictly follow procedures for managing the media. In particular, SAR workers should be briefed about procedures for releasing information to the media. These generally include referring reporters to the public information officer for information about the number and identity of victims and the progress of SAR operations.

10.3.6 Reception and Care of Victims

Reception and care of victims is a common demand in disasters. Emergency authorities provide short-term support in the form of food, accommodations, and limited medical care. To avoid confusion, the following sections use the terms *reception center* to refer to a location where evacuees are registered. We use *mass care facility* to indicate places that provide food and temporary accommodations for victims. Sometimes reception and mass care are located in a single facility. However, many areas, especially host counties for hurricane evacuations, establish a reception center on the major evacuation route. From there, evacuees are directed to the nearest available mass care facility.

During an incident, emergency responders should observe the EOP's designation of the reception and care coordinator. This group, usually the Red Cross or Salvation Army, operates the reception center and mass care facilities. Many large municipal fire departments operate technical rescue units, emergency medical services, and ambulance transportation. In these cases, the transportation of uninjured victims to reception and care facilities is straightforward. Of course, search and rescue, transportation, and reception and care are coordinated with the EOC. Personnel should follow predetermined decision protocols for determining the number and location of reception and care centers. In small scale disasters such as localized floods, centers within a single jurisdiction can be opened as needed. In large scale disasters such as major hurricanes, planning studies are used to estimate the number of people who will evacuate (Texas Governor's Division of Emergency Management, 2004). These evacuation data can be combined with estimates of the percents of evacuees who typically use mass care facilities, which has a historical average of 15% (Mileti, Sorensen, and O'Brien, 1992).

One issue in the operation of reception centers is registering evacuees and allocating them to mass care facilities. Some communities use laptop computers with simple databases for this. Evacuee registration provides a link to the population monitoring function, especially accountability and casualty assessment. It also enables separated family members to find each other. It provides accurate counts for feeding and sleeping facilities. The EOP should contain a procedure for assessing whether victims need clothing and sleeping facilities. It should address how to handle the logistics of fulfilling these needs. Feeding demands also require attention. The places for service, cooking arrangements, and food storage and transportation arrangements are big challenges. Sanitation in bathroom and shower facilities must also be considered. More specific arrangements are required for the presence of children in these facilities. Games, toys, and nursery facilities are commonly provided. The arrival of family pets can be a problem. Experience has shown that not allowing pets discourages needy families from using shelters. Some facility managers arrange to have evacuees' pets housed in local animal protection facilities. Finally, victims need and want frequent updates about the status of the incident, response activities, and the condition of evacuated areas.

10.3.7 Emergency Medical Care

This function varies significantly across jurisdictions. In some cases, fire departments house the emergency medical services (EMS) function as well as the ambulance function. In other cases, these functions are all provided by different organizations. Medical care for victims of major disasters is provided by three components of the emergency response:

- ▲ Emergency personnel in the field.
- ▲ The network of local hospitals.

- The National Disaster Medical System (NDMS). The NDMS is a system of military aircraft equipped to sustain treatment and move patients anywhere in the United States.

At the scene, victims receive medical intervention in a chain of care that continues to hospital emergency rooms and on to definitive care. Casualty assessments are aimed at appropriately distributing and managing patients. Medical management at an incident scene serves four functions:

- Triage.
- Medical treatment.
- Mental health support.
- Patient transportation to definitive care.

The objective of triage is to sort victims into categories so emergency responders can make the most effective use of medical treatment to save the maximum number of lives (Auf der Heide, 1989). Emergency medical personnel use triage tags to indicate patient treatment classification. The most common categories are

- Dead.
-  Catastrophic.
- Urgent.
- Minor.
- Critical.

The triage tag, which identifies the injury type and treatment administered in the field, is the initial patient tracking system. In incidents with many victims, triage may be indicated initially by marking the priority on the patient's forehead with a felt pen. A triage tag is then attached to the patient as soon as feasible.

The IMS medical branch establishes treatment areas away from the immediate threat area to begin treatment at the scene. Medical treatment addresses patients' basic life support needs. Treatment might also involve administration of an antidote in the case of chemical hazards. It might involve administering potassium iodide in the case of radiological hazards. Areas should be designated near treatment areas to serve as collection points for patients' transportation to hospitals. Treatment personnel oversee patients in such zones to decontaminate them (if necessary), monitor their physical conditions, and deliver any needed continuing care.

It is also important to attend to the mental health needs of victims and their families. This would be particularly true in terrorist incidents. Many fire departments maintain behavioral health units to respond to such needs. The Red Cross and other voluntary associations have regular staff and volunteers with mental health specialization. Behavioral health support to victims and their families is likely to be needed both during operations at the scene and later. At an incident scene, behavioral health personnel can be located at decontamination lines and treatment areas. They can also be located at the transportation branch's staging area. Behavioral health units can also be deployed to receiving hospitals to support hospital professionals in caring for short-term victim needs, including debriefings. If mass care facilities are opened, behavioral health personnel can provide similar services at those locations. After the incident, the behavioral health units can serve as referral resources to victims and families. They can link those in need with appropriate community resources including medical or mental health care.

Disaster victims may be moved from the scene to either receiving hospitals or mass care facilities. This depends on their medical assessments. In chemical, radiological, or biological incidents, only patients who have been decontaminated should be transported unless severe threats to life safety arise. This reduces the load on decontamination teams at hospitals. It also reduces the probability that a health facility will itself become contaminated. Victims are transported in a variety of vehicles, depending upon victim condition and medical need. The options include ambulances and multiple occupancy vehicles. If the situation is particularly urgent, helicopters can be used. If a patient's injuries are severe, or the local hospital system is overloaded, patients can be moved into the NDMS directly from the scene.

Hospital disaster response is guided by each institution's disaster plan. These plans address six issues:

1. Internal and external hospital security.
2. Lock-down procedures.
3. Decontamination.
4. Tracking for walk-in patients.
5. Decisions to treat patients inside the facility and/or in treatment areas outside the hospital.
6. Triage for walk-in patients.

A relevant dispatch center notifies receiving hospitals that a disaster is in progress. Hospital disaster plans usually require that receiving hospitals go to *lock-down status*. They secure all doors to control access to the facility. They also notify hospital staff and physicians of the emergency. By following their procedures, hospitals determine the need to mobilize out-of-hospital areas for mass casualty incidents. Hospitals can open prearranged onsite areas to handle the patient overload. In addition, the jurisdictional EOC can establish off-hospital site medical care facilities. These are commonly known as **medical aid stations.** Factors in the decision to treat outside the hospital include the:

- ▲ Number of victims.
- ▲ Nature of injuries.
- ▲ Types of treatment/antidote administration required.
- ▲ Potential for victims to contaminate hospital facilities.

You may have to manage a chemical or radiological incident. If you do, you should work through the EOC to ensure that drugs, antidotes, and equipment are moved to the receiving hospitals. The EOC pharmaceuticals representative monitors pharmaceutical needs. This representative obtains additional drugs and resupply through EOCs links to local pharmacies, drug distributors, and the National Pharmaceutical Stockpile.

In addition to patients transported from the scene, hospitals should expect "walk-ins" or self-referred patients. They might transport themselves or be transported to hospitals by bystanders (Auf der Heide, 1989). Hospitals must decontaminate, triage, and treat such patients. The possibility of victim contamination may require hospitals to outfit their personnel in personal protective equipment. This might include face shields, gloves, overalls, and respiratory protection (Goetsch, 1996). If the demand exceeds hospitals' decontamination capacity, you might assign jurisdictional resources to support them.

The hospital medical staff determines patient treatment needs. For some types of chemical or radiological exposures, appropriate care may not be available in the local area. In such cases, hospital staff will refer the patient to the NDMS for transportation to facilities where appropriate treatment is available. Each hospital must determine its patient capacity. When a hospital reaches its maximum patient load, any additional arriving victims are transported to other receiving hospitals. If all area hospitals are full, victims will be transported to the NDMS receiving area for transport to other locations.

In most communities, the establishment of morgues and the handling of dead in disasters are regulated by law (Hershiser and Quarantelli, 1976). EOPs normally specify the location of permanent and temporary morgues. They specify the procedures for moving the dead to morgues and maintaining records of the bodies. Finally, they specify procedures for claiming bodies. In the U.S., the county medical examiner's office typically performs the morgue function. This includes

- ▲ Receiving human remains.
- ▲ Safeguarding personal property.
- ▲ Identifying the deceased.

- ▲ Preparing and completing case file records on each deceased.
- ▲ Photographing, fingerprinting, and collecting DNA specimens as appropriate.
- ▲ Producing death certificates.
- ▲ Coordinating and releasing remains for final disposition.

The medical examiner is responsible for handling remains. He or she assumes an important role in the *chain of custody* for evidence related to the prosecution of criminal acts. The number of fatalities may exceed the capacity of the morgues. Then the capacity can be expanded as long as the additional facilities have adequate security, utilities, and access to transportation. Alternate sites must accommodate the rapid mobilization of multiple examination stations. This includes partitioning areas for:

- ▲ Receiving bodies.
- ▲ Decontaminating bodies, as needed.
- ▲ Examining/autopsying the bodies.
- ▲ Conducting toxicological chemical laboratory examinations.
- ▲ Assigning the bodies for disposition, including issuance of a death certificate.

Most medical examiners' offices maintain a permanent force of vehicles. They also have personnel to move deceased victims to the morgue. In mass casualty incidents, you should plan to add to that transport capability.

Biological incidents require the expertise of public health agencies. These agencies usually have little or no history of working with emergency managers because they play a limited role in other types of disasters. Local public health departments monitor clinics and hospital records for evidence of epidemics. They conduct scientific investigations aimed at biological agent identification and control. They also specify preventive measures for exposed populations. It is a good idea to have local public health personnel in the EOC during any large-scale disaster. They can provide expertise regarding two special powers that are only granted to public health departments.

The first of these special powers is the decision to administer *mass prophylactic measures*. Prophylactic measures are disease preventing measures such as immunizations. A state or local health department has the authority to order the administration of mass prophylactic measures if there is an epidemic. This is a medical decision that is based on the type of agent, the efficacy and availability of the available medications, and the time available. Public health officials will obtain the needed drugs or vaccines from local wholesale stocks. If this is not sufficient, they can get more from a local government cache or the National Pharmaceutical Stockpile.

The first of these special powers is the decision to quarantine. The definition of quarantine varies among the states, but it generally refers to the confinement of citizens or property due to a public health threat. The Centers for Disease Control (CDC) distinguishes between *isolation* and *quarantine*. Isolation is the confinement of symptomatic patients. Quarantine is the confinement of asymptomatic exposed individuals (who might or might not be infected with the biological agent). Whatever distinction your state makes, it is your public health authorities who must determine the area and timing of implementation. Police will conduct any evacuation that is necessary to implement quarantine. In most states, citizens under a quarantine order may legally be removed from their homes by force. They are then transported to mass care facilities and confined there, if necessary. Police then maintain a perimeter and oversee access control. Just as public health authorities are the only legal source of a quarantine order, they are the only ones that can rescind it.

10.3.8 Hazard Exposure Control

This function seeks to reduce people's hazard exposure to a level that is *as low as reasonably achievable* (ALARA). This is a very important consideration that should be adjusted for the importance of the objectives for which exposure will be incurred. According to the ALARA principle, it is appropriate for an emergency responder to enter an unstable building to extricate a group of severely injured victims trapped inside. In this case, the potential for saving many lives offsets the risk to the emergency responder. The ALARA principle does not support risking an emergency responder's life by entering an unoccupied building just to remove some valuable papers. Emergency personnel can achieve the ALARA objective in three ways:

▲ Minimizing the amount of time spent in hazardous areas (time).

▲ Staying as far away from hazard sources as necessary (distance).

▲ Insulating themselves from the hazard (shielding).

As noted earlier, incident zone access control protects people by keeping them out of the risk area. This reduces their time in the risk area to zero. Hazard exposure control is the reason why you might want to limit the amount of time residents spend in the risk area to salvage property. Moreover, provision of police escorts enables you to rapidly communicate any unexpected increase in hazard level. You can then order a quick return to safety. Keeping people out of the risk area also uses distance for protection. If you establish access control points far enough away, you can minimize the risk to the public and also to the emergency responders who staff those access control points. Finally, it is mostly emergency response personnel who use shielding. They protect themselves from the hazard when they use personal protective equipment. Air filters or bottled

Figure 10-2

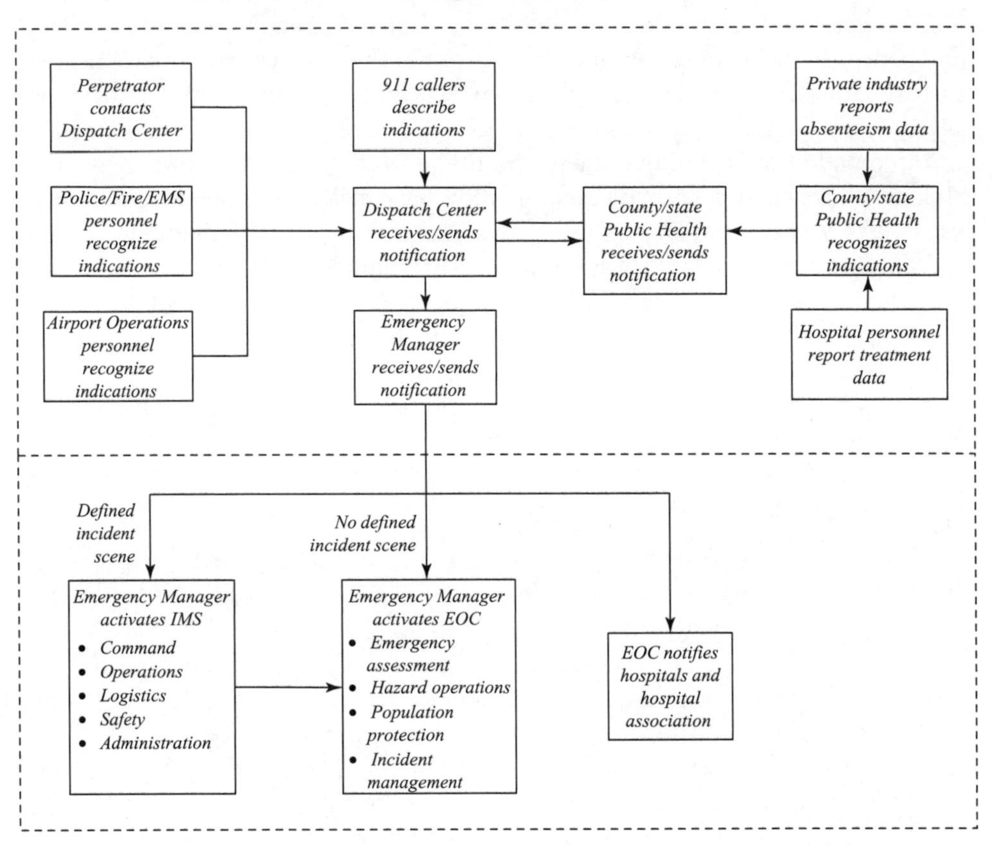

Local notification process for a chemical, biological, or radiological (CBR) incident.

air, chemical protective suits, gloves, and boots all provide shielding against exposure to chemical, biological, or radiological (CBR) agents (see Figure 10-2).

10.3.9 Environmental Surety

This is an issue whenever a hazard agent, most likely CBR, leaves a dangerous residue in the air, water, or soil. Hazmat incidents involving even small quantities of some CBR agents can create major contamination problems. A variety of hazards can cause hazmat releases as a secondary hazard. Finally, a terrorist attack could involve a deliberate hazmat release. In all of these cases, population exposure after reentry to the impact area could cause long-term health consequences. Thus, many communities use hazmat response teams to test for, abate, and monitor such contamination.

The achievement of environmental surety is complex. It might be addressed at many stages of incident management. However, it becomes a major priority in the later phases of response and often continues into disaster recovery. A contaminated area might also be a crime scene. Then the goal of reducing contamination to a safe level must be balanced with the goal of recovering and preserving evidence. The FBI must be involved in threat assessment and evidence collection throughout all phases of this process.

Some incidents have one or more contaminated areas. This is very likely to be true where hazmat technicians have been operating. The incident commander usually insures that all personnel and equipment in the hot and warm zones are decontaminated. In addition, all other personnel and equipment that might have been exposed should be decontaminated. After decontamination, hazmat technicians should collect samples of runoff from the decontamination corridors and then shut them down following local SOPs. This runoff water can be analyzed to ensure any unknown CBR agents are identified. It should also be tested to verify that any acids or caustics have been neutralized.

For all contamination, site threat assessment and remediation follows local procedures. In most cases, these are consistent with federal environmental guidelines. Many different agencies contribute to the site remediation effort. The state emergency management agency normally oversees the process. The county (or state) health department, the National Guard, and state transportation department can all support hazard monitoring by using their hazard monitoring devices. These agencies may also have specialized roles in the collection and analysis of samples for agent identification.

County or state environmental protection departments collect environmental samples to produce their initial assessments. In CBR events defined under the Comprehensive Environmental Response Compensation and Liability Act (CERCLA), these agencies coordinate with the EPA to implement the National Contingency Plan (NCP). The NCP coordinates environmental response, including:

- ▲ Site assessments.
- ▲ Consultation.
- ▲ Agent identification.
- ▲ Environmental monitoring.
- ▲ Environmental decontamination.
- ▲ Long-term site restoration.

Additional support is available from the U.S. Department of Health and Human Service (HHS) Health and Medical Services Support Plan.

Environmental samples of air, water, or soil—as well as unknown substances—are usually collected and packaged by hazmat technicians at the incident scene. In most cases, county or state resources are used to collect environmental samples and prepare them to be transported for laboratory analysis. This is done according to the instructions contained in the CDC SOPs for containerization.

FOR EXAMPLE

Communicating with the News Media during Search and Rescue

It can be very important for official sources to limit the news media's access to information transmitted during emergency response operations. Thirteen miners became trapped in a coal mine in West Virginia in early January 2006. During the search and rescue, reporters overheard a discussion between the SAR team in the mine and the authorities in the control center. Although no one is sure what exactly what was said, media members and family members thought they heard SAR team members report the miners were still alive. The news media immediately reported the story before seeking confirmation that it was correct. Later, rescuers found that twelve of the miners had died. The inaccurate report raised false hopes and added to the pain of the victims' families.

First responder teams communicate with the local FBI office regarding transportation to the laboratory conducting the analysis. For security, and to ensure a continuous chain of custody for the evidence, an FBI approved law enforcement office transports the material. Chain-of-custody paperwork is required unless otherwise determined by the FBI. Responsibility for the samples is transferred to the laboratory staff through the chain-of-custody paperwork. The laboratory is required to be a secure facility as determined appropriate by the FBI.

After samples are received, the laboratory tests the samples. Or, if needed, laboratory personnel refer the sample material to the CDC. Before transporting any unknown material, on-scene hazmat teams should assess it for stability. They should call the FBI for assistance if necessary. Transported samples should be accompanied by a hazard assessment. This should include the probability that the material is explosive or might release some gas when the container is opened.

SELF-CHECK

- Describe the major requirements of an effective evacuation protocol.
- List the four reasons to implement access control and security.
- Explain the morgue functions that the county medical examiner's office typically performs.
- Name the three ways that emergency personnel can achieve the ALARA objective.

10.4 Incident Management

Successful response to a community-wide disaster requires the local emergency response organization to mobilize rapidly. This, in turn, requires a predetermined concept of operations and its elaboration in the jurisdiction's EOP. The **concept of operations** is a summary statement of what emergency functions are to be performed and how they are accomplished. In almost every case, this requires coordination of many public sectors, private sectors, and nongovernmental organizations. It also requires specification of how extra-community resources will be mobilized and integrated into the response effort. This section discusses seven specific functions that are the core of the incident management function.

10.4.1 Agency Notification and Mobilization

Notification to the jurisdictional authorities comes from different sources, depending on the nature of the threat. Federal agencies usually notify a predetermined warning point. This warning point might be the local emergency manager or the police or fire department dispatch center. For routine emergencies, dispatch centers are the most common warning points. After the warning point is notified, it must notify other agencies and mobilize appropriate resources. The EOP should specify how people should be contacted. The principal emergency response agencies operate on a 24-hour basis. Relevant departments that do not operate on this schedule should always be available by having on-call duty officers. This notification process should end only when all of the parties that have a duty or capacity to respond have been informed. The aims of notification are to identify the organizations needed, alert them to begin their own activation processes, and prepare them to initiate the response. Thus, you should establish criteria for determining who is likely to initiate the notification process. They need to know who they should notify, which communications channels are available, and what information should be given.

This notification process differs among different types of hazards. In hurricanes, the notification process is "top down." When the National Hurricane Center detects a hurricane, it notifies state and local emergency management agencies. In turn, they notify other departments within their jurisdictions. Accidents at hazmat facilities have more of a "bottom up" notification process. The facility notifies the local jurisdiction, which notifies its departments before notifying state and federal agencies. Notification of a hazmat transportation accident is even more complex because the hazmat carrier does not routinely interact with a specific local jurisdiction. In most cases, the driver of the truck or crew of a train attempts to notify their company dispatcher and a local or state police office. In either case, it is likely that local or state police is the first agency at the scene. The first on scene will, in turn, notify other local and state agencies. If necessary, the lead state agency will make a link to the federal emergency response system.

Notification in terrorist incidents is especially complex because CBR agents might be detected in many different ways. First, a caller might inform the police or fire department. Or dispatchers might determine from their call screening protocol that such an agent has been released. Fire and police personnel responding to an apparently routine call might notice signs and symptoms of CBR exposure. County or state health departments could discover it during a routine screening of local employers for indications of increased absenteeism. Surveillance systems in hospitals and clinics could show symptoms consistent with CBR exposure.

10.4.2 Mobilization of Emergency Facilities and Equipment

A major step in emergency response is the activation of a jurisdiction's EOC. As facilities, EOCs vary. In more hazard-prone (and wealthier) communities, EOCs have full-time staff with extensive communications equipment, powerful computers, and sophisticated display screens. Such arrangements create stable, visible, ready locations for supporting emergency response operations. At the other extreme, many communities' EOCs are converted from conference rooms by hanging some status boards on the wall and installing a few additional telephones. This does not mean that only an expensively designed EOC is adequate. In fact, even a very basic EOC can be effective. It must be based on careful analysis of the functions that will be performed there (Lindell et al., 1982). It is better to have a modest facility that matches the EOC's design to its function than it is to build a large expensive facility that provides inadequate support.

There are often many facilities in a community that are called EOCs. One facility might serve as the jurisdiction's EOC. However, many fire and police departments also have their own departmental EOCs. Fire departments usually locate their EOCs in, or adjacent to, their dispatch centers. Some police departments do the same, whereas others have stand-alone technical operations centers. It is also common for public works and transportation departments to maintain their own departmental EOCs. However, each EOC focuses on managing the response of its own organization. Departmental EOCs accept directives from the jurisdictional EOC, call upon their own SOPs, and dispatch their own resources.

EOCs all perform essentially the same function. They are the hub of the emergency information processing within the jurisdiction. They are also the hub between the jurisdiction and external sources of assistance. The EOC requests data, receives it, and processes it. The EOC then uses this information to coordinate the community's emergency assessment, hazard operations, and population protection actions. Figure 10-3 describes the information flow in a typical EOC (Lindell et al., 1982; Perry, 1995). The communications team requests and receives data from three principal sources. The first type of source is state and federal hazard detection agencies such as the National Weather Services (NWS).

Figure 10-3

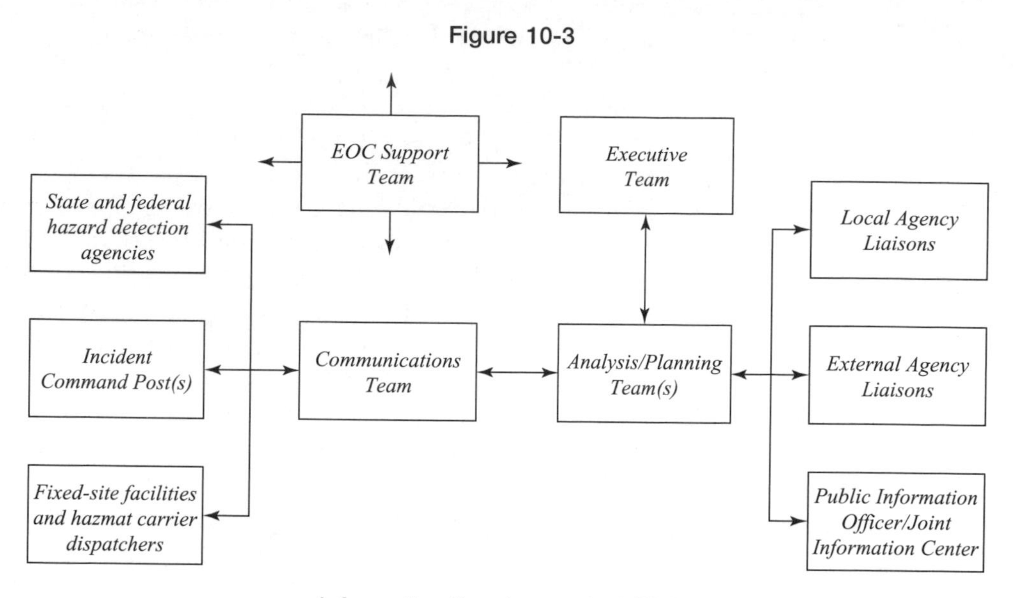

Information flow in a typical EOC.

The second type of source is one or more local incident command post. The third type of source is a local fixed-site facility or hazmat carrier's dispatch office. These organizations provide hazard-monitoring data.

The communications team receives information about the status of the situation. They also receive updates about the status of the emergency response and field teams' needs for additional resources. The communications team passes on the data to one or more analysis/planning teams. These teams process the data to predict likely future states of the hazard and environment. They also examine the implications of current and future environmental conditions for the population protection function. These analysis/planning teams typically address the planning, logistics, and finance/administration activities defined under the Incident Command System (ICS)/Incident Management System (IMS). The analysis/planning teams present the results of their analyses to the executive team and recommend further emergency assessment, hazard operations, and population protection actions. The executive team reviews the analysis/planning teams' recommendations and acts on them. The EOC support team periodically briefs the analysis/planning teams, the executive team, the public information officer (PIO), and the agency liaisons. In turn, the PIO uses this information to prepare press releases and briefings. The PIO briefs the news media in a joint information center located near the EOC. The agency liaisons communicate the information that is relevant to their agencies. This includes requests for the activation of additional resources and the dispatch of mobilized resources to staging areas. At the municipal level, the agency liaisons are representatives from police, fire service, emergency medical services, and public works. There are also representatives from public and private utilities, the Red Cross, and Salvation Army. In addition to these organizations, there might be representatives of higher levels of government in the EOC. Some agencies, such as FEMA, are at all major disasters. Others are only present for certain types of disasters. The need for organizations to be represented in the EOC depends on the nature of the threat. It also depends on the network of governmental resources needed to respond to that threat.

The EOC support team provides administrative and logistical support for the EOC. It also documents the status of the incident over time and the actions taken in response to those conditions. In addition to the EOC coordinator, the EOC is also staffed by a variety of officers. They are charged with oversight and coordination of the emergency at hand. The officers in charge of the communications team and the analysis/planning teams frequently report to the EOC coordinator during an emergency even though they usually report to someone else during normal operations. The reason for this is that the EOC officers play roles that are very different from those of the agency liaisons. Instead of representing agencies, the EOC staff members perform functions that serve the needs of the jurisdiction as a whole. In addition, the EOC coordinator ensures that the EOC is fully functional. This includes maintaining computer hardware and software, display devices, and communications equipment. It also includes establishing duty schedules for EOC personnel (12 hours on/12 hours off) to ensure continued staffing in a lengthy incident.

10.4.3 Communication/Documentation

This function has a much higher profile in the EOC than at the incident command post because communications is crucial to the EOC. The communications team collects, displays, and records data on environmental conditions, casualties, and damage to property and the environment. For example, the communications team obtains weather data from the NWS in hurricanes and floods. During incidents involving unknown agents, agent identification is done on-scene. The incident command post relays the outcome to the EOC. In some incidents, the analysis/planning teams might not have adequate expertise in the EOC. In such cases, the communications team might contact specialists elsewhere to obtain the necessary information.

It sometimes happens that small jurisdictions assign untrained office staff to serve as emergency communicators. These communicators maintain uninterrupted telephone contact with other response organizations. If clerical staff must be used, then they must be trained. Experience has shown that if they are unable to understand the messages they receive, they will probably record them incorrectly. In many ways, an inaccurate message may be more misleading, and ultimately more dangerous, than no message at all.

Documentation is an important task of all EOC staff rather than the sole task of a single unit. Thus, the communications team must maintain communication logs. These logs record who called into or out of the EOC, when the call took place, who participated in the call, and the content of the call. The analysis/planning teams must always update the situation status and resource status. The PIO must document media inquiries and responses to them. Agency liaisons must document the inquiries they receive and the responses they make. The EOC coordinator may choose to establish a documentation unit. This unit collects information for the after-action report that assesses the strengths and weaknesses of the emergency response. Documentation must also be established for activities that take place at the incident scene and at other locations outside the EOC. Indeed, these will be the only locations where activities need to be documented when the EOC is not activated. In addition, the incident command post should save incident action plans and other documents generated by the ICS.

Documentation is easier when you have forms that meet the needs of all the agencies. Normal documentation systems are designed for normal circumstances. However, they may be time intensive or otherwise unsuited for use in emergencies. You should work with your department of finance and administration to create a documentation system. In addition, federal reimbursements for expenses incurred during presidential disaster declarations may demand different documentation processes. You should ensure those documents meet these standards as well.

The EOC staff should not try to record every single message transmitted everywhere. However, it should follow basic procedures to ensure there are records of the most important response actions. The basis for those response actions must be documented. Moreover, the resources used in the response actions must be documented. This means keeping information on the status of the incident and the response. It also includes information on the timing and effectiveness of operational decisions. Such data are needed in the short run to adapt response strategies to disaster demands. However, these data are also needed in the long run to support after-action assessments. These assessments provide feedback to improve the management of future disasters. They also provide the basis for defense against litigation by any parties who believe they were adversely affected by the actions of the emergency response organization.

10.4.4 Analysis/Planning

The analysis/planning teams assess the current status of the situation and projects its future status. For example, analysis/planning teams might use an NWS forecast of an approaching weather front to anticipate a change in wind direction that will require evacuations in neighborhoods that previously were considered to be safe. These teams may activate technical specialists in specific areas, such as toxicology, and radiology, for CBR incidents. Analysis/planning teams also use data from hazard monitoring to examine anticipated changes in incident demands. This information allows them to determine if the incident commander needs additional resources for these new demands. Conversely, as the incident demands decrease, analysis/planning teams assess the opportunities for demobilization of response units that are no longer needed. Finally, analysis/planning teams are sometimes responsible for posting current data on situation status, resource status, and response actions to EOC status boards. This activity also requires recording such information for briefing oncoming shifts and for generating after-action reports.

10.4.5 Internal Direction and Control

This provides the answer to the question "Who is in charge?" There are three basic ways to answer this question. The CAO could be the emergency director, with or without an executive committee of department heads as advisors. This strategy has the advantage of maintaining continuity with the way the community is managed under normal conditions. However, it can have the disadvantage of overloading the CAO with decisions for which he or she has little training or experience. In addition, provisions must be made for an alternate to provide regular rest periods. In the second management strategy, the CAO delegates responsibility to the head of one of the disaster-relevant departments. The advantage of this strategy is that these officials have training and experience directly tied to response. However, they still need to consult the CAO if they need clarification on some policy issues. They also need to make arrangements for an alternate.

In the third strategy, the CAO delegates to the local emergency manager. This gives authority to the person who has the most disaster training and experience. However, this strategy requires someone else to assume the duties of EOC coordinator. Also, this could create conflicts in some jurisdictions. The problem here is that some jurisdictions establish emergency management as a division within another department. In that case, the local emergency manager would have to give orders to someone who is normally his or her superior.

The appropriate management strategy depends in part on the normal organizational structure. In addition, it could depend on the demands of the emergency. For example, the emergency manager serving as EOC coordinator could handle minor emergencies by:

- ▲ Consulting with EOC staff.
- ▲ Devising policy for immediate disaster response.

- ▲ Consulting with the CAO and agency heads for policy approval when appropriate.
- ▲ Implementing approved policy.

For major disasters, EOC management would shift to the CAO. The CAO would be supported by an executive team comprising the heads of the major disaster-relevant agencies. The EOC coordinator would alert this group to the need for policy decisions and implement their policies.

Internal direction and control involves three specific activities. These are management oversight, policymaking, and coordination with local agencies. The first activity, *management oversight* of disaster operations, involves monitoring the performance of field units to verify that they are responding in accordance with the EOP and agency SOPs. This does not mean that the EOC assumes tactical direction of on-scene operations. The incident commander is best positioned for such decisions. Instead, the EOC focuses on providing information and resource support to the incident commander. The EOC monitors the response to ensure that the disaster demands are being addressed. Often, these demands change over time. The demands of initial impact might decline and new demands may arise from secondary threats. For example, in large floods, the initial concern with evacuation and rescue gives way to public health concerns about sewer systems overflowing and contaminating drinking water systems. The changes in disaster demands cause changes in response operations which, in turn, produce changes in incident management activities. The EOC must continually monitor the incident commander's response to the threat. The EOC must continually review the need to redeploy resources to respond to the changing demands.

The second activity in internal direction and control is *policymaking*. When a disaster strikes, the EOP ordinarily serves as the framework for coordination. It defines the functional assignments for different response organizations. It defines the chains of command and lists available resources. It defines the procedures for requesting mutual-aid. If a disaster generates demands that are not adequately addressed in the EOP or SOPs, incident managers might need to improvise solutions. The executive team needs to authorize the improvised procedures.

The third activity is *coordination with local agencies*. This involves ensuring that emergency response agencies not reporting directly to the incident commander receive current information about the status of the incident and response. In addition, it involves giving each agency instructions to perform unplanned activities that evolve from the policymaking process.

10.4.6 External Coordination

This addresses the relationship between the local emergency organization and outside organizations. Many of these contacts are made through telephone calls, but an increasing number come through emails. There is considerable potential for web-based crisis information management software to support EOCs (Department of Justice, 2002). However, many of the available systems are incompatible with each other and some provide inadequate support to ICS/IMS (Hunt, 2005). Widespread adoption is likely to occur only if no- or low-cost systems such as CAMEO are distributed.

In a major disaster, state and federal agencies dispatch liaison personnel to local EOCs to coordinate operations. You need to provide space in or near the EOC for these personnel. In addition, the EOC must host visitors without disrupting response operations. EOC managers often fail to anticipate that visitors want to visit and be briefed. Some of these visitors have legitimate disaster related functions. Others have no purpose other than a desire to show concern. Unfortunately, the larger the scale of a disaster, the larger the number of VIP visitors is likely to be. They are difficult to turn away, so you must develop a procedure for hosting them. The EOC coordinator can assign a PIO to escort visitor tours that explain the emergency response operations. It is also helpful to designate a space near the EOC where these VIPs can receive special briefings and ask questions without getting in the way.

10.4.7 Public Information

The location from which public information is given depends upon the size of the incident. It also depends on whether emergency operations are taking place at a defined incident scene. In small, short duration incidents with a defined incident scene, the public information function is often based at the scene. In large incidents, or where there is no defined incident scene, public information is handled through the EOC. In major disasters affecting many communities or multiple levels of government, the site for public information will be a joint information center.

Incident managers should designate a broadly knowledgeable chief spokesperson who can call upon specialists to respond to specific questions. All other response personnel should direct media inquiries to the spokesperson. Conversely, the PIO must keep incident managers informed about the information demands of the news media and the public. The information demands of the media can be determined by the specific questions reporters ask. The information demands of the public can be determined by routinely monitoring the content of calls to the jurisdiction's rumor control center. Any questions that are asked repeatedly should be addressed in press conferences.

In addition, the PIO should schedule regular media briefings. Provisions should be made for the rapid preparation of graphic materials, such as maps, to be used in briefings. Such materials should describe the location of the disaster impact zone. It will help those conducting briefings in describing the response. The PIO facilitates all requests for media orientation tours. The PIO is responsible for providing appropriate personal protective equipment for the media at the incident scene and ensuring its proper use. Finally, the PIO advises the EOC

FOR EXAMPLE

Rapid Notification

Rapid notification of an emergency is critical as it allows time for population protective actions. Even if an accident occurs in a rural area or in the middle of the night, rapid notification is possible. In a spill of radioactive materials in southeast Colorado, the shipper was notified within one hour of the truck wreck by the local county sheriff's office (Hornsby, Ortloff & Smith, 1978). This was in spite of the fact that the accident took place in a rural area in the middle of the night. Also, the driver of the truck was pinned inside the truck cab.

when conditions have reached the point that the public information function needs to be moved from the incident command post to the EOC or from the EOC to the joint information center.

10.4.8 Administrative and Logistical Support

Administrative and logistical support is handled at the incident command post during minor events. However, it is transferred to the EOC during a community-wide disaster. This function comprises the logistics and finance/administration sections of ICS/IMS. The EOP should recognize that the emergency response organization, like all other organizations, requires support services. EOC staff need purchasing, accounting, and support staff. They need routine office equipment repairs, and office supplies such as printer paper. Moreover, the EOC can support the incident command post by assuming many of the logistics section's service and support branches. It can also assume much of the burden of the finance/administration section.

SELF-CHECK

- Define concept of operations.
- List the seven specific functions that are the core of the incident management function.
- List the objectives of agency notification.
- Describe the responsibilities of the PIO.

SUMMARY

As shown, the EOP and procedures developed during emergency preparedness impact the actions you take during emergency response. However, as no emergency can ever be predicted with accuracy, the demands of any specific incident can never be predicted with perfect accuracy. Therefore it is imperative that you learn to improvise effectively as when the need arises. Focus on the four basic functions: emergency assessment, hazard operations, population protection, and incident management, and understand what tasks are associated with each of them. It is imperative that these responses are carried out effectively to minimize casualties and damage.

KEY TERMS

Alert A class of a nuclear power plant emergency defined by the NRC and FEMA that involves substantial degradation of plant safety. Releases are expected to be well below EPA exposure limits.

Concept of operations A summary statement of what emergency functions are to be performed and how they are accomplished.

Emergency classification system A method of organizing a large number of potential incidents into a small set of categories. These categories link the threat assessment to the level of activation of the responding organization.

General emergency A class of a nuclear power plant emergency defined by the NRC and FEMA that involves substantial core degradation and the possibility of radioactive material escaping from the containment building. Releases might exceed EPA limits offsite.

Medical aid stations Off-hospital site medical care facilities.

Rapid damage assessment The first stage of damage assessment that provides you with immediate information about the magnitude of the impact. It defines the boundaries of the physical impact area and assesses the intensity of damage within that impact area.

Site area emergency A class of a nuclear power plant emergency defined by the NRC and FEMA that involves major failures of plant safety functions. Releases might exceed EPA limits onsite but not offsite.

Unusual event A class of a nuclear power plant emergency defined by the NRC and FEMA that involves potential degradation of plant safety. No releases are expected unless other events occur.

ASSESS YOUR UNDERSTANDING

Go to www.wiley.com/college/lindell to evaluate your knowledge of hazards, vulnerability, and risk analysis.
Measure your learning by comparing pre-test and post-test results.

Summary Questions

1. A hurricane warning indicates the possibility of hurricane conditions (sustained winds of at least 74 mph) within a designated section of coast within 36 hours. True or False?
2. Hazard operations measures are feasible for all hazards. True or False?
3. What do people need in an evacuation?
   (a) mass care
   (b) evacuation transportation support
   (c) evacuation traffic management
   (d) all of the above
4. Which of the following is true about EOCs?
   (a) EOCs are the hub of the emergency information processing within the jurisdiction.
   (b) EOCs are managed by a large group of people with different titles.
   (c) To be effective, an EOC needs to be large and expansive.
   (d) all of the above
5. In an Alert:
   (a) releases are expected to be well below EPA exposure limits.
   (b) releases are expected to be right at EPA exposure limits.
   (c) releases are expected to be higher than EPA exposure limits.
   (d) there have not been any releases of any hazmat.
6. Which factor is *not* used when defining emergency classes?
   (a) hazard generation
   (b) hazard transmission
   (c) community vulnerability
   (d) capability of the local emergency management team
7. A PIO should
   (a) rarely have meetings.
   (b) work in the field along with first responders.
   (c) provide appropriate personal protective equipment for the media at the incident scene and ensure its proper use.
   (d) only talk to VIPs.

Review Questions

1. What does threat detection include?
2. How can you establish guidelines for choosing hazard operations actions?
3. What might people need in order to implement protective actions?
4. Name and explain the three specific activities that internal direction and control involves.

Applying This Chapter

1. There is a train derailment at the edge of your town and a tank car of liquefied natural gas is burning. Next to it is a tank car filled with chlorine gas. The wind is currently blowing away from town. Why do you need to monitor weather conditions with this hazard?
2. A tornado has destroyed a major part of your city's industrial area. Why would you need to implement access control and security measures?
3. There has been an earthquake in the town for which you are the local emergency manager. You immediately see the need for heavy rescue. What steps do you take to directly address this need?
4. The roof of the local convention center collapsed during a major exhibition. How will you handle this mass casualty incident?
5. What would you say to the press if you were the PIO responsible for the communication after an explosion at a local factory where you were not sure if and how many fatalities occurred?

YOU TRY IT

Chemical Facility Accident

You are working with the staff of a nearby chemical facility to develop an emergency classification system. What are the major issues you need to address?

Evacuation

You need to evacuate low-lying areas of your city in anticipation of a major flood. What steps do you take to be sure everyone is warned and can evacuate?

Support Team within the EOC

There is a wildfire approaching your city and you have activated your EOC. What tasks should the EOC support team perform?

11

DISASTER RECOVERY

Managing the Process

Starting Point

Go to www.wiley.com/college/lindell to evaluate your knowledge of disaster recovery.
Determine where you need to concentrate your effort.

What You'll Learn in This Chapter

- ▲ How communities function before a disaster strikes
- ▲ Steps in the recovery process
- ▲ Three basic components of household recovery
- ▲ What businesses need for to recovery
- ▲ Government impact on recovery
- ▲ The role of hazard insurance in disaster recovery
- ▲ Local government's role in recovery
- ▲ Roles of recovery operations plans

After Studying This Chapter, You'll Be Able To

- ▲ Compare and contrast how households, business, and governments normally function with how they function during recovery
- ▲ Examine the recovery process
- ▲ Differentiate between the three components of household recovery
- ▲ Examine ways to help businesses recover
- ▲ Analyze how the government helps in recovery
- ▲ Identify the impact of hazard insurance
- ▲ Examine how local governments aid in recovery
- ▲ Analyze the function of a recovery operations plan

Goals and Outcomes

- ▲ Outline normal operations for your community
- ▲ Prepare plans to ensure an effective and speedy recovery after a disaster
- ▲ Support households in their recovery process and identify ways your organization can assist in the recovery
- ▲ Support businesses after a disaster and assist them through recovery
- ▲ Manage government assistance effectively
- ▲ Propose a plan to increase the awareness of hazard insurance
- ▲ Support local government efforts in recovery
- ▲ Create and develop a recovery operations plan (ROP)

INTRODUCTION

Recovery differs from activities that take place during other phases of the emergency management cycle. Through effective planning, you can see households, businesses, and communities through the recovery process. Households, businesses, and communities can also receive assistance from other sources including state and federal governments, hazard insurance, and charities.

In this chapter, you will examine how to manage the recovery process. First, you will look at how communities normally function before a disaster strikes. Then, you will assess the recovery process: how businesses and households recover, how the government helps in recovery, and how to design a recovery operations plan. You will also look at how to increase hazard insurance awareness. Finally you will analyze the local government recovery functions that are critical to recovery and how to implement them quickly and effectively.

11.1 The Routine Functioning of U.S. Communities

Before we examine recovery, let's examine how communities function before a disaster strikes. First, a **community** is a specific geographic area. It is frequently considered to be a town, city, or county with a government. A community has two additional elements. These are psychological ties and social interaction (Poplin, 1972). Psychological ties involve a sense of shared identity that arises from common goals, values, and behavioral norms. This leads "insiders" to distinguish themselves from "outsiders" (Lindell and Perry, 2004). Moreover, insiders interact more frequently with each other than they do with outsiders. Many of these interactions involve the exchange of money for goods and services. However, some interactions are based on affection and social support.

Communities have basic units-households, businesses, and government agencies. Each social unit has people and resources. Households supply labor to businesses in exchange for money. In turn, households pay money for goods and services from private suppliers and government services. In addition, households interact with friends, relatives, neighbors, and coworkers. Businesses use the labor they receive from households to produce goods or services, which they then sell to their customers. As is the case with households, businesses use the money they obtain from customers to pay suppliers, infrastructure, and government. *For-profit* businesses provide goods and services for a fee. The government provides them in exchange for taxes. However, there are also *nonprofit* nongovernmental organizations (NGOs) that provide goods and services at or below cost-and sometimes free. The steady flow of money in exchange for goods and services is known as *cash flow.*

In a free market economy, government establishes broad rules within which people can freely exchange resources. For example, certain goods (e.g., heroin) and services (e.g., prostitution) are unacceptable and illegal. Government provides some services that the private sector cannot or will not provide at acceptable cost. Units with more social, economic, and political power can force less powerful units to accept less favorable outcomes.

These basic units act in cooperation, competition, and conflict (Poplin, 1972; Thomas, 1992).

- ▲ **Cooperation** refers to activities that result in mutual benefit. An example is when a supplier provides a good or service to a customer in exchange for money.
- ▲ **Competition** exists when two parties strive toward a goal that only one can achieve. In fair competition, the parties use legitimate methods. For example, two businesses compete to sell a product to customers on the basis of quality and price.
- ▲ **Conflict** occurs when one party attempts to directly frustrate the goal achievement of another. For example, one business might attempt to use its greater resources to force its suppliers to refuse to serve its competitor.

Many social institutions, such as schools and churches, promote agreement on values and legitimate methods of goal achievement by socializing their members. Complete consensus is never reached. Political institutions exist to resolve differences.

Households, businesses, and government agencies have *human assets.* Human assets include the intelligence, physical abilities and the personality characteristics of the people who live or work in homes, businesses, and for the government. These characteristics combine with time and effort to produce *labor* (Schneider & Schmitt, 1986). In addition, there are physical assets. These include land, buildings, equipment, furniture, clothes, vehicles, crops, and animals. These physical assets are also known as *goods.* Finally, social units have *financial (capital) assets* such as cash, stocks, bonds, savings, and insurance. In many cases, these assets were obtained through loans, mortgages, and credit card debt. Loans, mortgages, and debts are known as *financial liabilities.* However, most assets generate *income* from employment, rental of physical assets, interest or dividends. This income must be balanced against expenses for *consumption,* and *production,* as well as *investment* in additional assets. Finally, social units vary in the resources they have. Ethnic minority, aged, and female-headed households frequently have fewer resources than other social units. Similarly, small businesses and small political jurisdictions also have fewer resources. This makes it difficult for them to withstand the impacts of a major disaster.

11.1.1 Household Activities

Households engage in a variety of activities over the course of the day. Some activities such as sleeping and eating are essential. By contrast, other activities such as attending cultural events and singing/dancing are highly discretionary. Discretionary activities can be reduced or eliminated when the need arises. Moreover, some activities are age or gender related. For example, adult males are more likely to be the household members involved in yard work and car repair, whereas adult females are more likely to be the ones involved in shopping and child care. In recent years, it is increasingly likely for adult males and females to be involved in work. If a disaster causes household members to change their normal patterns of daily activity, this can cause psychological and economic distress and social conflict.

11.1.2 Business Activities

Businesses produce a wide variety of goods and services. Some of the industries generate goods and services that are sold to customers outside the community. These industries define the community's *economic base*. The economic base model identifies the amount of goods and services from exports, internal investment, and consumption (Chapin and Kaiser, 1985). More money is available for internal investment and consumption when exports exceed imports. A *multiplier effect* is set in motion when money that is received from the sale of exports is spent inside the community. As a result, urban areas receive between $1.50 and $2.50 in induced local income for every dollar of revenue from exports (Blair and Bingham, 2000). In general, mining, manufacturing, wholesale and retail trade, banking and finance, and high quality service facilities (e.g., nationally renowned medical clinics) are major contributors to an economic base. However, there can be exceptions to this rule (Chapin and Kaiser, 1985).

These facts have implications for disaster recovery. First, some communities have weak economic bases. They have low exports, low investments, and high internal consumption. These communities need outside help to recover from a disaster. Second, industries that produce exports should receive immediate attention in the disaster aftermath. They can stimulate local investment and consumption. This will spread the recovery to other industries.

11.1.3 Government Activities

The governments of most local communities perform a variety of functions that cannot be performed by the private sector. Each function is assigned to a subunit called an agency or department. All of the departments report to the jurisdiction's chief administrative officer (CAO). The CAO might be a mayor, city manager, or chair of the county board of supervisors. One department or agency is emergency management. Other examples include planning, law enforcement, and public works.

FOR EXAMPLE

Savings Rate

The normal routines in households affect disaster recovery as well. Different economists define savings in different ways, but all agree that the amount of money Americans save every year is very low. Most households save somewhere between 0 to 4% of their incomes. When their property is damaged or destroyed in disaster, many need financial assistance.

SELF-CHECK

- **Define community.**
- **Define conflict.**
- **Define competition.**
- **Explain how the savings rate affects recovery.**

11.2 The Recovery Process

Recovery begins when the emergency has been stabilized so there is no longer a threat to life and property. Recovery ends when the community has recovered from the disaster. Most people's goal is to have their lives resume exactly as they were before the disaster. People assume that they must rebuild the buildings and infrastructure as it was. It is now understood that restoring the community to its previous status can reproduce the hazard vulnerability that led to the disaster.

Most American communities recover quickly from disasters. However, the fact that communities *as a whole* recover does not mean that specific neighborhoods recover rapidly or even at all. It does not mean that specific businesses can maintain or resume operations. It is important to know which population and economic sectors will have the most difficulty recovering from disaster. This enables you to help with technical and financial assistance when it is needed. You can also monitor recovery, and encourage households and businesses to adopt mitigation measures that reduce their hazard vulnerability.

Disaster recovery is both physical and social. Disaster recovery includes actions taken to cope with casualties. Households must find emotion-focused strategies for dealing with the death of affective support from loved ones. They must also have problem-focused strategies for coping with the loss of resources needed to earn an income, manage the home, and rear children. Injuries can add emotional and financial strain. Similarly, businesses must cope with the loss of employees who might be dead, injured, or overwhelmed with caring for families and friends.

Disaster recovery includes coping with property damage. Households must repair minor damage and rebuild destroyed property. Businesses and government agencies must repair commercial and industrial structures. They repair critical facilities such as hospitals, police stations, and fire stations. They must also repair infrastructure such as water, sewer, electric power, fuel, transportation, and telecommunications.

One of the most difficult parts of recovery is restoring social routines and economic activities. The process of "getting back to normal" involves restoring people's psychological stability. It also involves learning positive lessons from the experience. Interacting with friends is another part of returning to normal. People also need to return to full-time employment and receive pay equal to what they received before the disaster.

Unfortunately, "normal" is what got the community in trouble in the first place. For example, when cities allow too much development in floodplains, "normal" is not a sustainable goal. A disaster resilient community learns from its experience. It looks for inadequate designs, construction methods, and construction materials. It identifies and fixes the buildings, infrastructure, and critical facilities that have these inadequacies. Finally, it recognizes which households, businesses, and government agencies have inadequate resources or lifestyles that make them unable to respond or recover from a disaster.

A disaster resilient community learns how to use the disaster as a focusing event that changes people's behavior. After a disaster, people should realize that they are at risk. They should consider what actions they can take to protect themselves and their property. They should discover what hazard adjustments are best for their community. A disaster resilient community supports sustainable development policies. It mobilizes the government and demands that effective policies be implemented.

11.2.1 Facilitating Conditions for Disaster Recovery

Disaster recovery includes many activities. Some of them happen at the same time. Others occur in sequence. At any one time, some households might be engaged in one set of recovery activities while others are engaged in other recovery activities. Some households might be fully recovered months or years before others. There might be households or businesses that never recover at all. Scholars agree that disaster recovery occurs in phases, but they disagree on the number and names of phases. However, all of them agree disaster recovery has at least two phases—short-term recovery and long-term recovery.

Communities must be able to identify and respond to specific problems. This ability helps them recover more rapidly. As a local emergency manager, you should work with local government agencies to respond to demands. Disaster recovery is easier if you can anticipate the biggest demands and plan how to deal with them before disaster strikes (Rubin, et al., 1985). Planning before a disaster does not eliminate the need to improvise after a disaster but it does make the recovery more manageable (Kreps, 1991). Just as you must anticipate disaster demands and plan your emergency response, you should anticipate disaster demands and plan your disaster recovery.

Schwab et al. (1998) emphasized the need to engage in predisaster planning for postdisaster recovery. Planning for recovery is the best way to become aware of recovery demands. There are short-term decisions such as where to locate evacuees. There are long-term decisions such as how to finance reconstruction, where to allow rebuilding, and where to rebuild public infrastructure. You should establish many of your community's recovery policies at the same time as you engage in emergency preparedness, comprehensive planning, and mitigation planning (Schwab et al., 1998).

Developing preimpact plans for disaster recovery is important because there will not be much time for recovery planning after disaster strikes. Preimpact recovery plans allow the community to incorporate mitigation measures into disaster recovery. Preimpact recovery plans also help elected and appointed officials resist pressure to return the community to "normal" after a disaster. By developing disaster resilience, communities can minimize disaster impacts. They can strengthen their ability to recover without assistance. They can help all population segments and economic sectors recover.

11.2.2 Disaster Recovery Functions

Recovery involves a network of tasks that that need to be performed by community subunits. As path A in Figure 11-1 indicates, affected households go through a sequence of steps to housing recovery. This process takes them through emergency shelter, temporary shelter, temporary housing, and permanent housing (Quarantelli, 1982).

As path D indicates, affected businesses pass through a slightly different sequence because they can suspend operations (represented as a dashed line) until they find a temporary operating location. As path B indicates, households and businesses need infrastructure such as water/wastewater, electric power, fuel (e.g., natural gas), transportation, and telecommunications before they can resume normal operations. Finally, path C is especially important because disaster assessment and a federal disaster declaration are preconditions for federal financial aid.

Figure 11-1

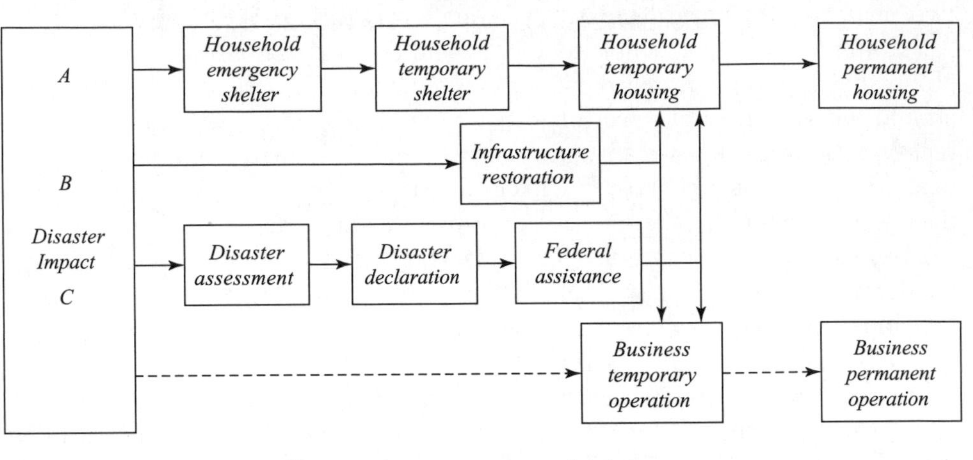

The recovery management process.

This aid allows the most severely stricken communities to rebuild public infrastructure and for households and businesses to rebuild damaged or destroyed buildings. To explain this figure more completely, the following sections examine household recovery, business recovery, infrastructure restoration, and the disaster declaration process.

FOR EXAMPLE

When Normal Might Not Be Desirable

It is human nature to want to return things to the way they were before a disaster. However, returning to "normal" also means repeating the same mistakes that led to the devastation caused by the disaster in the first place. In the case of Hurricane Katrina, there is debate about whether New Orleans should allow homes that are below sea level to be rebuilt. Some argue these homes will be safe if the levees are raised to protect against a Category 5 hurricane. They point to the extensive (and very expensive) flood control works in the Netherlands. Others argue that the Dutch have little choice. Much of their country is below sea level so they must protect what they have. They don't want to spend large amounts of federal funds to solve what they view as a local problem.

SELF-CHECK

- Explain where recovery begins and ends.
- Explain why it is difficult to return to social routines and economic activities immediately after a disaster.
- Explain why communities must identify and respond to specific problems that arise during recovery.
- Explain the differences between short-term and long-term decisions in postdisaster recovery.

11.3 Household Recovery

There are three basic components to household recovery (Bolin and Trainer, 1978). These are housing recovery, economic recovery, and psychological recovery. All three of these involve obtaining resources to recover. However, households must invest a lot of their time to obtain most of these resources. This means household members will experience major disruptions in their normal activities.

11.3.1 Housing Recovery

Households typically pass through four stages of housing recovery following a disaster (Quarantelli, 1982).

▲ **Emergency shelter** is the first stage. Emergency shelter is an unplanned location that is intended only to provide protection from ordinary weather conditions of temperature, wind, and rain. For example, some families sleep in their cars after earthquakes (Bolin and Stanford, 1991, 1998).

▲ **Temporary** shelter is the second stage. This includes food preparation and sleeping facilities that are sought from friends and relatives or are found in hotels or motels. Mass care facilities in school gymnasiums or church auditoriums are a last resort.

▲ **Temporary housing** is the third stage. Temporary housing allows victims to reestablish household routines in nonpreferred locations.

▲ **Permanent housing** is the last stage. Permanent housing reestablishes household routines in preferred locations.

The use of emergency shelter usually peaks on the day of the disaster and declines rapidly thereafter. However, this decrease does not immediately increase

Figure 11-2

It takes time for disaster victims to complete the paperwork to receive aid or insurance compensation.

occupancy rates for permanent housing. Indeed, the proportion of the affected population in permanent housing can also decline after a disaster. This is because additional households are forced to move out of damaged homes that are condemned by authorities. For example, it took nine days after the Whittier Narrows earthquake for mass care facility occupancy to reach its peak (Bolin, 1993).

Sites for temporary shelter include homes of friends and relatives, commercial facilities such as hotels and motels, and mass care facilities such as Red Cross shelters (see Figure 11-2). During Hurricane Lili, 53% of the evacuees stayed with friends and relatives, 30% stayed in commercial facilities, and 3% stayed in mass care facilities (Lindell et al., 2005). The average percent of people that stay in mass care facilities is 15%, but ranges from 1% to over 43% (Mileti et al., 1992). The location where a household seeks temporary shelter is predictable. Households will stay with friends and relatives if these have undamaged homes and live nearby. Households with higher incomes who lack nearby friends and relatives with undamaged homes seek commercial facilities. Lower income households in such conditions have no alternatives to mass care facilities.

Areas with large minority populations can pose problems because of their "unconventional" household structures. Some households are multigenerational, including grandparents, parents, and children (Bolin, 1993; Yelvington, 1997). Others involve multinuclear kinship, where siblings and their families live with each other. Still other households involve multinuclear friendship, where immigrants from the same country share a residence. These complex household structures create problems in identifying a single *head of household* to whom the authorities can issue an assistance check. In addition to the normal reluctance to seek mass shelter and housing, some disaster victims hesitate to approach authorities because they have no immigration documents (Yelvington, 1997).

Sites for temporary housing also include homes of friends and relatives, rental houses and apartments, and trailer parks. Some of these sites are in or near the stricken community. Others are hundreds or even thousands of miles away. Lack of alternative housing within an acceptable distance of jobs leads households to leave the area. The population loss after Hurricane Andrew was 18% in South Dade County, 33% in Florida City, and 31% in Homestead (Dash, Peacock, and Morrow, 1997). Other households remained in severely damaged units—or even condemned units—without electric power or telephone service for months (Yelvington, 1997).

The loss of housing in a disaster can be extremely problematic in a tight housing market. After Hurricane Andrew, housing availability dropped to 1.6% from 5.5% a year earlier. The housing scarcity drove up rents by 15 to 20%, which priced low-income victims out of the market (Yelvington, 1997). Even when victims find temporary housing, they can still take a long time to return to permanent housing. In one working class neighborhood after Hurricane Andrew, the average length of displacement was 95 days. The percent of returnees was still only 62% nearly a year after the disaster (Morrow, 1997).

Households encounter many problems during reconstruction. One problem is high prices for repairs. Another problem is poor quality work. Contract breaches also occur (Bolin, 1993). The rebuilt structures do benefit from improved quality and hazard resistance. This is especially true for public housing (Morrow, 1997). However, few victims think the improvements are worth the inconvenience they experienced.

Lower income households tend to have higher hazard exposure. This is because they live in more hazard-prone locations. They have higher physical vulnerability. This is because they live in structures that were built according to older, less stringent building codes. The builders used lower quality construction materials and methods. The homes have not been well maintained (Bolin and Bolton, 1983). Because lower income households have fewer resources for recovery, they also take longer to transition through the stages of housing. Sometimes they remain for extended periods of time in severely damaged homes (Girard and Peacock, 1997). They are sometimes forced to accept temporary housing as a permanent solution (Peacock, Killian, and Bates, 1987). There might still be low-income households in temporary housing even after high-income households all have relocated to permanent housing (Berke, Kartez, and Wenger, 1993; Rubin et al., 1985).

11.3.2 Economic Recovery

Insurance coverage varies by hazard agent. Some studies have shown 86% coverage for a tornado (Bolin and Bolton, 1986) and 25% coverage for an earthquake (Bolin, 1994). Risk area residents are likely to forego earthquake insurance because they consider premiums to be too high and deductibles too large (Palm, Hodgson, and Blanchard, 1990). Income, education, and occupational status are all correlated with earthquake insurance purchase (Bolin, 1993).

Households lacking adequate insurance coverage must use other strategies for coping with their losses. These include (Bolin 1993):

- Obtaining small business administration (SBA) or commercial loans.
- Seeking Federal Emergency Management Agency (FEMA) or NGO grants.
- Withdrawing personal savings.
- Failing to replace damaged items.

SBA loans are difficult for households because they involve long-term debt that takes many years to repay (Bolin, 1993). FEMA grants are also difficult because they require households to meet specific standards. One requirement is proof from the victims that they are residents of the disaster impact area. However, there can be problems in registering people who evacuated or were rescued without identification (Yelvington, 1997). Relaxing the standards for loans and grants might seem more humane, but this would create other problems. It would allow the chronically homeless and out of area construction workers to obtain services intended only for disaster victims. In turn, the resulting resentment toward "freeloaders" could generate pressure to reduce services for legitimate victims.

The economic recovery of some households takes place quickly, but others take a long time to recover. For example, 50% of households affected by the Whittier earthquake reported complete economic recovery at the end of the first year. However, 21% reported little or no recovery even at the end of four years (Bolin, 1993). Those who had a larger income and greater savings recovered more quickly. Those who had a large family or badly damaged home or had to move more times took longer to recover (Bolin, 1993). In some cases, the problems in economic recovery are due to the loss of permanent jobs that are replaced only temporarily or not at all (Yelvington, 1997).

There are differences in the rate of economic recovery among ethnic groups. For example, African-American households (30%) lagged behind Caucasian households (51%) in their return to preimpact economic conditions eight months after the 1982 Paris, Texas tornado (Bolin and Bolton, 1986). However, the factors affecting recovery were similar for both ethnic groups. Economic recovery was faster for households with fewer members, higher income, and more years of education. Those with higher incomes were also more likely to have insurance. In addition, African-American household recovery was negatively related to primary group aid and the number of household moves. Larger households were less likely to use disaster assistance, have adequate insurance, or receive adequate aid (see Figure 11-2). Larger households made more moves. Higher socioeconomic households made fewer moves. The overall effect is that large poor households are doubly handicapped in their economic recovery.

11.3.3 Psychological Recovery

Few victims develop major psychological problems from disasters. Instead, most people experience mild distress. For example, Bolin and Bolton (1986) found negative impacts such as lack of patience (38%) and strained family relationships (31%) after the Paris, Texas tornado. However, victims also experienced positive impacts. These included strengthened family relationships (91%), decreased importance of material possessions (62%), and increased family happiness (23%).

Researchers have also examined public records in their search for psychological impacts of disasters. However, there were no long-term trends of births, marriages, deaths, and divorce applications due to Hurricane Andrew (Morrow, 1997). Domestic violence rates remained constant for about six months after the hurricane but did increase about 50% for nearly two years after that. In all, only 12% of households affected expressed a need for counseling (Morrow, 1997). Most victims simply experience the normal grieving process for the losses resulting from the disaster. However, they may face many frustrations during the recovery process. Many have to interact repeatedly with public (governmental) and private (e.g., insurance companies) bureaucracies. To help people cope with the aftermath of disaster, you should work with local mental health agencies to ensure an organized mental health referral system will be available. This system should give extra attention to:

- ▲ People with preexisting mental conditions.
- ▲ Those who have witnessed the death or severe injury of loved ones.
- ▲ Single female heads of household.
- ▲ Children.
- ▲ Emergency responders involved in difficult search and rescue operations.
- ▲ Medical personnel handling heavy work loads.

Perhaps the best summary of psychological recovery is "the recipients of [mental health] services are normal people, responding normally, to a very abnormal situation" (Gerrity and Flynn 1997, p. 108). The majority of victims and responders recover quickly from the stress of disasters without help. Those who lose their homes and loved ones are likely to experience the most distress. The appropriate strategy for helping victims and first responders deal with their grief is not to intervene in a heavy-handed way. Disaster recovery specialists should provide information about sources of material support. Mental health professionals should facilitate victims' involvement in social and emotional support groups.

11.3.4 Sources of Household Recovery Assistance

Households can be characterized in terms of three modes of disaster recovery, although few households fit exclusively into a single category (Bolin and Trainer, 1978):

- ▲ Autonomous recovery.
- ▲ Kinship recovery.
- ▲ Institutional recovery.

In autonomous recovery, a household receives no help from others. The success of this recovery mode depends on the household's available human, material, and financial resources. Human resources are available to the extent the household members have come through the disaster alive, uninjured, and with confidence in their ability to recover. In addition, autonomous recovery depends on the degree to which members can continue to generate income from employment, rental of assets, and interest or dividends from financial assets. Moreover, it depends on the degree to which material resources are available. This includes the extent to which its possessions are undamaged or can be restored. Autonomous recovery also depends on being able to receive adequate compensation from insurance. In some cases, autonomous recovery also depends on the degree to which creditors accept delayed payments on liabilities such as loans, mortgages, and credit card debt. Finally, it depends on members reducing consumption.

Kinship recovery relies on the help of friends and family. Kinship recovery depends on other family members' willingness to help the disaster victims. In addition, it depends on other family members' ability to help. The ability to help depends on what assets other family members have and their ability to share them. Obviously, kinship recovery is more difficult when other family members are poorer or when they live far away from the victims.

Institutional recovery relies on governmental and nongovernmental organizations. Such organizations are constrained by organizational policies that determine whether victims qualify for aid and, frequently, if they can prove their claims of disaster losses. Significantly, institutional recovery also depends on victims' ability to devote the time required to travel to assistance centers and prepare and process applications for assistance.

Table 11-1: Household Recovery Problems (by Ethnic Group)

Problem perceived to be large	*Anglo-Americans*	*African-Americans*	*Hispanics*	*Total sample*
Dealing with mortgage companies about insurance money	68	49	68	64*
Dealing with building inspectors	52	38	76	63*
Living in damaged home	59	63	59	60
Neighborhood conditions	55	60	39	47*
Living in temporary quarters	45	61	38	46*
Dealing with insurance companies	33	26	48	40*
Dealing with contractors	38	18	45	37*
Unemployment	11	29	30	25*
Household finances	14	40	20	22*
Neighborhood crime	34	23	16	22*
Transportation	2	28	17	16*
Job relocation	7	21	17	15
Dealing with agencies	11	20	13	15
Behavioral problems with children	19	18	10	14
Family violence	17	11	5	9*
Gain of member(s)	14	0	4	5*
Loss of member(s)	4	0	13	4

Adapted from Morrow, 1997.

*Difference between highest and lowest percentage statistically significant at $p < .05$.

Some aspects of household recovery are relatively similar across ethnic groups, but others reveal distinct differences. For example, Table 11-1 shows Anglo-American, African-Americans, and Hispanics experienced similar levels of frustration in some areas. However, most of these commonalities were for problems that did not occur often. These are listed at the bottom of the table. By contrast, there were significant differences in frequent problems. For some of these problems, the Anglo-Americans reported the greatest frequency of frustration.

FOR EXAMPLE

Recovery Takes Time

Households must invest time to obtain the resources they need to recover. This includes time to find and purchase alternate shelter, clothing, food, furniture, and appliances to support daily living (Yelvington, 1997). Time is also needed to file insurance claims, apply for loans and grants, and search for jobs. The time required for these tasks is increased by multiple trips to obtain required documentation (Morrow, 1997). FEMA provides telephone registration, but its value can be undercut if there is a loss of telephone service after a disaster or if there are not enough operators to avoid long waits. There will also be increased travel time if cars, street signs, traffic signals, and landmarks are destroyed. Public transit may not be available for weeks. Adding to the time burden is increased cost for many items due to supply scarcities. Finally, victims need skills and self-confidence to cope with the disaster assistance bureaucracy (Morrow, 1997).

For other problems it was the Hispanics that experienced the greatest frustrations. In general, however, African-Americans had the highest level of frustration with more problems than either of the other two groups.

SELF-CHECK

- Explain why lower-income households tend to have higher hazard exposure.
- Name the other strategies that households who lack adequate insurance coverage must use to cope with their losses.
- Name the three modes by which households recover from disasters.
- Identify the similarities and differences across ethnic groups in household recovery from disasters.

11.4 Business Recovery

Several studies have examined the ways in which individual businesses prepare for, are disrupted by, and recover from these disasters. Older, larger, and more financially stable businesses are more likely to adopt hazard adjustments. So are businesses in the manufacturing, professional services, and finance, insurance and real estate sectors.

Disasters disrupt business operations in a variety of ways. It is easy to understand how direct physical damage can force businesses to shut down. In addition, disruption of infrastructure such as water/sewer, electric power, fuel, transportation, and telecommunications also forces businesses to shut down.

Small businesses are more vulnerable than large businesses for two important reasons. First, they are more likely to be located in nonengineered buildings that become damaged. Second, they are less likely to have hazard management programs to reduce this physical vulnerability. Thus, in this respect, small businesses are equivalent to the most physically vulnerable households—ones that are poor, female-headed, or members of ethnic minorities. Small businesses also face increased costs to repair structures and replace contents. At the same time, these businesses could lose customers if they relocate. Three years after the Whittier earthquake, 50% of destroyed commercial space and 100% of damaged commercial space had been replaced (Bolin, 1993). In the meantime, however, a number of businesses were forced to relocate. These businesses were located in the old central business district and that was mostly unreinforced masonry structures. Because Whittier is located in within the Los Angeles metropolitan area, local residents could readily obtain the goods and services they needed from undamaged businesses in adjacent communities. Thus, by the time the damaged space was available for reoccupancy, it was leased by new tenants. The former tenants did not have the resources to wait that long to reopen.

A disaster can affect businesses in even more subtle ways. One indirect effect is population dislocation. If many households move away, this decreases the revenues of the community's businesses. Even if households remain in the community, a disaster affects their discretionary income. This loss of discretionary income can weaken the demand for many products and services. People will spend more money on building supplies and less on movies and restaurants. Disasters can also increase competitive pressure from large outside businesses that recognize a major new market for reconstruction materials. These factors cause many small local businesses to fail in the aftermath of a disaster, especially if they were only marginally profitable beforehand (Alesch & Holly, 1996; Alesch et al., 2001). Indeed, businesses can produce business failures long after the event. This is especially true if the community was already in economic decline before the event (Bates and Peacock, 1993; Durkin, 1984; Webb, Tierney, and Dahlhamer, 2002).

There are also differences among business sectors in their patterns of recovery. Wholesale and retail businesses report significant sales losses. However, manufacturing and construction companies often show gains following a disaster (Durkin, 1984; Kroll et al., 1990; Webb, Tierney, and Dahlhamer, 2000). Moreover, businesses that serve a large market tend to recover more rapidly than those that only serve local markets (Webb et al., 2002). Small businesses have been found to experience more obstacles than large firms. Compared to their large counterparts, small firms are more likely to depend on local customers. They also lack the financial resources needed for recovery. Finally, small businesses lack access to governmental recovery programs (Alesch and Holly, 1996; Alesch et al., 2001; Dahlhamer and Tierney, 1996, 1998; Durkin, 1984; Kroll et al., 1990). Thus, business sector and business size can be seen as indicators of operational vulnerability. These indicators are equivalent to the demographic indicators of social vulnerability in households.

Businesses' hazard vulnerability explains the changes a disaster causes in production, sales, and profits. In particular, four cases can be used to illustrate firms' variation in their postdisaster sales levels. The first case is defined by businesses in the impact area that have minimal hazard vulnerability. Professional services are a likely example. These businesses experience only small decreases in sales after disaster impact and return quickly to their predisaster levels. The second case consists of businesses that also are in the impact area, but have moderate vulnerability. For example, large manufacturers experience a larger initial drop in sales levels and their recovery takes a longer time. Tourism oriented businesses may also suffer major initial losses. It will take them some time to return to their prior level of profitability because people are often afraid to visit the disaster area for a period of time.

The third case consists of businesses that experience initial sales losses because they are inside or near the impact area. However, they later experience an increase in demand for their products/services during the disaster aftermath. Recovery-related businesses in the building construction, materials, and hospitality industries tend to follow this pattern. The last group comprises recovery-related businesses located just outside the impact area. They avoid any initial losses because they are undamaged. In addition, they can reap gains in the aftermath of the disaster because they are close enough to the impact area to benefit from the reconstruction.

These principles can be seen in data from business recovery in two communities affected by Hurricane Andrew (Dash et al., 1997). Homestead had a larger population, a higher per capita income, and a higher average home value than Florida City. Homestead had population that was 42% Anglo-American and 35% Hispanic. Florida City was 61% African-American and 37% Hispanic. Florida City is slightly farther from the point at which the hurricane eye made landfall. However, there was no initial difference in the hurricane's impact on the two city's businesses. The overall commercial property loss after the hurricane was 29% in Homestead. It was 32% in Florida City. Table 11-2 describes the business impacts of the hurricane.

Overall, there were significant differences in the two communities over the next year. For example, total sales volume declined 83% in Florida City

Table 11-2: Changes in the Number of Businesses, Employees, and Sales Volume after Hurricane Andrew

	Businesses change (%)		*Employees change (%)*		*Sales volume change (%)*	
Industry	Florida City	Home-Stead	Florida City	Home-Stead	Florida City	Home-Stead
Agriculture	−71	+4	−92	+74	−93	+66
Construction	0	−20	+12	−20	+12	−59
Manufacturing	0	−12	−67	−19	−59	−32
Transportation/ communication	−50	+9	−100	+4	−26	+51
Wholesale trade	−60	−4	−50	+6	−84	+57
Retail trade	−64	−2	−84	+16	−84	−5
Finance/ insurance/real estate	−20	0	−59	−1	−32	−32
Business services	−63	+6	−94	−5	−65	−14
Professional services	−45	−3	−73	+16	−69	+1
Public administration	−50	+38	−69	+7	n/a*	n/a*

Adapted from Dash et al., 1997.

*Sales volume is not applicable to public sector organizations.

but only 1.1% in Homestead. However, inspection of Table 11-2 reveals that there are distinct differences from one industry to another. The impact depends on whether one examines the change in the number of businesses, the number of employees, or sales volume. For example, Florida City shows dramatic declines for agriculture on all three indicators. However, there is no change or even modest increases in construction. By contrast, Homestead showed a slight increase in the number of agricultural businesses. It shows significant increases in the number of agricultural jobs and sales volume. Moreover, it experienced significant declines for all three indicators in construction. This is almost the opposite pattern of Florida City. These differences in business impacts indicate that local authorities should be aware of the health of local businesses before a disaster strikes. They should also monitor these businesses' economic status in the disaster's aftermath to determine if government intervention is needed.

FOR EXAMPLE

Business Closings Due to Service Interruptions

Tierney (1997) reported that extensive lifeline service interruption after the 1993 Midwest floods caused a large number of business closures in Des Moines, Iowa. The closings occurred despite the fact that the physical damage was confined to a small area.

SELF-CHECK

- Discuss the subtle ways in which a disaster can affect businesses.
- Identify the differences among business sectors in their patterns of recovery.

11.5 The Role of State and Federal Governments in Recovery

State and federal agencies can play significant roles in disaster recovery. However, the burden most frequently falls on local governments. This is because only about 1% of all disasters receive presidential disaster declarations (PDD). Local governments should prepare to undertake a variety of functions during a disaster recovery process. They must understand that they might not receive any aid from higher levels of government for minor disasters. In smaller disasters, the state governments can still play significant roles in assisting the local governments. In large events, the PDD opens up a broad range of programs for relief and reconstruction. In these events, the state plays a coordinating role between federal and local governments. The additional funding a PDD can bring provides the federal government with a major opportunity to influence state and local behavior during the recovery period. Disaster response may be mostly over before the PDD is granted. However, federal assistance is certainly welcome when it finally arrives. The Recovery Function Annex of the Federal Response Plan of January 2003 is available on the Department of Homeland Security (DHS) website (www.dhs.gov/dhspublic) and lists 71 federal disaster recovery programs that are administered directly by the DHS or by dozens of other federal and volunteer organizations.

The lead recovery agency at the federal level is FEMA, which was placed in the Department of Homeland Security in 2002. Other federal agencies might be called upon when a PDD is granted. These agencies include the Small Business

Administration, the U.S. Army Corps of Engineers, and the Economic Development Administration. Each of these agencies has specific disaster recovery programs that it funds.

The National Response Plan provides for the establishment of disaster field offices (DFOs) in the vicinity of the disaster. Emergency response teams (ERTs) are located in the DFOs. These include an operations section, which coordinates federal, state and voluntary efforts. The ERT operations section has a human services branch that is responsible for many tasks including:

- ▲ Needs assessment.
- ▲ Establishment of disaster recovery centers.
- ▲ Initiation, coordination, and delivery of recovery programs authorized by the Stafford Act.
- ▲ Managing DHS and state grant programs.

Finally, there is an infrastructure support branch that deals with restoration of public utilities and other infrastructure services. There is also a deputy field coordinating officer for mitigation. This officer coordinates with the infrastructure support branch to promote mitigation and preparedness activities.

The main types of programs providing recovery assistance are the individual assistance, infrastructure support, and hazard mitigation grant programs. Individual assistance is available to households through the Temporary Housing Assistance program, individual and family grants, disaster unemployment assistance, legal services, special tax considerations, and crisis counseling programs. Individuals and businesses can receive aid through the SBA Disaster Loans program. This program can provide help with repairs to housing, businesses, and economic losses. In the past, many loan programs have been inaccessible to low-income households. Low-income households tend to rent rather than own their housing. They fail to qualify for loans because of their low incomes and lack of collateral. The Individual and Family Grant program was intended to fill the need for those whose needs were not being met by the SBA loan program, private insurance, or NGO assistance. However, the amounts awarded tend to be small.

Public assistance programs are offered through the infrastructure support branch. They are targeted at state and local governments, certain nonprofit organizations that provide emergency services, and Indian tribes. These programs support the repair or replacement of public facilities damaged by disaster.

Assistance provided has increased in importance since the passage of the Disaster Mitigation Act of 2000 (DMA 2000). This legislation requires local governments to identify potential mitigation measures that could be incorporated into the repair of damaged facilities. They must do this to be eligible for pre- and postdisaster funding. This policy encourages local government to engage in

FOR EXAMPLE

FEMA's Performance during Hurricane Katrina

Due to FEMA's performance during Hurricane Katrina in 2005, which was widely considered as unacceptable by the public and government leaders, several politicians have called for FEMA to be completely dismantled and have suggested that a new organization take its place. Some have argued that FEMA should not be dismantled but be a separate entity removed from the Department of Homeland Security, as it was prior to the creation of Homeland Security. As of this writing, it is unclear as to what the future of FEMA will be.

mitigation activities such as hazard mapping, planning, and development of building codes. Other activities supported by DMA 2000 include development of training and public education programs, establishing reconstruction information centers, and assisting communities to promote sustainable development.

State governments vary widely in the level of attention and resources they devote to planning for and implementing recovery. Some states have established programs that provide assistance to households and local governments for recovery if they do not receive a PDD. Some states have created state disaster funds to provide this assistance. States also designate departments to provide help. States can fund these programs through the creation of state disaster funds. However, only about half the states have done so. Typically, state legislatures have appropriated funds after disasters on the basis of need. Another type of disaster fund is a disaster trust fund. This creates revenue by dedicating a percentage of sales taxes or other revenues to the fund.

SELF-CHECK

- Name the tasks that the human services branch of the ERT operations section is responsible for.
- Explain why many loan programs have been inaccessible to low-income households.

11.6 The Role of Hazard Insurance

Hazard insurance is a preimpact recovery preparedness action. Theoretically, it could completely replace current programs of disaster relief if everyone paid insurance premiums according to their homes' hazard exposure and structural vulnerability. In addition, hazard insurance could decrease government workload and expense. It would do this by shifting part of the administrative burden for evaluating damage to insurance companies in the private sector. Finally, hazard insurance defines the terms of coverage in advance. This would reduce opportunities for politicians to increase benefits after disasters. The desire to appear to be generous creates a temptation to vote for "pork barrel" projects. The problem is that generous aid for uninsured victims angers those who had the foresight to purchase insurance in advance. This causes people not to want to purchase hazard insurance.

Unfortunately, the potential contribution of hazard insurance remains to be fully achieved. There are many difficulties in developing and maintaining a sound hazard insurance program. The National Flood Insurance Program has made significant strides over the past 30 years. However, it continues to require operational subsidies. One of the basic problems is that those who are most likely to purchase flood insurance are, in fact, those who are most likely to file claims (Kunreuther, 1998). This problem of *adverse selection* makes it impossible to sustain a market in private flood insurance. The federal government has tried to solve this problem. They have required flood insurance for structures located in the 100-year flood plain that are purchased with federally-backed mortgages. Unfortunately, homeowners frequently allow their policies to lapse after the first year. The program has no effect on those who purchase their homes without a mortgage or have paid off their mortgages.

FOR EXAMPLE

Hazard Insurance and Disaster Recovery

The lack of universal hazard insurance creates significant differences in disaster recovery. Some homes are rebuilt soon after a disaster because their owners have high quality insurance coverage. Other homes take much longer because they are only partially insured. In some cases, the homeowners lack *any* insurance. They may not be able to afford quality insurance or were denied access to it because of racial discrimination (Peacock and Girard, 1997).

SELF-CHECK

- Describe hazard insurance.
- Name some problems with flood insurance requirements.

11.7 Local Government Recovery Functions

After a disaster, local government needs to perform many tasks very quickly, and many of these must be performed at the same time. This makes it just as important to plan for disaster recovery as for emergency response (Schwab et al., 1998). There is almost never a clearly defined line between response and recovery. Some sectors of the community might still be in response mode while others are already into recovery. Some organizations will be carrying on both types of activity at the same time. This means that there is little time to plan for recovery after the response has begun. By planning for recovery before disaster strikes, resources can be allocated more effectively and efficiently. This increases the probability of a rapid and full recovery. A lack of planning can also increase the probability of conflicts arising due to competition over scarce resources.

Local government must perform specific tasks during disaster recovery. Some of these tasks involve restoring services it performed before the disaster. In addition, local government must rebuild any critical facilities that were damaged or destroyed. Finally, local government must perform its regulatory functions regarding land use and building construction. During recovery, these two functions require rapid action under a heavy workload. Special provisions are required to expedite the procedures for reviewing and approving the redevelopment of private property.

In preimpact recovery planning, a community must overcome three major misconceptions about recovery. The first misconception is that the recovery can be improvised after the response is complete. In fact, a timely and effective disaster recovery requires a significant amount of data collection and planning. Postponing data collection and planning until after the response is over delays recovery. The second misconception is that there is ample time to collect data and plan the recovery during emergency response. It is true that some recovery-relevant data must be collected during the response. However, an assessment of "lessons learned" from the disaster impact should be used to guide a recovery process that has already been designed before the disaster strikes. Finally, the third misconception is that the objective of recovery should be to restore the community to the conditions that existed before the disaster. As noted earlier, this simply reproduces the community's existing disaster vulnerability.

You should establish a recovery/mitigation committee before disaster strikes. The committee should assign each recovery function to a specific organization, develop a ROP, and acquire any necessary resources. Finally, the committee should conduct the training and exercises needed to ensure the ROP can be implemented effectively.

11.7.1 The Recovery/Mitigation Committee

The recovery/mitigation committee can be an important part of the recovery process. It should be established before a disaster. Personnel who are designated to serve on this committee should include a chairperson and a lead agency. The lead agency should probably be the local planning department. The planning process should begin when the community's CAO publishes a planning directive. The recovery/mitigation committee chairperson should establish a planning schedule. Many government agencies should participate in the committee. For example, the directors of planning, building construction, and public works should be included. In addition, you should include representatives from utility companies, businesses, churches, charities, and neighborhood groups.

The committee should use the community hazard vulnerability analysis to identify the locations within the community that have the highest hazard exposure. The committee should work with the rest of the community to formulate a vision of the type of disaster recovery it intends to implement. Next, the committee should develop a ROP. The plan should integrate the likely disaster impacts, community goals, and public and private sector capabilities within the community. In addition, the ROP should identify external sources of assistance and their loan/grant requirements. These should be integrated into a comprehensive disaster assistance program. The committee should develop a financial plan for responding to the disaster. Business interruption caused by a disaster decreases the community's tax revenues. Finally, the committee should establish agreements with NGOs and CBOs for support in recovery. These groups provide financial and in kind support, as well as legal and technical assistance. After a disaster strikes, the committee should ensure that organizations implement the ROP.

11.7.2 Envisioning a Community Recovery Strategy

The recovery/mitigation committee needs to work with the community to develop a shared vision of disaster recovery. The short-term recovery following a major disaster can generate an economic boom. State and federal money flows into the community to reconstruct damaged buildings and infrastructure. These funds pay for construction materials and the construction workforce. To the extent that the materials and labor are acquired locally, they generate local revenues. In addition, the building suppliers hire additional workers. The workers spend their wages on places to live, food, and entertainment. Unless there are undamaged towns nearby, this money is spent within the community.

Communities must also consider the long-term economic consequences of disaster recovery. What will happen after the reconstruction boom is over? They can attract new businesses if they have a skilled labor pool and good schools. Other assets include low crime rates, low cost of living, good housing, and environmental amenities such as mountains, rivers, or lakes (Blakely, 2000). Communities can also enhance their economic base if they can attract businesses that are compatible with the ones that are already there. Such firms can be identified by asking existing firms who are their suppliers and distributors. These new firms might be attracted by the newer buildings and enhanced infrastructure.

If a disaster stricken community does not already have such assets, they can invest in four components of economic development—locality development, business development, human resources development, and community development. Locality development involves enhancing a community's existing physical assets. This can be done by improving roads or establishing parks on river and lakefronts. Business development involves efforts to retain existing businesses or attract new ones. Although it is not easy, this can be accomplished by working with businesses to identify their critical needs. In some cases, this might involve establishing a business incubator that allows start-up companies to obtain low cost space and share meeting rooms. Human resources development involves the development of a skilled workforce. Finally, *community development* involves using nongovernmental organizations, community-based organizations, and local firms that hire residents of the community whose incomes are below the poverty level.

11.7.3 The Role of Nongovernmental Organizations and Community Based Organizations

The role of groups such as the American Red Cross and Salvation Army is widely publicized. The role of local churches and service organizations is increasingly recognized. All of these organizations provide housing, food, clothing, medicine, and financial help to disaster victims. The *existing* government social service agencies are supplemented by nongovernmental organizations. NGOs *expand* their membership to perform the tasks they are expected to perform during disaster recovery (Dynes, 1974). By contrast, existing community-based organizations *extend* themselves beyond their normal tasks to perform novel activities. In addition, there are situations in which organizations cannot successfully meet the recovery needs of disaster victims. In such cases, government agencies, NGOs, and CBOs form an **unmet needs committee.** This is an *emergent* organization that is designed to serve those whose needs are not being addressed.

FOR EXAMPLE

Bringing Businesses Back

Some businesses might be reluctant to move into a disaster impact area, or even to a community that has been affected by disaster. A program for developing small businesses, affordable housing, community health clinics, and inexpensive child care can help to eliminate some of what new businesses might consider to be the risks of relocating to the community.

SELF-CHECK

- List several tasks that local government must perform during disaster recovery.
- Explain the purpose of a recovery/mitigation committee.
- Identify what the American Red Cross, Salvation Army, and local churches often provide after a disaster.
- Explain the function of an unmet needs committee.

11.8 Developing a Recovery Operations Plan

The demands of disaster recovery imply that specific functions be performed. Four principal disaster recovery functions are disaster assessment, short-term recovery, long-term reconstruction, and recovery management see Table 11-3).

11.8.1 Disaster Assessment

According to the disaster impact model, disaster assessment should include both physical and social impact assessment. Therefore, physical impact assessment should involve assessments of casualties and damage. However, casualty assessment is such an important task that it is performed during the emergency response. The social impact assessment should examine the psychological, demographic, and economic impacts of disaster. These can all be addressed together in *victims' needs assessments*. Finally, community agencies need to determine if they need to make changes to the ROP, building code, or other local ordinances. A report that describes the "lessons learned" from the disaster can guide these changes.

Table 11-3: Disaster Recovery Functions

Disaster assessment	
Rapid assessment	Victims' needs assessments
Preliminary damage assessment	"Lessons learned"
Site assessment	
Short-term recovery	
Impact area security and reentry	Emergency demolition
Temporary shelter/housing	Repair permitting
Infrastructure restoration	Donations management
Debris management	Disaster assistance
Long-term reconstruction	
Hazard source control and area protection	Infrastructure resilience
Land-use practices	Historic preservation
Building construction practices	Environmental remediation
Public health/mental health recovery Economic development	Disaster memorialization
Recovery management	
Agency notification and mobilization	Public information
Mobilization of recovery facilities and equipment	Recovery legal authority and financing
Internal direction and control	Administrative and logistical support
External coordination	Documentation

Damage Assessment

There are three basic types of damage assessment, and the first two of these are also done during emergency response (FEMA, 1995). The first type, rapid **assessment**, identifies the areas affected by the disaster. It also assesses the severity of the physical impacts so you can determine the need for lifesaving activities. Rapid assessment should be completed within one to three hours after impact. This allows you to determine where there are buildings requiring search and rescue operations. It also allows you to determine if there is a potential for secondary hazards. Rapid assessment also provides information about the status of infrastructure and critical facilities. A rapid assessment is performed by available police, fire, and public works personnel. Additional data can be provided from private sector organizations that own or operate lifelines and critical facilities.

Preliminary damage assessment is the second type of assessment. It produces counts of destroyed, severely damaged, moderately damaged, and slightly damaged structures. This level of assessment should be completed within 3 to 4 days. The data from the preliminary damage assessment is used to support requests for state and federal disaster declarations. A preliminary damage assessment is performed by having local government personnel perform a *windshield survey.* Inspectors do this by driving along all of the streets in the impact area. As the name suggests, they do not get out of their cars unless roads are blocked. Inspectors tally counts of damaged structures, with residential structures being classified by income levels and structural categories. Buildings can then be given a red, yellow, or green tag depending on the level of damage and occupant safety, with red-tagged buildings being unsuitable for occupancy. A preliminary damage assessment should also include estimates of percentages of households with insurance coverage.

Finally, a **site assessment** is meant to produce detailed estimates of the cost to repair or replace each affected structure. This information is used to support requests for federal assistance to the owners of the damaged property. It includes estimates of losses to residential, commercial, industrial and public property. Site assessments require technically trained personnel for multi-story structures such as apartment buildings. These include architects, structural engineers, and building inspectors who can usually be drawn from city staff. Additional technical personnel might be recruited from other local organizations or called in from outside the community. Skilled construction professionals can be supplemented by volunteers who can conduct site assessments for most single family residences if they have been trained in the use of well designed checklists. A site assessment might take weeks to complete. These methods of damage assessment can be compared to the procedures of cost estimation that are used in routine construction projects, as shown in Table 11-4.

Table: 11-4: Types of Post-Disaster Damage Assessments

Damage assessment	Routine construction cost estimation
Rapid damage assessment	
Preliminary damage Assessment	
Site assessment ⟶	Preliminary cost estimate
	Detailed cost estimate

In preparing for the necessary damage assessments, staff from local government departments should be assigned to damage assessment teams (DATs). All DAT members should be trained in a common assessment procedure. This accelerates the process and generates results that are comparable across all DATs within the community.

Victims' Needs Assessment

The psychological, demographic, and economic impacts of disasters can be evaluated using a **victims' needs assessment.** Preparation for victims' needs assessments should begin during the preimpact recovery planning process. The first step is to identify the community's vulnerable population segments as part of the social vulnerability analysis. These may be defined as specific locations and neighborhoods or types of households and businesses. The local jurisdiction should assign staff to victims' needs assessment teams (VNATs) and supplement them with staff from other organizations. These supplementary staff should be assigned by contract with NGOs and CBOs. They should be trained together with the government staff in methods of victims' needs assessment.

The lower the savings rate, the higher the need for public assistance. Unfortunately, the savings rate in the U.S. has been extremely low for the past decade. Consequently, VNATs should be prepared to find large numbers of households and businesses needing assistance. In addition to housing needs, VNATs should also be prepared to identify households' needs for employment. Households will also need other economic assistance and have psychological needs. VNAT team members need preimpact training. VNAT team members need to know the availability of local, state, federal and NGO disaster recovery programs. In turn, this enables them to accurately diagnose victims' needs and refer them to the appropriate recovery programs.

Lessons Learned

The recovery/mitigation committee must establish evaluation procedures to ensure lessons are learned *and applied* to improving the community's resilience. Therefore, it should establish a "lessons learned" subcommittee. The subcommittee should establish procedures for studying the event. The recovery/mitigation committee should use the assessment to determine how the community should modify its land-use plan, building code, and other community operations. Other issues to be considered should include infrastructure location and replacement, as well as the capital improvements program. The delivery date of the report should be set early in the recovery process, perhaps 30 days after the disaster. That way, its recommendations can be incorporated into the recovery process.

11.8.2 Short-Term Recovery

The first benefit of the disaster assessment is to guide the short-term recovery. This comprises eight tasks that are performed in the immediate aftermath of a disaster to take care of community members' immediate needs and to prepare for long-term reconstruction.

Impact Area Security and Reentry

Security must be maintained in the impact area. This is to ensure that residents do not return before it is safe to do so. This is also to assure people that their property is being protected from looting. You must have procedures for residents' reentry. There is a need to provide for temporary reentry to remove essential items. There is also a need for permanent reentry for people to return to live in their homes. In both cases, conditions must be safe enough to allow people to enter. This might require that you demolish severely damaged buildings and remove heavy debris. In addition, proper identification is needed to assure that only residents or authorized reconstruction personnel are allowed in. Finally, you must establish basic criteria before people can return home. Functioning transportation and sewer systems are especially important criteria for reentry. It is possible to allow people to return before electric power is available because some people have their own generators. Whatever criteria your community sets, they should be established ahead of time. If the disaster has had a regional impact, reentry should be coordinated with nearby communities.

Temporary Shelter/Housing

The majority of evacuees prefer to stay temporarily in the homes of friends and relatives. Among those whose friends and relatives are either too far away or are themselves victims, the more affluent choose commercial facilities. Poorer households, usually 10-25% of the evacuees, tend to stay in mass care facilities (Mileti et al., 1992).

Mass care facilities must accommodate differences due to age, ethnicity, and physical limitations. Mass care facilities make it difficult to accommodate household differences. These differences include behaviors such as personal sanitation, privacy, child rearing, and hours and loudness of social interaction. Mass care facilities also place increased demands on time for other tasks. This reduces time for childcare and can result in loss of control over children. Lack of personal space and privacy consistently generate tensions among those in these facilities (Yelvington, 1997). Operation of mass care facilities can be complex after major disasters. In such cases, there will be a need for many multilingual volunteers to assist in multiethnic communities. You need enough people to provide continued staffing. There are likely to be thousands of volunteers in the first few weeks. However, there are likely to be dramatic drops in volunteerism after the second week (Yelvington, 1997). Crowding and stress make it important to maintain transparency in making decisions about shelter operation. You need to establish procedures for coping with predisaster homeless people, construction workers, and others who do not qualify for housing (Bolin, 1993).

There is a movement from temporary shelter to temporary housing. "Doubling up" eventually causes friction in interpersonal relationships. Commercial facilities are a drain on family finances. Mass care facilities are crowded, noisy, and lack the privacy to which people are accustomed. The number of displaced households may be less than the vacancy rate for affordable housing within commuting time of jobs. When this happens, the existing housing market can accommodate the relocation. If the rental rates are high, or if they are far away from jobs, government can increase temporary housing by bringing in mobile homes.

The ROP should recognize that the need for temporary housing grows in importance when there is a limited amount of affordable housing. The number of displaced households will be compounded by those evicted from undamaged homes because they lost their jobs and could not make payments. In a major urban area exposed to large scope disasters, temporary housing could be provided by thousands of mobile homes. If trailer parks are established, local officials should try to reduce social friction by locating people close to friends and family.

Businesses also need temporary operating locations when their normal locations have been damaged or destroyed. Many small businesses have customers who are loyal enough to travel an extra distance. However, loyalty does have its limits. The government might need to allow temporary business operations in parking lots or other space that is close to the displaced businesses' normal locations. The ROP should also identify sites for temporary housing and businesses. They may be needed for as much as a year. Some of these businesses include hotels/motels and restaurants. These are needed to provide places where emergency workers and construction crews can live and eat while they are rebuilding damaged or destroyed structures.

Infrastructure Restoration

There are often many households and businesses that cannot resume normal functioning simply because of the lack of water or power. Thus, there is a need to inspect and repair any damage to pipelines and electric power lines. Inspection and repair might also be needed for streets, bridges, street signs, and lights. Critical facilities such as hospitals, police stations, and fire stations must be quickly repaired. However, a community's public infrastructure is also served by other facilities such as water treatment plants, transit bus barns, public works equipment yards, and government offices. There is also privately operated infrastructure that includes electric power stations, television and radio facilities, and telephone switching facilities. An inventory of these facilities should be available from the hazard vulnerability analysis.

Resources are limited during the recovery period, so choices must be made among conflicting priorities. There are likely to be conflicts among residential, commercial, and industrial segments of the community for priority in infrastructure restoration. Worse yet, there may be conflicts among residential neighborhoods, among commercial sectors, and among industries. This is the reason the preimpact recovery plan must include priorities. There should also be links to damage assessment procedures that allow the recovery managers to adapt the predetermined infrastructure restoration priorities to the needs of the situation.

Debris Management

Natural disasters and explosions can destroy many structures. This produces an enormous amount of debris that must be removed. Debris management should designate temporary sites for sorting recyclable from nonrecyclable materials. Nonrecyclable materials should be moved to permanent sites for disposal. Debris management is complicated when criminal evidence must be gathered systematically. You might need to do this for investigations of accidents or when the site is a possible crime scene. In such cases, debris removal is likely to be delayed. Temporary sorting sites are needed to separate out material evidence from debris.

Emergency Demolition

Some structures will be damaged severely enough to pose a threat of collapse. Procedures are needed to rapidly assess their stability. You need to determine if they should be reinforced and rebuilt or demolished. This assessment clearly requires competent structural engineering assistance. Historic preservationists should also be consulted if the building has cultural significance (Donaldson, 1998). This process is more efficient if historic structures have been surveyed and inventoried before disaster strikes. Postimpact damage assessment procedures should be developed to avoid unnecessary demolition of damaged historic structures (Kariotis, 1998; Kimmelman, 1998). The ROP should establish policies that include criteria for demolition of severely damaged structures. Owners should be adequately notified. In addition, the procedures should contain samples of the demolition contracts. These contracts require legal counsel to assure that the process respects personal property rights.

Repair Permitting

The ROP should contain criteria for determining which structures are eligible for reoccupancy. This is based on the percent of damage to the different elements of the building such as the foundation, wall, and roof systems, exterior and interior walls, floors, plumbing, electrical systems, and HVAC systems. The large number of requests for building repair permits can overwhelm a local code enforcement department (Schwab, et al., 1998). In preparation for this, the permit office staff should be able to call on staff from other communities and businesses. In addition, the ROP should establish a permit process that includes a 10-day moratorium on minor repairs. There should be a 30-day moratorium on permits for substantial repairs involving 50% or more of the pre-impact property assessment. This allows time for the city to acquire enough staff to evaluate the properties and areas involved. Of course, exemptions may be needed for reconstruction of critical facilities. The process should be streamlined as much as possible. The streamlined process should be continued for a limited time period that has been defined in the ROP.

ROPs should anticipate the possibility that developers will purchase many damaged homes to replace them with apartments. For example, one city established a five-month moratorium on applications for construction of new apartments. It also established restrictions on new buildings. This allowed a Design Review Board to exclude building designs that were incompatible with the existing neighborhoods.

Every disaster produces complaints about building contractors. Thus, the ROP should address the need to monitor and register out-of-area contractors. You may need to provide contract advice to owners of damaged property. Care should be taken to ensure that regulation of contractors does not prevent groups such as Habitat for Humanity from using volunteer labor.

Donations Management

Disasters produce an outpouring of goods from households and businesses outside the area. There is usually a substantial amount of useful material in these donations but there also is a lot of junk. Donations take personnel to sort through the donations. Even useful items must be sorted. Another problem with donations is that an influx of goods lessens the need to buy from local businesses, thus threatening their revenues. Thus, financial donations are preferable to material donations. Since material donations will inevitably arrive, you need to have procedures to manage them. You must establish a staging area outside the impact area. Here, donations can be received, sorted, and prepared for delivery.

Disaster Assistance

During recovery, people often need to contact multiple agencies within a short period of time. The large number of people visiting the agencies and the small number of staff results in long lines. In some disasters, these problems are made worse by offices moving from one location to another. It is important for you to provide "one-stop shopping" so victims can resolve all of their needs at a single location. It is also important that the location be accessible by public transportation. Additional staff must be recruited and trained to minimize victims' processing delays.

FOR EXAMPLE

Money Is Better

In the aftermath of a disaster, people donate all kinds of items. Donations of women's formal gowns have been recorded. After hurricanes in the South, donations of heavy wool coats were received. Even useful donations such as canned food must be sorted and labeled. New York City received so many items after the terrorist attacks of 9/11, that they stored the excess in several warehouses for future disasters. New York City still hasn't distributed all of these donated items.

11.8.3 Long-Term Reconstruction

A disaster opens a window of opportunity to change hazard management policy. The recovery/mitigation committee should have already assessed the community's hazard exposure, physical vulnerability, and social vulnerability. In addition, it should be well prepared with suggestions for ways in which to reduce future risks by integrating hazard mitigation into recovery (Schwab et al., 1998; Wu and Lindell, 2004). Finally, the committee should identify sources of funding for mitigation.

Hazard Source Control and Area Protection

The recovery/mitigation committee should have examined hazard source control and area protection mitigation strategies before a disaster strikes. Of course, these mitigation strategies are not feasible for some hazards. If these strategies are implemented, there may be additional growth in the protected area as people overestimate the protection these mitigation measures provide. You can avoid the increases in vulnerability by making changes in the land-use and building construction practices within the affected areas.

Land-Use Practices

Long-term reconstruction planning provides an opportunity for implementing some of the changes in land use policies that were developed during the preimpact recovery planning process. This is also a perfect time to reexamine the community's existing land-use plans. You can also use this time to pass new ordinances that will reduce hazard exposure. Alternative land uses can reduce the total population and property at risk. This can be accomplished through purchase of private property and development rights. Public facilities and other infrastructure should be located away from hazardous areas. Road width and access regulations might also need to be established or revised at this stage. Lot restrictions can be used to reduce population density. Landscaping and vegetation requirements can be established to reduce potential for flooding, landslides, or fires. The ROP should provide guidance on the reconstruction of **nonconforming uses**, which are structures that do not meet the zoning requirements for their geographic areas. Usually these are older structures that were constructed prior to the establishment of the current zoning requirements and, thus, are "grandfathered."

Building Construction Practices

The ROP should also address the implementation of new mitigation requirements such as elevating structures located in floodplains. Other portions of the building code can reduce the physical impact of a disaster on those structures located in risk areas. These include increasing disaster resistance of building structures and increasing the resistance of their "soft spots" such as windows. In addition to addressing new code requirements, the ROP should also address the building construction process. For example, it should address the regulation of out of area contractors to minimize the perennial problem of work that is paid for but never performed. Communities need to provide fair regulation of volunteer construction laborers to balance the legitimate interests of local contractors against the needs of the community for rapid provision of affordable housing for low-income residents (Peacock and Ragsdale, 1997).

Public Health/Mental Health Recovery

Most natural disasters in the United States have had minimal public health consequences. This is because the country has few endemic diseases that are likely to increase after a disaster. Contrary to many people's beliefs, dead bodies are a public health threat only if those who died had communicable diseases when they were alive. Death itself does not generate disease. Waterborne illnesses are a problem if survivors drink from, wash food in, or bathe in water sources that have been contaminated. Water may be contaminated by raw sewage or chemical spills. Of course, such exposures can be avoided by having survivors use bottled water or by evacuating the impact area until water is safe. Disease must also be controlled in areas where pests harbor diseases. For example, mosquito control has become increasingly important. Mosquito-transmitted diseases, such as West Nile virus, have become increasingly prevalent.

Natural disasters produce minimal mental health consequences. Few victims use formal psychological services after a disaster (Gist and Stolz, 1982). The two most prominent problems are material resource loss (Freedy et al., 1992) and disruption of social networks (Kaniasty and Norris, 1995). Material resource loss is addressed by the programs for housing and economic recovery. However, mental health professionals can facilitate the recovery process by acting as victim advocates. Other recommendations include designing community interventions to provide social support (Salzer and Bickman, 1999).

Some mental health professionals have concluded that the failure to seek formal psychological counseling is a potential threat to the mental health of victims and even first responders. In connection with the latter, Mitchell (1983) developed a system called the *Critical Incident Stress Debriefing* (CISD). Despite its widespread popularity, there is no evidence of CISD's effectiveness (McNally, Bryant & Ehlers, 2003). One problem seems to be that establishing a rigid schedule for victims to discuss traumatic events disrupts their ability to control the alternation between psychological phases of active processing and avoidance (Pennebaker and Harber, 1993).

Economic Development

The ROP should provide guidance on the economic development of disaster stricken areas. Redevelopment should have been planned during the process of envisioning the community recovery strategy. In communities that are highly dependent on tourism, active promotion is needed to assure potential visitors that all facilities are back in operation.

Infrastructure Resilience

During recovery you can decrease the vulnerability of infrastructure. In most cases, roads and bridges can be strengthened. Aboveground lines can be placed underground to reduce their vulnerability to wind and ice. In some cases, pipelines for infrastructure can be rerouted to reduce vulnerability. However, most of these lifelines must pass through high hazard exposure areas at some point. All of these lifelines are critical to a community's resilience. Planning should strengthen infrastructure to decrease its vulnerability.

Historic Preservation

Recovery is the time to determine how to protect historic buildings from future disasters (Cliver, 1998). The federal government has funds, as do many states, for the preservation of historic buildings. However, the community must recognize the value of these structures. They must invest time and money into their preservation (Alfaro, 1998).

Environmental Remediation

Hazmat spills are an increasing problem during natural disasters. The process of cleaning up oil and chemical spills could take months (Lindell and Perry, 1996, 1997; Showalter and Myers, 1994). In most cases, such work is performed by specialized contractors hired by the government. Efforts should be coordinated with local personnel from the department of public health, land-use planning, or fire/hazmat response.

Disaster Memorialization

Another part of recovery is memorializing the victims. When there is a significant loss of life or damage to a community's historic buildings, the sense of loss can be tremendous. Communities frequently derive some solace from building a memorial. These disaster memorials can play an important part in the recovery of a community's sense of identity and pride. They should be considered when a community has suffered a traumatic event. They must be planned and developed in a carefully designed, transparent, and participatory process. Loved ones of victims will all have different ideas as to how their loved ones should be remembered. For example, the memorial to the New York victims of the 9/11 terrorist attacks has come under constant criticism from survivors and the design has been revised many times due to this criticism.

11.8.4 Recovery Management

To manage disaster recovery, you need to perform many of the same types of tasks as you do during the incident management function of the emergency response phase. You do not need to establish the kinds of special procedures for agency notification and mobilization that were used in the emergency response. However, you need to mobilize facilities for donations management, debris management, and disaster assistance. If you are in a community with many victims who lost their homes, and not enough available housing, you might need to develop one or more mobile home parks. You need to identify appropriate sites that have suitable zoning and access to utilities.

You need to work with other agencies to maintain internal direction and control because many recovery tasks require multiagency coordination. Fortunately, the recovery process typically involves the same tasks that agencies perform as part of their normal duties. This makes it easy to allocate recovery functions to agencies.

There is a need for external coordination, especially in presidentially-declared disasters. This is because people from state and federal government agencies are involved. There should be a clear understanding of which agencies should address which problems. However, local agencies need to understand the restrictions of different state, federal, NGO, and CBO programs.

There is also a need for public information, especially to inform disaster victims about recovery policies and procedures. However, there is also a need to inform other citizens about the progress of the recovery. Thus, the ROP should describe the procedure for providing public information. The procedure should describe which agencies are the sources of each type of information, what will be the general content of their messages, and what channels they will use. General information can be distributed through the mass media. Brochures can be targeted at individuals and groups located in vulnerable zones. Telephone hotlines can be useful for answering questions about recovery. Public meetings should be held frequently.

The recovery/mitigation committee needs to obtain legal authority for a wide range of short term recovery actions. These include a development moratorium, temporary repair permits, demolition regulations, and zoning for temporary housing (Schwab et al., 1998). In addition to ensuring adequate legal authority, the recovery/mitigation committee must identify financial tools for achieving mitigation objectives. Financing can be obtained by *directing Community Development Block Grant funds* to mitigation activities, *establishing special assessment districts,* and *charging impact fees* for new development—especially when it is in a hazard prone area.

During the recovery period, the pace of operations decreases so the management of specific emergency response and recovery functions does not need to be focused at incident scenes or centralized in the EOC. Thus, the activities performed by the planning, logistics, and administration sections within the IMS are gradually dispersed back to the jurisdiction's normal departments. Nonetheless, special provisions are required to support the additional staff generated by obtaining mutual aid personnel from other jurisdictions and volunteer personnel such as architects and engineers used as building inspectors. Moreover, records accumulated by the finance section must be available to provide a justification for expenditures on disaster recovery and hazard mitigation that are reimbursable by state and federal agencies. Documentation is needed to learn from the events. Maintaining an event log provides the committee with information to revise the ROP. In addition, this provides justification for expenditures on disaster recovery and hazard mitigation. These are reimbursable by state and federal agencies. Finally, documentation provides legal counsel with the information that might be needed to defend against any lawsuits.

SELF-CHECK

- Explain the differences among rapid assessment, **preliminary assessment,** and **site assessment.**
- Define **victims' needs assessment.**
- Define **nonconforming uses.**
- Describe the role of mass care facilities.

SUMMARY

During this chapter, you examined how communities function before disaster strikes. Once disaster strikes, recovery takes center stage as many organizations and people come together to rebuild. As shown, recovery involves its own set of planning and implementation. Recovery involves households, businesses, and communities. Working together and understanding the obstacles can aid in the recovery process. It is important to utilize the state and federal governments, hazard insurance, and charities to assist people and groups to get the community back to its normal patterns of social functioning. The American Red Cross, Salvation Army, local churches, and volunteers play a big role to get people back to their normal state.

KEY TERMS

Community A specific geographic area that is frequently considered to be a town, city, or county with a government. A community also has stronger psychological ties and social interaction among its members than with outsiders.

Competition The effort of two parties striving toward a goal that only one can achieve. In fair competition, the parties use legitimate methods.

Conflict The opposition that occurs when one party attempts to directly frustrate the goal achievement of another.

Cooperation Activities that result in mutual benefit.

Emergency shelter An unplanned location that is intended only to provide protection from ordinary weather conditions of temperature, wind, and rain.

Nonconforming uses Structures that do not meet the zoning requirements for their geographic areas.

Permanent housing Housing that reestablishes household routines in preferred locations.

Preliminary damage assessment Damage assessment that produces counts of destroyed, severely damaged, moderately damaged, and slightly damaged structures.

Site assessment Damage assessment that is meant to produce detailed estimates of the cost to repair or replace each affected structure. This information is used to support requests for federal assistance to the owners of the damaged property. It includes estimates of losses to residential, commercial, industrial, and public property.

Temporary housing	Housing that allows victims to reestablish household routines in nonpreferred locations.
Temporary shelter	Housing that includes food preparation and sleeping facilities that are sought from friends and relatives or are found in hotels, motels or mass care facilities.
Unmet needs committee	An emergent organization that is designed to serve those whose needs are not being addressed.
Victims' needs assessment	An evaluation of the psychological, demographic, and economic impacts of disasters on victims.

ASSESS YOUR UNDERSTANDING

Go to www.wiley.com/college/lindell to evaluate your knowledge of disaster recovery.
Measure your learning by comparing pre-test and post-test results.

Summary Questions

1. Which of the following is not considered a human asset?
   (a) intelligence
   (b) retirement account
   (c) physical abilities
   (d) personality characteristics
   (e) All the above are human assets.
2. Disaster recovery is both physical and social. True or False?
3. What is the average percentage of people who stay in a mass care facility following a disaster?
   (a) 5%
   (b) 15%
   (c) 25%
   (d) 49%
4. Large businesses are more likely to adopt hazard adjustments than small businesses. True or False?
5. What percentage of all disasters receive presidential disaster declarations?
   (a) 1%
   (b) 5%
   (c) 10%
   (d) 15%
6. The federal government requires flood insurance for structures located in the 100-year flood plain that are purchased with federally-backed mortgages. This has solved the problem of flood victims with no insurance. True or False?
7. Existing CBOs *extend* themselves beyond their normal tasks to perform novel activities. True or False?
8. There is little time to plan for recovery after the response has begun. True or False?
9. During the recovery period, the pace of operations increases. True or False?

Review Questions

1. What is the economic base model?
2. Why is it important to develop preimpact plans for disaster recovery?
3. What are the four stages of housing recovery following a disaster? Explain each stage.
4. Why are small businesses more vulnerable than large businesses?
5. What are the main types of federal government programs that provide recovery assistance?
6. What is the problem with giving aid to uninsured disaster victims?
7. What is the role of nongovernmental organizations and community-based organizations after a disaster strikes?
8. What are the three major misconceptions about recovery that a community must overcome in preimpact recovery planning? Explain each misconception.
9. What are the four principal disaster recovery functions?

Applying This Chapter

1. What is the economic base of your community?
2. You have been asked to develop preimpact plans for disaster recovery in your area. Describe the plans you would develop.
3. If a tornado struck your community, are there mass care facilities that are prepared for a disaster? Explain the solutions you would recommend to a town that doesn't have mass care facilities.
4. You live in a town made up of independent, small businesses. The town's chamber of commerce has asked you to outline how small businesses can minimize their vulnerabilities to disasters. What do you include in a presentation on this issue?
5. Interview your local Red Cross chapter director to find out if there proper plans in place should a disaster strike. If not, what steps need to be taken to ensure proper preparedness?
6. Your community has been the target of a terrorist attack. What extra steps do you take around the impact area because it was the site of a terrorist attack?
7. You are the emergency manager for New Orleans. After Hurricane Katrina, many homes were severely damaged and were not safe to enter. You want to demolish the structures and remove the remaining debris. Homeowners, however, wanted to return to get some of their personal

effects before the structures are demolished. Would you allow the homeowners to return? Why or why not?

8. Does your local government have a recovery operations plan? If not, pick a hazard and identify the area of your community that is most vulnerable to that hazard. Identify the disaster recovery functions that need to be implemented if a major disaster strikes.
9. Who do you need to involve in forming your local recovery/mitigation committee?

YOU TRY IT

Hazard Insurance

You are the local emergency manager for a town and attend a town meeting to discuss mitigation strategies. A man stands up and tells you that he lives in a flood plain. He asks you if he should purchase flood insurance. How do you respond and why?

Home Repairs

What common complaints do homeowners have about repairs after a disaster and why? What can you do to ensure the homeowners aren't being taken advantage of?

Disaster Assistance

You are the emergency manager for New Orleans. Thousands of victims need financial assistance, medical care, housing, jobs, and some way to transport their children to school. Unfortunately, to get all of this accomplished, victims have to go to six different agencies and travel miles from one place to another. How do you make the process easier for the victims?

12

EVALUATIONS

Improving Performance

Starting Point

Go to www.wiley.com/college/lindell to evaluate your knowledge of evaluations.

Determine where you need to concentrate your effort.

What You'll Learn in This Chapter

▲ How to evaluate personnel performance

▲ Importance of evaluating organizations such as the local emergency management agency (LEMA) and the local emergency management committee (LEMC)

▲ How to measure performance in drills, exercises, and incidents

▲ How to determine if training and risk communication programs work

After Studying This Chapter, You'll Be Able To

▲ Analyze an individual's performance

▲ Examine and critique a LEMA's and LEMC's performance

▲ Prepare and implement effective drills and exercises

▲ Analyze risk communication programs to determine their effectiveness

Goals and Outcomes

▲ Write an employee performance appraisal

▲ Design ways to improve specific weaknesses of an organization such as a LEMA or LEMC

▲ Compare and contrast reaction criteria, learning criteria, behavior criteria, and results criteria

▲ Design risk communication and training programs based on evaluation of previous programs

▲ Evaluate a training program

INTRODUCTION

Any organization is only as good as its people. For people to improve and grow professionally, they must receive constructive feedback on a regular basis. Whether you give them or receive them, formal performance appraisals are an important aspect of people's jobs. You can use the ideas from individual performance appraisal to determine if your LEMA and your LEMC are effective. You generally produce these evaluations by reviewing people's performance over an entire year. However, you sometimes need to evaluate the performance of individuals, teams, or entire organizations during drills or exercises that take place over a few hours.

This chapter discusses how to assess individuals, teams, or organizations. It shows you how to conduct your own performance appraisals and how to prepare drills and exercises to assess the performance of an emergency response organization.

12.1 Evaluating Personnel Performance

Performance appraisals contribute to the performance of any organization. They provide a systematic review of an individual employee's performance on the job (Cascio, 1998; Schmitt and Klimoski, 1991). Appraisals serve four functions:

- **Development** focuses on improving an employee's *ability* to do a job. In this context, performance appraisal can be used to guide decisions about training, reassignment, or termination. Training keeps the job constant but changes the person by giving them new skills. Reassignment keeps the person constant but changes the job. The change can be lateral, moving the employee to another job with the same level of responsibility. Reassignment can also be a promotion or a demotion. Termination is ending the employee's job.
- **Reward** focuses on improving a person's *motivation* to do the job. Appraisals should have clear criteria that provide guidance to the employee about what is important to the organization. In addition, a good appraisal process can guide rewards. Rewards improve productivity and satisfaction. Rewards also decrease turnover.
- **Internal research** lets you know what skills your employees have. Internal research also lets you know if you were looking for the right qualities when hiring for the position. This function is usually performed by organizational psychologists. This is a concern of big organizations with large human resource departments.

▲ **Legal protection** is achieved when an organization conducts performance appraisals. Appraisals must be completed according to generally acceptable procedures. You must also retain documentation of the review.

There are four principal questions that need to be addressed in the performance appraisal process.

1. When should the appraisal happen?
2. Who can do the appraisal?
3. What should be evaluated?
4. How should the performance appraisal be done?

An appraisal should be conducted at least once a year. A formal appraisal is different from informal feedback about job performance, which should be frequent. You can also link a performance appraisal to the task cycle. That is, if a person works on a lengthy project, an appraisal should be conducted soon after project completion.

As to *who evaluates,* consider who has objective information about the employee's performance. Also consider who understands the goals of the organization. You should also consider who can give raises and who can assist in training. Typically, the immediate supervisor knows the goals and job descriptions and has observed the employee's performance. In addition, supervisors want to maintain reward power and can assist in training. These are the reasons why supervisors are the most common evaluators. However, there are others who can make valuable contributions. For example, an employee's peers are valuable sources of information. They often have more accurate information because they have more frequent contact with the employee. As a consequence, they observe a more representative sample of behavior. However, they might give an employee a low assessment if they think this will improve their own standing in the organization. As a result, peer appraisals are more appropriate when there is a trusting, noncompetitive atmosphere. You can also implement procedures to prevent competitive behavior from influencing ratings (Kane and Lawler, 1977). Those who work for the employee being evaluated are also good sources of information. These subordinates see aspects of an employee's behavior that are not seen by supervisors or even by peers. Subordinates are often concerned about sharing negative information about their bosses. They are worried that their bosses will find out and retaliate. Finally, employees themselves are good sources of information because they have more information about their performance than anyone else. This is especially true when they work independently. However, self-evaluations are often too positive. Employees tend to attribute their successes to their own efforts and their failures to other people and conditions in the workplace. In many organizations, the supervisor and employee both rate the employee's performance. They seek to reconcile their different ratings through discussion. Where possible, feedback is sought from peers, subordinates, and even an organization's customers and suppliers. In the case of LEMAs, this might include personnel from other agencies in the LEMC.

As to *what should be evaluated,* you should seek to rate performance on data that meet three conditions. First, the data must be available within the time period in which the appraisal is being conducted. Second, the data must be relevant to job performance. Third, the data must be comprehensive. Data availability can be a problem if a person works on projects that take years to show results. For example, a risk communication program might easily take more than a year to conduct and even longer to produce measurable changes in households' emergency preparedness. This is why performance must sometimes be evaluated on intermediate results rather than final outcomes.

Also pay attention to the relevance of the evaluation criteria. It is obvious that an employee's choice of music or office decoration does not affect job performance. However, it can sometimes be difficult to distinguish what is personally distasteful from what is actually disruptive. It is common to see evaluations that reflect factors other than performance. For example, two different employees of a LEMA might be given what seems to be the same assignment—implement a risk communication program in a hazard-prone neighborhood. Suppose, however, that one is assigned to a middle class neighborhood with long-term homeowners and an active community council. The other is assigned to a working class neighborhood where there is high turnover among renters and they do not have a neighborhood organization. The first employee is likely to have a much more successful risk communication program than the second. However, the difference in the two employees' performance is due to factors that are beyond their control.

All parts of the job should be measured. In many cases, the short-term impacts of a person's behavior are quickly noticed. However, the long-term effects of performance are not easily recognized. For example, an emergency manager might face short-term pressure to meet with police and fire personnel to ensure the emergency operations plan (EOP) is updated. However, the need to meet with land-use planners to develop a preimpact disaster recovery plan is easily overlooked. You should ensure that both the short-term and long-term aspects of the job are evaluated. If it is not possible to perform all aspects of the job in a single year, set explicit goals to focus on some activities in one year and the rest in later years.

As *to how should it be done*, your human resource department typically has a set of performance appraisal criteria that have been devised for all civil service jobs. Typically, these instruments separately address *task* and *interpersonal* performance. In turn, task performance is frequently broken down into motivational and ability components. Typical performance appraisal categories include:

- **Time and project management:** Understands own job description and the function of the unit; well organized; sets and adjusts priorities in response to job impediments; delegates as appropriate; follows through on objectives; consistently produces work of a quality and quantity that is consistent with organizational needs.

- **Resource and knowledge management:** Understands budget processes relevant to the position; uses allocated resources wisely; understands and follows organizational procedures relevant to daily job operations; knows and uses sources of additional information and assistance as needed.
- **Decision making and problem solving:** Identifies problems, collects information, and weighs viable options; makes decisions and follows through.
- **Innovation:** Generates new ideas.
- **Personal management:** Initiates activity without waiting for directions from superiors; seeks additional responsibility; recognizes mistakes and adapts to them; perseveres until projects are completed.
- **Change management:** Accepts and supports new methods of job performance and organizational procedures.
- **Interpersonal skills:** Works well with supervisors, peers, subordinates, and customers; manages conflict effectively.
- **Communication:** Able to speak and write clearly but is diplomatic in dealing with others.
- **Quality of work life:** Demonstrates respect for individual differences, contributions, and family related responsibilities of others; supports and promotes organizational diversity initiatives.

It is difficult to remember all the behavior that each employee displays over the year. Take time at the beginning of each year to review the criteria that is used to evaluate performance. Periodic review of the evaluation criteria helps you notice examples of effective and ineffective behavior, and interpret these when they occur. Take time during the year to identify and record each employee's typical level of performance. It is a good idea to encourage employees to keep their own job diaries as well. You should note any instances of good performance and poor performance. At the end of the year, use this information to rate each employee's performance on each of the evaluation criteria.

Before the annual review, give employees copies of the performance appraisal form and ask them to review the past year. The supervisor and the employee should then rate the employee on each of the criteria using a numerical scale such as the one listed in Table 12-1.

Both the supervisor and the employee should prepare to explain the reasons for their ratings in terms of specific examples. Supervisors often are required to give a written explanation for ratings of 1 because low ratings lead to termination. Supervisors are also required to give explanations for ratings of 5 because high ratings lead to promotion. Supervisors may not be required to give detailed explanations for ratings 2-4. However, the more frequent and specific the feedback you give to employees, the more likely it is they will improve.

Table 12-1: Sample Performance Appraisal Rating Scale

	Performs far below job requirements	*Performs below job requirements*	*Performs job requirements adequately*	*Performs above job requirements*	*Performs far above job requirements*
	1	2	3	4	5
Supervisor					
Employee					

Schedule a private meeting with each employee. Open the meeting with positive achievements in the past year. You also need to have a frank discussion of specific performance shortcomings. The objective is to describe actual incidents of performance. It is important for supervisors to focus on behavior, which can be changed. Do not focus on personality characteristics, as they are virtually impossible to change. The supervisor should emphasize that good performance is rewarded and poor performance must be corrected. Performance change can be accomplished by training if the problem is a correctable lack of ability. Change can also be accomplished by withholding rewards if the problem is a minor lack of motivation. If the problem is either an uncorrectable lack of ability or motivation, performance change can be achieved by transferring or firing the employee.

It is important to avoid sending the wrong message. If you have an overall positive appraisal, then do not dwell on the negative aspects of a performance appraisal. Unfortunately, dwelling on the negative is quite common. This is easy to understand because most supervisors want to save time by focusing on what needs to be fixed rather than "waste" time talking about what is being done well. Nonetheless, praise is important because it lets subordinates know good work is being noticed and will have positive consequences. Recognition of good performance is especially important when there is little difference between the smallest and the largest salary increases within the organization. Of course, it is even better if good performance is recognized throughout the year.

To conduct a review:

Step 1: Schedule a private meeting.

Step 2: Discuss the employee's accomplishments.

Step 3: Discuss the areas where the employee could improve.

Step 4: Allow employees an opportunity to explain their self-ratings, particularly if they can provide a reason for those ratings.

Step 5: Allow time for a full discussion, listening carefully to what employees have to say and focusing on behavior and performance and not personality characteristics.

FOR EXAMPLE

Salary Freezes

In 2002, there was a record number of salary freezes in effect. Sixteen percent of all employees reported that their company had decided not to give any raises that year. Unfortunately, salary freezes are a popular way to keep operating expenses level in a company and within state and local budgets. Even if you are faced with a salary freeze, it is still important to give an employee an annual review. The employee and the organization can only improve with feedback. If you can't give an outstanding employee a raise, you can look at other forms of compensation. Paid time off is one example of a reward that does not increase costs.

Step 6: Collaborate with the employee in setting specific and measurable objectives for the coming year.

Some objectives should be performance-oriented, such as, completing tasks or projects. Others should be developmental. For example, an employee can set a goal to take a specific training course. In addition, objectives should be set only if they can be accomplished within the period of performance. Thus, the objective "Get an emergency operations plan *approved* by the end of the year" should be revised to "Get an emergency operations plan *submitted* by the end of the year." This is because the employee can't control the approval process, which might take longer than the end of the year to complete. Setting objectives is an important way of showing high-performing employees how they can obtain promotions. Just as important, it is a way to keep poor-performing employees from giving up altogether. A good development plan, based on clear objectives, shows them how they can achieve better performance ratings.

SELF-CHECK

- Define **development** and **reward.**
- List other forms of compensation you can offer if you do not have enough money to give an employee a raise.
- Define **legal protection** and **internal research.**
- Give three reasons why employee performance appraisals are important.

12.2 Evaluating the LEMA and the LEMC

Periodic evaluations help organizations in the same way they help employees. Identifying specific weaknesses can lead to suggestions for improved performance. There are general principles that one could use to evaluate any organization. For example, setting goals and reviewing achievement annually is a general principle that applies to any organization. In addition, we have very specific standards for evaluating the performance of emergency management agencies.

12.2.1 General Principles of Organizational Evaluation

You should work with the other members of the LEMA and LEMC to set specific, measurable objectives that can be accomplished within the period of performance. These objectives should be developed collaboratively because such goals gain greater commitment than goals imposed by a supervisor. The goals for the LEMA and LEMC should differ from each other for two reasons. The first reason is that LEMA and LEMC are different organizations with different responsibilities. The second is that your control over the allocation of resources in the LEMC is more limited than your control over the LEMA.

To evaluate the LEMA,

- ▲ Assess the hazard vulnerability analysis.
- ▲ Assess the hazard mitigation program.
- ▲ Assess the emergency preparedness program.
- ▲ Assess the recovery preparedness program.
- ▲ Assess the community hazard education program.
- ▲ Review the capability shortfall identified in previous years.
- ▲ Review the multiyear development plan that was designed to reduce the capability shortfall.
- ▲ Revise goals, if needed, based on the LEMA's current capability.
- ▲ Set specific milestones (objective indicators of task performance) for each quarter of the year to determine if the LEMA is making progress at a satisfactory rate throughout the year.
- ▲ Assign tasks to the personnel who are most qualified to perform them.

Evaluating performance of the LEMC is somewhat more complex, but follows basically the same procedures as are used for evaluating the LEMA. Each LEMC subcommittee should identify the specific tasks that must be accomplished to make progress in its functional area, including

- ▲ Hazard vulnerability analysis.
- ▲ Planning, training, and exercising.

- ▲ Recovery and mitigation.
- ▲ Public education and outreach.
- ▲ LEMC management.

In some cases, this leads subcommittee members to set an objective to acquire the resources needed to perform a task. Or the objective might involve actual task performance. For example, a hazard vulnerability analysis committee might first set an objective of acquiring software. They may later set an objective of getting someone trained to use the computer to conduct HVAs.

As is the case with the LEMA, the LEMC should use a collaborative process to set specific, measurable objectives. The objectives must be ones they can achieve within the performance period. The subcommittees should coordinate their objectives with each other. This can be done either through the executive committee or in general meetings of the LEMC. Once all of the subcommittees have set objectives, they should monitor their performance informally throughout the year. They should also review their performance at the end of the year. The executive committee should then meet with senior elected and appointed officials to discuss the LEMC's achievements during the previous year.

12.2.2 National Fire Protection Association Standard 1600

The National Fire Protection Association (NFPA) Standards Council established a Disaster Management Committee in 1991. The committee developed standards for preparedness, response, and recovery. They did this for the entire range of disasters. The committee's first product, *NFPA 1600 Recommended Practice for Disaster Management,* was published in 1995. Standards are reviewed and revised on a five-year cycle. The committee preparing the 2000 edition determined that the scope of the original effort was too narrow. With representatives from FEMA, International Association of Emergency Managers (IAEM), and the National Emergency Management Association (NEMA), the committee adopted what they called a "total program approach." This approach is consistent with the principles of comprehensive emergency management and integrated emergency management systems. The new standard was written to cover both public and private sector organizations. They also included business continuity programs in their scope. Business continuity planning first began in the private sector as a means of incorporating disaster planning and consequences into business plans. Business continuity planning is based on federal government continuity planning. The federal government has plans to ensure the continued delivery of government services following nuclear attacks (Perry and Lindell, 1997). The standard issued in 2000 was renamed the *Standard on Disaster/Emergency Management and Business Continuity Programs.* The third edition, issued in 2004, retained the same name as the second edition (National Fire Protection Association, 2004).

The current version of NFPA 1600 defines an **entity** as a public or private sector organization that is responsible for emergency/disaster management or continuity of operations. The standard requires the organization to have:

- ▲ A documented emergency management program.
- ▲ An adequate administrative structure.
- ▲ An identified coordinator.
- ▲ An advisory committee.
- ▲ Procedures for evaluation.

The program must address 14 elements identified in Table 12-2.

The Laws and Authorities element must address the legislation, regulations, directives, and industry standards that authorize the emergency management program.

Table 12-2: Emergency Management Program Elements

Element	*Title*
1	Laws and Authorities
2	Hazard Identification, Risk Assessment, and Impact Analysis
3	Hazard Mitigation
4	Resource Management
5	Mutual Aid
6	Planning
7	Direction, Control, and Coordination
8	Communications and Warning
9	Operations and Procedures
10	Logistics and Facilities
11	Training
12	Exercises, Evaluations, and Corrective Actions
13	Crisis Communications and Public Information
14	Finance and Administration

The Hazard Identification and Risk Assessment element must identify the:

- ▲ Hazards to which the organization is exposed.
- ▲ Probability of extreme events that would adversely affect the organization.
- ▲ Potential physical and social consequences of those events.

The Hazard Mitigation element requires the entity to develop a strategy to eliminate hazards or limit their consequences.

The Resource Management element requires the entity to:

- ▲ Identify the resources (personnel, facilities, equipment, materials and supplies, and money) that are needed.
- ▲ Inventory the resources that are currently available.
- ▲ Define the resulting shortfall.
- ▲ Address the role of volunteers.
- ▲ Create and participate in mutual aid agreements as a means of enhancing resources available.
- ▲ Establish performance objectives relative to each threat identified in the vulnerability analysis for personnel, equipment, apparatus, and facilities.
- ▲ Calculate resources in terms of quantity, response times, and capabilities.

The Mutual Aid element requires the entity to:

- ▲ Identify the need for resources from other entities.
- ▲ Establish formal agreements for requesting those resources.
- ▲ Refer to these agreements in relevant sections of the entity's emergency management plans.

The Planning element requires the entity to develop:

- ▲ A strategic plan.
- ▲ An emergency operations/response plan.
- ▲ A recovery plan.
- ▲ A hazard mitigation plan.
- ▲ A continuity plan.

The standard specifies the content of each type of plan. In addressing common plan elements, NFPA 1600 requires specifying roles and responsibilities. Roles must also be defined for external organizations that will participate in mitigation, preparedness, response, and recovery activities.

The Direction, Control, and Coordination element requires the entity to establish the authority for response and recovery operations. This specifically includes adopting an incident management system and assigning functional responsibilities to specific organizations within the entity.

The Communications and Warning element requires the entity to develop and test the equipment and procedures needed to activate the emergency response organization.

The Operations and Procedures element requires the entity to develop the procedures needed to respond to the hazards identified in the Hazard Identification, Risk Assessment, and Impact Analysis element. These procedures must include:

- Situation assessment and resource assessment.
- Transition from response to recovery operations.
- Continuity of operations.

The Logistics and Facilities element requires the entity to:

- Develop systems for identifying, obtaining, and delivering needed resources to response and recovery personnel.
- Handle unsolicited donations.
- Establish a primary and alternate emergency operations center.

The Training element requires the entity to:

- Identify training needs.
- Design and implement a training program.
- Document the delivery of training to all emergency response personnel.

The Exercises, Evaluations, and Corrective Actions element requires the entity to:

- Conduct periodic evaluations of the plans and procedures including tabletop, functional, and full-scale exercises.
- Ensure corrective action regarding any identified deficiencies.

The Crisis Communications, Public Education, and Information element requires the entity to establish procedures for disseminating relevant information to the news media and the public during pre-, trans-, and postdisaster phases of operation.

The Finance and Administration element requires the entity to develop fiscal procedures to ensure decisions can meet the time constraints of emergencies yet conform to accepted accounting standards.

NFPA 1600 provides a program that can be adapted to any level of government or private industry. The program helps you manage all hazards. You can use this program while working with multiple organizations. This reflects FEMA's all-hazards approach. The standard shows that the goal is to create a disaster resilient community. The standard includes response operations, planning, and recovery. There is, however, a *distinct and explicit* emphasis on employing effective land-use practices, codes and regulations, strategic community protection, and sustainable urban design.

NFPA 1600 is important to the emergency management profession for several reasons. First, it was issued by a respected and established authority. The government and major emergency managers respect and understand the importance of NFPA standards. Second, NFPA 1600 is important for its use in program assessment. The standard provides a model that can be used in self-assessment and also by external evaluators. Furthermore, NFPA 1600 can serve as a basis for planning either to create a program or to enhance an existing program so that it meets the standard. The government must allocate limited resources. When you must make arguments to expand, or change your programs, you can use NFPA 1600 as the authoritative basis for your position.

12.2.3 State Capability Assessment for Readiness Program

In 1997, FEMA and the National Emergency Management Association released the state **Capability Assessment for Readiness (CAR) program.** The CAR program describes a self-assessment process for state emergency management agencies (SEMAs) to:

- ▲ Evaluate their readiness to mitigate hazards.
- ▲ Prepare and respond to emergencies.
- ▲ Recover from disasters.

The CAR program consists of standards that begin, at the highest level, with 13 emergency management functions (EMFs) adapted from NFPA 1600. The CAR program is more specific than NFPA 1600. The CAR program divides each EMF into attributes. Attributes are broad performance criteria. Each attribute is further divided into characteristics. Characteristics are specific performance criteria. This results in a checklist of specific, objective, measurable items that each state can use to assess its capabilities. For example, Table 12-3 shows that the *Laws and Authorities* element is divided into ten attributes. Further, Table 12-4 shows division of attribute 1.1 into seven specific performance criteria.

Table 12-3: Attributes for EMF 1.0 of *Laws and Authorities*

Attribute	*Content*
1.1	The State Emergency Management Program/responsibility is legally established in State law.
1.2	Trust Fund legislation has been enacted by the State.
1.3	Legal authorities supporting regulations for Continuity of Government (COG) activities exist in State law.
1.4	The State supports the establishment of legal authorities for local emergency management jurisdictions.
1.5	The State complies with the requirements of the National Environmental Policy Act (NEPA) and other environmental laws.
1.6	The State complies with the National Historic Preservation Act (NHPA) requirements.
1.7	State law enables the use of statewide or local codes or ordinances for the purpose of mitigating hazards.
1.8	The State complies with applicable Civil Rights statutes.
1.9	State legislation is enacted for a State Dam Safety Program that includes all criteria outlined in the National Dam Safety Program.
1.10	The State complies with the requirements of the Emergency Planning and Community Right to Know Act.

For each attribute or characteristic, the state is evaluated on a 1 to 5 scale, where:

- 5 is fully capable.
- 4 is very capable.
- 3 is generally capable.
- 2 is marginally capable.
- 1 is not capable.
- N/A is not applicable.

The scores on all the characteristics are then added to create an overall score for that attribute. The scores on all the attributes within an EMF can then be added to create an overall score for that EMF. This produces a profile of the state's strengths and weaknesses.

Table 12.4: Characteristics for Attribute 1.1 of *Laws and Authorities*

Characteristic	*Content*
1.1.1	A legal basis for the emergency management program exists in State law.
1.1.2	The process for the Declaration of a State Proclamation of Emergency or Disaster exists in State law.
1.1.3	The State has adopted an executive order or other mechanisms for coordination among State agencies.
1.1.4	Development of Mutual Aid agreements, including specific provisions (e.g., liabilities, responsibilities, participants, review process), is supported by State law.
1.1.5	Legal authority for evacuations (e.g., hurricane, HAZMAT, etc.) is defined.
1.1.6	The State has adopted an executive order or other mechanism for the establishment of continuity of operations plans in all State agencies involved in disaster response and recovery operations.
1.1.7	A strategy addressing needs for legislative and regulatory revisions has been developed.

SEMAs can complete the assessment process and submit the results to FEMA. FEMA can then take all the information from the state assessments to create a national evaluation. SEMAs can also use their assessments to develop strategic plans. FEMA and state officials can use the results to identify action items that should be addressed in future years under the Emergency Management Performance Grant program.

12.2.4 The Emergency Management Accreditation Program

The Emergency Management Accreditation Program (EMAP) is also based on NFPA 1600. EMAP is closer to NFPA 1600 than CAR because it includes the requirements for program management. However, EMAP does differ from NFPA 1600 in some ways. First, it was written specifically for state and local emergency management agencies. EMAP accreditation can be obtained by LEMAs, as well as SEMAs. Second, the EMAP accreditation process is more elaborate than the CAR assessment. Once you submit an application, you have 18 months to conduct a self-assessment of your compliance with EMAP's 54 standards. The self-assessment begins with a letter from a jurisdiction executive stating a commitment to receiving accreditation. An accreditation manager is then selected to coordinate the accreditation application. You may be chosen for this job. One of your duties will be to develop the self-assessment plan. You will also assign responsibilities to each agency within your jurisdiction. You will set a schedule for the self-assessment process.

The self-assessment requires a proof of compliance record for each standard. You must identify the documents, interview sources, or observations that prove compliance. In addition, you must explain how these support the claim of compliance. If your proof is based on written documentation, you must also include copies of all relevant documents. You must also have documentation for interviews and observations.

After completing the self-assessment documents, you will submit them to the EMAP commission for review. If the EMAP commission judges your materials to be satisfactory, it will schedule an on-site assessment. You will be assigned an assessor team. During the on-site assessment, the team receives an orientation. You will give team members a tour of major facilities and give them any additional information they request. Members of the team will examine the written plans, procedures, and memoranda that were listed as proof of compliance. Team members contact interviewees listed on the compliance record to obtain independent verification of the information in the application. Team members inspect facilities, equipment, materials, and supplies. After they have collected all the information they need, the team conducts an exit interview. At this time, they describe their preliminary findings.

After the team finishes its visit, it then describes its activities and reports its findings to the EMAP review committee. This occurs at a meeting that the applicant is allowed to attend. Based on this information, the EMAP review committee recommends that the applicant be accredited, conditionally accredited, or denied. This recommendation is then sent to the EMAP commission. If accredited, your jurisdiction is issued a certificate that is valid for five years. Nonetheless, accreditation must be maintained during that time period. To do this, you must prove continuing compliance with the EMAP Standard, provide documentation

FOR EXAMPLE

Florida

Florida was the first state to apply and receive EMAP accreditation. As Florida is especially vulnerable to hurricanes and other coastal hazards, the state dedicates a lot of resources to hazard mitigation and disaster response. Arizona, North Dakota, and Washington D.C. have also received EMAP accreditation.

of compliance, and file an annual report with the EMAP commission. If accreditation is denied, you are informed of the reasons why. Further information about EMAP can be found at www.emaponline.org.

SELF-CHECK

- Define **entity.**
- List the elements of the **emergency management program.**
- Define **State Capability Assessment for Readiness (CAR) program.**
- Identify who can obtain the EMAP accreditation.

12.3 Evaluating Drills, Exercises, and Incidents

Evaluating drills, exercises, and incidents is similar to evaluating employees or LEMAs. However, there are also some significant differences. Performance is measured over a relatively short period of time. In drills, performance is measured over a period of minutes. In exercises and incidents, performance is measured over a period of hours to days. This shortened time period makes evaluation easier because there is less performance to evaluate. However, task performance is measured much more intensively during drills and exercises. You must observe the performance of many people. Finally, incidents have the potential for generating lawsuits. This is especially true for incidents that involve the loss of life or extensive destruction of private property. In turn, these lawsuits can stifle the free exchange of information needed to learn from experience and improve.

12.3.1 Drills

When planning to conduct drills, the first task is to specify clearly what objectives will be tested. Typically, drills are used to test people, facilities, and equipment on tasks that are *difficult, critical,* and are *performed infrequently.* The first two conditions are important because they make failures in task performance likely and escalate the consequences when failure does occur. The third condition is important because long time intervals between opportunities for task performance cause people's skills to decay and their equipment to deteriorate.

Drills usually involve one or a few people who must perform a specific task in response to a hypothetical scenario (see Figure 12-1). A **task** is "a distinct work activity carried out for a distinct purpose" (Cascio, 1991, p. 190). A task might

Figure 12-1

Table-top exercises are an essential tool for improving performance.

have many steps or elements. These steps are defined as "the smallest unit[s] into which work can be divided without analyzing separate motions, movements, and mental processes involved" (Cascio, 1991, p. 190). In a hazmat emergency response, a technician is assigned the task of plugging a hole in a container that is leaking a toxic liquid. This task has multiple steps. The steps must be performed in the following order:

Step 1: Obtain a briefing about site conditions.

Step 2: Don personal protective equipment.

Step 3: Verify radio operability.

Step 4: Collect repair tools and materials.

Step 5: Enter the hot zone.

Step 6: Perform the repair.

Step 7: Return to the warm zone for decontamination.

Step 8: Participate in a debriefing.

A drill is conducted by a **controller,** the person who provides the information from the scenario. Drills are relatively simple, so the same person can serve

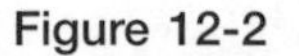

Figure 12-2

Drills usually involve one or a few people performing tasks that are difficult, critical, and which are performed infrequently.

as the **evaluator.** This is the person who observes the player's performance and notes any deviations from the EOP or its procedures. The evaluator must have skills that meet or exceed that of the person being evaluated. In addition, the evaluator must identify any facilities, equipment, and job performance aids that are needed for the task. Frequently, an evaluator randomly selects one or more individuals as the principal or alternate performer in a position to be tested. The player is asked to "walk through" each step in that task. Some tasks have a significant mental, as well as physical, component. In these cases, players might be asked to "think aloud" as they proceed through the scenario (Ericsson and Simon, 1993). To do this, players must:

1. Identify the information they need.
2. Describe the way they are processing the information.
3. Give the final judgment or decision they have made.

12.3.2 Functional Exercises

A functional exercise differs from a drill by involving more people. This makes functional exercises more comprehensive than drills. In addition, the scenario is usually more complex because it involves more tasks, equipment, and people. As a result, functional exercises test people's ability to perform both taskwork and teamwork.

- ▲ **Taskwork** is the ability to perform each separate step of the response.
- ▲ **Teamwork** is the ability to schedule tasks and allocate resources among team members to achieve a performance that is efficient, effective, and timely (McIntyre and Salas, 1995).

Unlike drills, **functional exercises** cannot combine the roles of controller and evaluator. These exercises sometimes require many controllers. These controllers will provide information to different teams of players, especially if the teams are in different locations. Several evaluators will also be needed to evaluate the different teams.

12.3.3 Table-top Exercises

A **table-top exercise** differs from a drill or functional exercise. A table-top exercise involves a group of senior personnel (see Figure 12-2). They are usually branch or departmental administrators and they serve as the directors of their functions. The scenarios for these exercises vary in their complexity. Some are as simple as open-ended questions designed to generate a discussion about a particular problem. For example, a table-top exercise might address the criteria for initiating an evacuation of local hospitals before a hurricane. The discussion would include a number of topics. These might include the resources available for providing transportation support to the hospitals. Other topics include ways to return ambulances and other emergency vehicles against the flow of evacuation traffic. These exercises involve an interacting group of people who are usually in a single room. Often, a single controller also serves as the evaluator.

12.3.4 Full-scale Exercises

A **full-scale exercise** simulates a community-wide disaster by testing multiple functions at the same time. It also tests the coordination among these functions. The complexity of full-scale exercises requires thorough planning of the scenario. It also requires coordination among the many controllers and evaluators. There also is a need for training the controllers. Controllers have to adjust if the players take actions during the exercise that deviate from the scenario conditions. The demand for controllers and evaluators might exhaust the available supply of qualified personnel. If so, outside personnel need to serve as controllers and evaluators. In this case, the outsiders need to be trained on the local EOP so they can see if the players are following it.

The magnitude of full-scale exercises varies. Small ones might provide a limited test of a few functions. For example, a single school might be selected to test the evacuation plan. However, large exercises conducted for nuclear power plants can involve thousands of players, and as many as 50 to 100 controllers and evaluators. Most full-scale exercises are unannounced. In addition, the exercise scenario and the time at which it will begin are unknown to the participants. Unannounced exercises prevent agencies from scheduling their best trained, and maybe their *only* trained, personnel for participation in the exercise. As a result, unannounced exercises provide a more accurate assessment of preparedness. This is especially true when exercises begin on the evening and night shifts.

Performing well in an unannounced exercise is a challenging goal. Achievable goals should be set first to build confidence and motivate people to improve. Poor performance can demoralize the participants and can be a public embarrassment. In addition, a large number of errors make it difficult to focus on a few tasks. It is also difficult to develop a consensus on how to improve with many errors. Thus, it is a good idea to work up to unannounced exercises by first verifying satisfactory performance in announced exercises.

FEMA organizes the development of an emergency exercise into eight steps:

1. **Needs assessment:** This includes addressing the community's primary and secondary hazards. In particular, the needs assessment focuses on testing the solutions that were implemented to correct past problems in the EOP, procedures, staffing, and training. It might also look at the effectiveness of new facilities, personnel, and equipment.
2. **Scope definition:** This includes the type of emergency, its location, the functions to be exercised, the members of the emergency response organization who will participate, and the type of exercise.
3. **Purpose statement:** This explains why the exercise is being conducted.
4. **Objectives:** The objectives should make clear who should take what action in response to which conditions and to what standard of performance.
5. **Narrative:** This is a specific description of relevant conditions, including a chronology of events. These include the initiating event, and contextual conditions in the physical and social environments.
6. **Major and detailed events:** A description of all major events requiring actions to meet the exercise objectives and of detailed events if they should initiate expected actions.
7. **Expected actions:** These are actions that achieve the exercise's stated objectives. These include assessment actions that obtain or verify information about the existence of environmental conditions or the performance of organizational actions. Other expected actions include preventive or corrective actions that reduce the magnitude of an event. Finally,

expected actions include protective actions that reduce the effects on people and incident management actions that consider alternative actions, make decisions, allocate resources, or coordinate the actions of responders and the public.

The key to effective exercises is the transmission of messages communicating specific events to exercise participants. Messages should indicate *what, how,* and to whom the message is directed. They should be designed so they generate the expected actions that meet exercise objectives.

12.3.5 Incidents

The evaluation of performance in an incident is extremely informative because it is unscheduled. This has the advantage of providing a realistic test of incident management tasks such as organizational activation and notification. These tasks are tested in a way they would not be in an announced exercise. The disadvantage of actual incidents as evaluations is that they are also uncontrolled. Both the size of the event and the response functions that are tested are matters of chance. Incidents also have no controllers or evaluators. Respondents must rely on their memories and any documentation to establish who did what, where they did it, when they did it, and why they did it. You must ensure there is adequate documentation of incident conditions and the response. Documentation is needed to support a critique.

12.3.6 Critiques

All three forms of exercises and incident responses benefit from an oral critique by the players, controllers, and evaluators (National Response Team, 1990). However, some full-scale exercises have so many participants that representatives must be selected from each unit. In an exercise critique, the discussions should

FOR EXAMPLE

TOPOFF

The federal government began running full-scale terrorist attack exercises in cities across the country during 1998. TOPOFF, which stands for *top officials,* is a national exercise that simulates several attacks simultaneously across several cities. *TOPOFF 1,* for example, took place in 2000. It simulated a radiological attack in Portsmouth, New Hampshire, and a biological weapons attack in Denver, Colorado. Another exercise, called *Dark Winter,* simulated a smallpox outbreak. That exercise indicated to officials that there wasn't enough small pox vaccine available. After learning this, the federal government quickly had enough smallpox vaccine produced and stored.

address whether the objectives were met. In an incident critique, the question is whether the response was consistent with the EOP and procedures. If there were deviations from the EOP and procedures, the participants should discuss why this occurred. In some cases, responders improvise actions that are better than the once prescribed by standard operating procedures (SOPs). In this case, you should consider revising the EOP or SOPs. In other cases, the improvised action will be worse than the procedure. In this case, you need to reassign personnel, improve training, or upgrade facilities and equipment. Document the critique results in a written report that includes an action plan. Write specific recommendations, assign responsibility for implementation, and schedule completion of each element of the action plan.

SELF-CHECK

- Define **drills and tasks.**
- Describe what a **controller** and **evaluator** do.
- Define **functional exercise.**
- Describe the difference between **taskwork and teamwork.**

12.4 Evaluating Training and Risk Communication Programs

The procedures for evaluating training and risk communication programs are different from the previous types of evaluations. Performance appraisals, drills, and exercises evaluate all, or at least most, of the people you want to know about. However, training and risk communication programs are usually administered to many more people than you can afford to evaluate. Consequently, we often test a subset (called a *sample*) of individuals from the larger group that received the training or risk communication program (called the *treatment group*). You can compare the performance of this sample from the treatment group to the performance of a group of people who did not receive the training or risk communication program (called the *control group*). There are many research designs that can be used to make scientifically rigorous comparisons between treatment groups and control groups (Cook, Campbell, and Peracchio, 1990; Schmitt and Klimoski, 1991). Despite their differences, procedures for evaluating training and risk communication programs are similar to the previous types of evaluations in one very important respect. In all cases, it is essential to define, in advance, what the criteria are for defining the success of the program.

FOR EXAMPLE

Results Criteria

It can be difficult to draw the correct conclusions from a single disaster. For example, a chemical spill might produce few casualties and damage. This could be because of specific event conditions. It could be that only a small quantity of chemical was released. Or, it could be that the wind was blowing away from populated areas. If we want to conclude that the organization responded effectively, we need to rule out these other possibilities.

The criteria for judging the success of training programs are often classified into four groups (Goldstein, 1993):

- **Reaction criteria:** Reaction criteria consist of trainees' opinions of the training program. This includes evaluations of the trainers, the facilities and equipment, the material, their enjoyment of the class, and their desire to take another class from the instructor.
- **Learning criteria:** Learning criteria are defined by performance on written tests or performance tests of skills addressed in the training program.
- **Behavior criteria:** Behavior criteria refer to trainees' ability to apply the new knowledge and skills to their jobs. This includes performance during drills, exercises, and incidents that take place after training is completed.
- **Results criteria:** Results criteria refer to the consequences of trainees' performance on the job. That is, did the training make a difference in the overall performance of the organization? Someone who can flawlessly demonstrate a skill that is never used on the job will probably never have an impact on the safety of the community.

The same four criteria can be used in assessing a community risk communication program. Reaction criteria are measured by reactions to the speaker, setting, communication medium, and message content. Learning criteria are measured by the beliefs about the hazard and hazard adjustments. Behavior criteria are measured by households' and businesses' implementation of hazard adjustments. Results criteria are measured by reductions in casualties, damage, and disruption from disasters. Reaction criteria are the easiest to collect. Learning and behavior criteria are more difficult to obtain. The infrequency of disasters makes it very difficult to collect results criteria.

SELF-CHECK

- Define **reaction criteria and learning criteria.**
- Explain how the procedures for evaluating training and risk communication programs are different from other types of evaluations.
- Define **behavior criteria and results criteria.**
- Explain the difference between a treatment group and a control group.

SUMMARY

As this chapter discusses, an organization is only as good as its people. For people to improve and grow professionally, you must give them constructive feedback on a regular basis. You can apply the same principles of feedback to determine if your LEMA and LEMC are performing effectively. As important as emergency management is, you must understand how important it is to hold people accountable to their jobs. To do this, you must be able to evaluate their performance. This chapter shows you how to conduct your own performance appraisals and how to prepare drills and exercises to assess the performance of an emergency response organization. Perform your job well by ensuring these evaluations take place on a regular basis.

KEY TERMS

Behavior criteria A standard for judging the success of a training program that refers to trainees' ability to apply new knowledge and skills to their jobs. This includes performance during drills, exercises, and incidents that take place after training is completed.

Capability assessment for readiness (CAR) program A state program that describes a self-assessment process for state emergency management agencies.

Controller The person who provides information from a scenario. Drills are relatively simple, so the same person can serve as the evaluator.

Development A function of an appraisal that focuses on improving an employee's *ability* to do a job. In this context, performance appraisal can be used to guide decisions about training, reassignment, or termination.

Drills A training exercise involving one or few people who must perform a specific task in response to a hypothetical scenario.

Entity A public or private sector organization that is responsible for emergency/disaster management or continuity of operations.

Evaluator The person who observes the player's performance and notes any deviations from the EOP or its procedures.

Full-scale exercise A training exercise that simulates a community-wide disaster by testing multiple functions at the same time. It also tests the coordination among these functions. The complexity of full-scale exercises requires thorough planning of the scenario. It also requires coordination among the many controllers and evaluators. There is also a need for training the controllers.

Functional exercise A training exercise that differs from a drill by involving more people. This makes functional exercises more comprehensive than drills. In addition, the scenario is usually more complex because it involves more tasks and equipment.

Internal research A function of an appraisal that lets an emergency manager know what skills his or her employees have. Internal research also lets emergency managers know if they were looking for the right qualities when hiring for the position.

Learning criteria A standard for judging the success of a training program that is defined by performance on written tests or performance tests of skills addressed in the training program.

Legal protection A function of an appraisal that is achieved when an organization conducts performance appraisals according to generally acceptable procedures.

Reaction criteria A standard for judging the success of a training program that consists of trainees' opinions of the training program. This includes evaluations of the trainers, the facilities and equipment, the material, their enjoyment of the class, and their desire to take another class from the instructor.

Results criteria A standard for judging the success of a training program that refers to the consequences of trainees' performance on the job. Results criteria evaluates whether the training made a difference in the overall performance of the organization.

Reward A function of an appraisal that focuses on improving a person's *motivation* to do the job. Appraisals should have clear criteria that provide guidance to the employee about what is important to the organization.

Table-top exercise A training exercise involving a group of senior personnel who are usually branch or departmental administrators and serve as the directors of their functions. Scenarios for these exercises vary in their complexity.

Task A specific activity carried out for a distinct purpose.

Taskwork The ability to perform each separate step of the response.

Teamwork The ability to schedule tasks and allocate resources among team members to achieve a performance that is efficient, effective, and timely.

ASSESS YOUR UNDERSTANDING

Go to www.wiley.com/college/lindell to evaluate your knowledge of evaluations. *Measure your learning by comparing pre-test and post-test results.*

Summary Questions

1. An appraisal should be conducted at least once a year. True or False?
2. Setting goals is not important to evaluations or organizations. It's the outcome of a goal that counts. True or False?
3. Which of the following exercises uses open-ended questions designed to generate a discussion about a particular problem and is usually conducted at higher levels in the organization?
   (a) drills
   (b) functional exercises
   (c) full-scale exercises
   (d) table-top exercises
4. Which of the following criteria are based on the performance on written tests or skills addressed in the training program?
   (a) reaction criteria
   (b) learning criteria
   (c) behavior criteria
   (d) results criteria
5. The drill controller can also be the drill evaluator. True or False?
6. Which of the following is comprehensive, involves many people, and involves many tasks?
   (a) drills
   (b) functional exercises
   (c) table-top exercise
   (d) taskwork
7. Which type of exercise involves a group of senior personnel only and includes a discussion of a number of topics?
   (a) full-scale exercise
   (b) drill
   (c) task
   (d) table-top exercise

8. In a performance evaluation, which of the following should be considered?
   (a) interpersonal skills
   (b) innovation
   (c) change management
   (d) all of the above

Review Questions

1. What four functions does an appraisal serve?
2. What are four of the typical performance appraisal categories?
3. How do you evaluate a LEMA versus a LEMC.
4. What are the differences between a drill, a functional exercise, a table-top exercise, and a full-scale exercise.
5. What are the steps into which FEMA organizes its emergency exercises?
6. What are the four criteria for judging the success of training programs?
7. What is the difference between a control group and a treatment group?

Applying This Chapter

1. One of your employees is a poor performer and always seems to have a negative attitude. Should you wait until the end of the year to discuss his performance during the annual performance appraisal your jurisdiction requires? If not, how do you conduct the review?
2. You have been asked to describe the criteria for the NFPA 1600 standard. The current version of NFPA 1600 defines criteria for what two types of programs?
3. As an emergency manager, you have been asked to write a report on the steps FEMA uses to organize full-scale exercises. Describe the steps you need to include in the report.
4. You have to evaluate the LEMC. What LEMC tasks do you evaluate and over what time period? Why? What type of objectives do you set in your evaluation?
5. You have two excellent employees who excel in all areas. Not only do they do their job well, but they consistently go above and beyond the call of duty for the organization. Unfortunately, there has been a salary freeze and you cannot increase their salary. Do you still perform a performance appraisal? Why or why not? Because you cannot raise their salary, what other things can you do to compensate the employees for their hard work?
6. The EMAP commission is going to perform an on-site evaluation of your organization. What information do you need to prepare for this assessment?

YOU TRY IT

9/11 Report

The 9/11 commission recommends in their report that the NFPA 1600 become the standard and be implemented in the private sector. What does NFPA 1600 require an organization to have?

Evaluating Exercises

You decide to put your jurisdiction through a full-scale exercise. The scenario is that a group of terrorists release a dirty bomb in the downtown area of your community. What are your criteria for judging the performance of the emergency response organization?

Evaluating the New Orleans Risk Communication Program

Hurricane Katrina caused billions of damage to property in the South and killed more than 1200 people. In the aftermath, the media focused a lot of attention on New Orleans. Knowing that the local authorities had warned the residents about the hurricane, what criteria would you use to judge their risk communication program?

13

INTERNATIONAL EMERGENCY MANAGEMENT

How Other Countries Manage Their Hazards

Starting Point

Go to www.wiley.com/college/lindell to evaluate your knowledge of international emergency management.
Determine where you need to concentrate your effort.

What You'll Learn in This Chapter

▲ The factors that cause variation in policy choices
▲ How a country's economic resources affect its emergency management programs
▲ How other countries responded to emergencies

After Studying This Chapter, You'll Be Able To

▲ Examine how the way a government is organized affects emergency management policy decisions
▲ Examine how the role of the military within a government affects emergency management strategies
▲ Assess the strengths and weaknesses of different countries' emergency management systems

Goals and Outcomes

▲ Work with people from other countries to develop an effective response to an emergency
▲ Assess a country's unique characteristics when developing an emergency management program
▲ Model responses to local emergencies based on case studies from successful responses to disasters in other countries

INTRODUCTION

You may be well aware of emergency management practices in the United States; however, you probably do not know how other countries approach emergencies. Countries are learning and borrowing policy from each other (Dolowitz and Marsh, 2000). Globalization has had two important effects. It has exposed all countries to an increasingly competitive economic system. Also, advances in communications have made it possible for policy-makers to communicate quickly and easily. Because of these changes, policy makers increasingly look beyond their national borders for ideas on how to address problems at home. In this chapter, you will the many variables that affect emergency response, as well as several examples of disasters from countries around the world.

13.1 Factors That Cause Variation in Policy Choices

Countries can be compared on the basis of many characteristics. These include regime type, political culture, modernity, and level of economic development (Barber, 1997; Inglehart, 1997). Yet, administrative structures are relatively similar across a broad range of countries. This is because the function of an administrative structure has an effect on the shape it takes (Peters, 1995). In addition, organizational models are frequently shared among groups of countries. For example, former colonies frequently have administrative structures that closely resemble those of their former colonial masters. Countries that share membership in a multinational organization such as the European Union (EU) frequently come to share forms of administrative organization in order to simplify cross-national cooperation. In disaster management, the influence of the United Nations has contributed to the use of common models, while encouraging countries to adapt these models to their own realities. Regional organizations such as the Asian Disaster Preparedness Center (ADPC) have also influenced the evolution of emergency management across a wide variety of nation-states. When comparing emergency management policies, the comparison must include *hazard vulnerability* and *local capability.*

13.1.1 Hazard Vulnerability

Hazard vulnerability varies according to hazard type and level of exposure. Frequently, the type of hazards and the level of exposure that affect a country influence the structure and the quality of its emergency management organizations. Countries with high levels of exposure have been described as having "disaster cultures." This enables them to adapt and respond to recurrent events. They can also show a high level of adaptation to particular hazards. For example, countries that face frequent typhoons have developed more sophisticated programs and policies than those that do not. Bangladesh has developed a system of evacuation platforms because of its very flat terrain. Taiwan has put money into typhoon warning research.

Hazard exposure and experience shaped programs in Central America after Hurricane Mitch in 1998. Prior to the arrival of this hurricane, most countries of the region had devoted little attention to emergency management. They had established basic "civil defense" programs. These were associated with the military and concentrated on disaster response. However, there was little or no attention paid to the connections between economic development programs and hazard vulnerability. After the hurricane, the governments of the affected countries began to change their national development programs. These emphasized the links between social factors, environmental degradation, and hazard vulnerability (Lavell, 2002).

13.1.2 Economic Resources

Emergency management is low on the priority list in poorer countries. Entire societies live on the brink of economic collapse, so more immediate problems take precedence. In the meantime, poverty and rapid urbanization generate large concentrations of vulnerable populations in high-risk urban areas. Both rich and poor countries have ignored the connection between the environment and human settlements. This neglect has delayed the development of a better understanding of how to develop land in a sustainable way. Ultimately, this has increased the number of disasters.

The quality of emergency management in a country is related to the amount of internal and external resources available. Many poor countries struggle with high levels of foreign debt. Often, this debt was incurred by earlier undemocratic regimes. To compound the problem, some of these countries devoted much of their national budget to the military. This left little money for education and health care. Emergency management was left far down the list. Sometimes this situation has been made worse by programs imposed by multinational lending agencies such as the World Bank and the International Monetary Fund (IMF).

Hazard insurance availability varies from one country to the next. Few countries have systems with market penetration as widespread as the U.S. National Flood Insurance Program (NFIP). In part, this is because participation in the NFIP is a condition for getting a federally-backed mortgage. Many foreign countries lack disaster insurance programs. If they do have one, premiums are usually too high so the majority of the population cannot afford them. In these countries, businesses might have disaster insurance, but few homeowners do. National governments might want to require homeowners to purchase hazard insurance. However, they cannot link hazard insurance purchase to mortgage approval, as in the United States. The reason is that fewer people in developing countries borrow money to buy a house. Typically, people first buy the land and then build the house one room at a time as they save enough money to purchase the construction materials. As a result, these governments lack the mechanism the U.S. government uses to intervene in the market.

Haddow and Bullock (2003) mention the availability of "specialized assets" as a factor affecting emergency management. These specialized assets may include items needed during response operations such as:

- ▲ Heavy equipment.
- ▲ Trained urban search and rescue (USAR) teams.
- ▲ Hazmat capabilities.
- ▲ Technical expertise such as geographic information systems (GIS).
- ▲ Training facilities.

Such resources are not available everywhere. They are often shared regionally through organizations such as the Caribbean Community's Caribbean Disaster Emergency Response Agency (UNISDR, 2002). USAR teams in particular are eager to participate in response efforts no matter where they occur. They can provide valuable assistance. They can provide on-the-job training. They can help with the difficult task of body retrieval. However, it is rare that they are able to arrive quickly enough to accomplish rescues during the critical first hours. This is due to the logistics of moving large numbers of people and their equipment. It is also due to legal and political problems with such movements. In some cases, fly-over rights have been denied to USAR teams. The entry of search dogs without the normal quarantine or veterinary procedures frequently causes problems. Thus, there is an increasing interest in development of local USAR teams.

13.1.3 Organization of Government

One of the most important issues is the degree of political centralization in a country. The control of policies, programs, and resources by the national level limits the ability of local governments to mount a rapid emergency response. It also makes it difficult for them to develop mitigation programs. Like police and fire protection, emergency management is a service that is delivered over a dispersed area. Consequently, it benefits from a significant degree of decentralization. This allows local governments to manage the service delivery (Peters, 1995, p. 161).

Too much emphasis on large disasters tends to lead to over-centralization. This is because it is assumed that there is a need for coordination over the large areas affected. Unfortunately, this also requires communication of information through multiple layers of government. This delays response and recovery operations. In reality, small frequent events cause more deaths and economic losses.

When small events occur, local governments can respond better if they do not need to wait for instructions and resources from the central government. Thus, empowering local governments and their populations to deal with local events is very effective. However, decentralization is frequently resisted by the national governmental authority because it reduces central control. Frequent, small events also point to the connection between patterns of development and hazard vulnerability. Increased recognition of this connection can lead to calls for more public participation in national goal-setting. This also threatens a status quo that benefits the elite.

The location of emergency management agencies in all levels of government is related to their effectiveness. It is also related to the emphasis given to different aspects of hazard vulnerability. An agency may be charged with responding to disasters. But, it will have difficulty in quickly finding and delivering the needed resources if it has a low status in government. Emergency managers benefit when they receive input from scientific agencies. However, they are often isolated from such input by their location in government.

Countries vary in the degree to which their emergency management agencies are staffed by professional emergency managers. Some countries have agencies with high political profiles and adequate resources. Those agencies are able to attract and keep well-qualified and dedicated personnel. Few countries have an adequate supply of well-trained emergency management professionals. This situation is of great concern, so some countries are developing training programs at their universities. For example, Istanbul Technical University in Turkey has recently developed a multidisciplinary program in emergency management.

13.1.4 Quality of the Built Environment

The quality of a country's infrastructure and housing affects the level of its disaster exposure. So do the quality of the business and industrial installations. These factors also affect the type of emergency management program it needs. For example, the better the construction, the less need there is for urban search and rescue teams after earthquakes. Similarly, good roads make it easier to evacuate large numbers of people. However, it does not necessarily follow that countries with higher quality infrastructure have lower hazard vulnerability. Countries with large numbers of high-rise buildings have an increased need for highly developed firefighting capabilities. This is also true for those with large chemical manufacturing installations.

13.1.5 Civil Society

Civil society includes all groups that are independent of the government. It includes religious groups, civic clubs, political parties, and other groups with

FOR EXAMPLE

Civil Society and Disasters

Civil society is often strengthened during response and recovery. This is especially true when government agencies prove inadequate to the task. Emergent organizations then rise to take on intractable problems. Such organizations were created in Mexico City after the 1985 earthquake (Velázquez, 1986). This also occurred in Kobe, Japan, after the Great Hanshin earthquake of 1995 (Shaw and Goda, 2004). In addition, existing organizations are strengthened as they extend their missions to take on new disaster-related tasks (Dynes, Quarantelli, and Wenger, 1990).

specific interests. Public opinion of their government's abilities affects the degree of trust they have in their government's emergency management efforts. When the public is well-informed and has strong beliefs in their rights, they are likely to demand competence. Nongovernmental organizations (NGOs) and community-based organizations (CBOs) are organizations that civil society uses to change governmental priorities or supplement weak powers with its own capabilities. True civil society groups are not organized by government agencies. They have grass roots organizations that emerge independently of government. These groups meet specific needs such as flood mitigation in a local watershed. They can exert substantial influence. They can even contribute to processes of regime change. For this and other reasons, governments may be wary of strengthening civil society.

13.1.6 The Role of the Military Emergency Management

The armed forces are involved in emergency management to some degree almost everywhere. The military has a high degree of organization. In some countries that is enough to differentiate it from other agencies. In addition, the armed forces usually have more resources needed for disaster response. These resources include communications, transportation, fuel, power, water, shelter, health care, and food. Equally important is the large number of strong, young men who are organized into groups and used to taking orders.

There is another reason for the strong influence of the military in emergency management. This has to do with the nearly universal roots of emergency management in civil defense. This includes the training and equipping of non-military personnel to repel invaders. In many countries, the military retains a strong influence on emergency management organizations. This usually leads to a strong emphasis on command and control models of disaster response that have little or no role for civilian input. Moreover, little attention is given to other needs such as disaster recovery and hazard mitigation. In other countries, the military is part of the emergency management system. However, it is under civilian control. In such cases, it cannot respond unless its presence has been requested. This option is preferred, if only to avoid overdependence on the military and risking a slide into an authoritarian government during a period of national weakness. The term "civil defense" is frequently used. However, this does not always imply a historic link to the national armed forces (Nikoluk, 2000).

13.1.7 The Role of International Organizations

Countries vary widely in their approach to foreign aid. This includes policies on when and how to send help to disaster areas or become involved in mitigation projects. Some countries have adopted a reactive approach. They confine themselves to offering assistance with search and rescue or postdisaster cleanup. By contrast, others have adopted a more developmentalist perspective. They assist in the formation of intergovernmental institutions and programs. Major goals in this approach are to reduce the incidence of disasters and to increase poor countries' ability to respond to emergencies.

Many international institutions are devoted to promoting improved emergency management practices. Noteworthy among them is the United Nations. The United Nations has an organization called the International Strategy for Disaster Reduction (UNISDR). It carries out the goals of the UN International Decade for Disaster Reduction. Regional institutions are also working on improvements. The UNISDR, the government of Japan, the World Meteorological Association, and the Asian Disaster Reduction Center issued an online publication, *Living with Risk: A Global Review of Disaster Reduction Initiatives,* in 2002. This resource, available at www.unisdr.org/eng/about_isdr/bd-lwr-2004-eng.htm, contains information on emergency management worldwide.

In addition to the United Nations, there are regional groups involved in emergency management. The Organization of American States has supported the development of disaster-resistant schools, hospitals, and road networks through its *Natural Hazards Project.* This is a division of the Unit for Sustainable Development and Environment. **La Red** is a network of Latin American social scientists that publishes scholarly work on disasters in the region. Their work highlights the issues of social vulnerability and sustainable development. The **Pan American Health Organization** is the regional office of the World Health Organization. It emphasizes retrofitting hospitals and strengthening public health programs. It also publishes an influential newsletter called *Disasters: Preparedness and Mitigation in the Americas.* It can be found at *www.paho.org/english/dd/ped/newsletter.htm.* Other regions around the world have similar groups. One example is the Asian Disaster Preparedness Center, which is based in Bangkok, Thailand. Organizations like these provide mutual aid, promote regional discussions, and share technology.

SELF-CHECK

- Explain why emergency management is low on the priority list in poorer countries.
- Explain why the degree of political centralization in a country is an important issue.
- Define **civil society.**
- Describe the role of the military in society and emergency management?

13.2 Examples of International Emergency Management Programs

The following sections of this chapter offer examples of emergency management programs and practices in a selection of countries. These examples show the wide variety of problems and solutions facing the varied populations of the globe as they confront hazardous environments.

13.2.1 Preparedness: Landslide Evacuation in São Paulo, Brazil

Brazil is a republic whose states have strong political powers. It also has wide regional variation in topography, soils, vegetation, and climate. Historically, the population centers have been in the mountainous coastal region. The interior plains states experienced increasing development only after construction of a new capital city, Brasília, in the 1960s. Brazil is fortunate in its lack of exposure to major earthquakes or hurricanes, two of the most catastrophic natural disasters in South America. However, it does experience frequent natural disasters of other types. These include floods, droughts, landslides, and wildfires. There are also significant technological hazards due to a high degree of industrialization in the central and southern regions. Brazil has a significant nuclear industry as well.

Emergency management in Brazil has developed over time to reflect the structure of the government and its hazards. The 1988 Constitution mandated "planning and promotion of defense against public calamities, especially drought and floods" (Ministério da Integração Nacional, 1999). The state of São Paulo created its State Coordinator of Civil Defense after several major disasters in the

1970s, including floods, landslides, and fires in high-rise buildings (Marcondes, 2003). This office developed the Plano de Prevenção da Defesa Civil (Civil Defense Preparedness Plan [CDPP]) as part of an effort to meet the goal of natural disaster reduction during the International Decade for Natural Disaster Reduction (IDNDR). It also established regional and municipal emergency management offices. These offices vary in quality and some municipalities have yet to establish their civil defense committees.

Landslides are the most common cause of deaths from natural disasters in Brazil. Many factors contribute to this situation. These include environmental factors such as varied topography, high levels of rainfall, and soil types that are prone to slipping. Social factors, which are even more important, include high poverty levels, rapid urbanization, and lack of adequate housing in safe areas. Another problem is a lack of education about the causes of natural disasters (Macedo, Ogura and Santoro, 2002).

The CDPP was developed as a cooperative effort among technical agencies, state and local governments, and local emergency management professionals. It is based on the monitoring of watersheds. They are monitored for rainfall levels, soil saturation, and weather forecasts. The first step in this plan was hazard mapping and risk analysis of the state. This was undertaken by the Geological Institute of the State of São Paulo and the Institute for Technological Research at the University of São Paulo. They determined that the Serra do Mar is the most vulnerable region of the state. This is a mountain range that follows the coastline and is undergoing rapid urbanization. Consequently, landslide risk areas in the Serra were mapped. Local governments were involved in the risk analysis to collect data on landfills and dumps, cuts, surface drainage patterns, and landslide scars. They also collected data on the number and location of houses (Scachetti, no date).

The Southern Hemisphere's summer rainy season is from December through March. During this time, emergency managers in the Serra do Mar activate a landslide monitoring system by entering the *observation* stage. During this stage, rainfall is closely monitored by computerized data collection stations distributed throughout the region. The rainfall data are continuously transmitted by satellite to the state Coordinator's Office. When rainfall reaches a critical level, the *attention* stage is activated. During this stage, field observers search the area for ground instability features. These include cracks at the upper part of slopes and bulges at the bottom of slopes. Other signs of slope instability include color and quantity changes of spring water, small rockfalls, small ground failures, and cracks in the walls of houses. When these signs are detected, the *critical* stage is entered. During this stage, inhabitants of the houses at immediate risk are evacuated. If conditions persist or deteriorate, declaration of an *emergency* stage requires all inhabitants of the risk area to be evacuated. The plan includes a significant public education component, with the State Coordinator of Civil Defense Office and municipal governments producing a wide variety of print and video materials.

These include graphic and verbal descriptions of ground instability features. They also include lists of what to do and who to call when the signs of a dangerous situation are detected. Figure 13-1 shows a sample of the public education materials developed by CEDEC.

The CDPP, which was instituted in 1988, has been highly successful in reducing the number of deaths due to landslides in the Serra do Mar. There were 48 deaths in two cities alone (Cubatão and Ubatuba) in 1986. There were 39 deaths in Cubatão and Santos during 1987. During the first year of the system's operation, 17 people died. There were 9 deaths in 1996 and 7 deaths in 1995, but the rest of the years from 1989 to 1999 saw 0–3 deaths. The system has been so successful that it is being implemented in another landslide-prone region, the Serra da Mantiqueira in the Paraíba Valley (Ridente et al., 2002).

13.2.2 Restructuring Emergency Management in New Zealand

New Zealand is exposed to floods, earthquakes, tsunamis, and cyclones. There is a popular image of New Zealanders as sheep farmers, but the population is 85%

Figure 13-1

Raios! O que faço?!

- Nas tempestades, nunca fique no mar ou em piscinas;

- Fique longe de árvores, postes ou linhas de energia elétrica pois podem conduzir raios;

- NÃO fique em lugares descampados (praias, campos de futebol, quadras, etc.) ou altos;

- Mantenha-se longe de estruturas metálicas, como varais ou trilhos;

- Não segure objetos pontiagudos ou metálicos.

NA SUA CASA:

- NÃO fique próximo a tomadas, canos, janelas e portas metálicas, nem use telefone ou aparelhos ligados a tomadas.

- Verifique se sua casa ou prédio tem sistema de pára-raios, obrigatório para prédios maiores de 30 m e terrenos com mais de 1.500 m2.

SOBRE O PÁRA-RAIOS:

- Não protege aparelhos eletrônicos, portanto é melhor desligá-los nas fortes tempestades;

- É importante sua manutenção a cada seis meses, ou após o equipamento ter sido atingido por um raio;

Coordenadoria Estadual de Defesa Civil

Moro em um morro... tem perigo?

Não desmate, principalmente em morros, pois sem árvores, vegetação ou grama, ficam sujeitos a desmoronamentos.

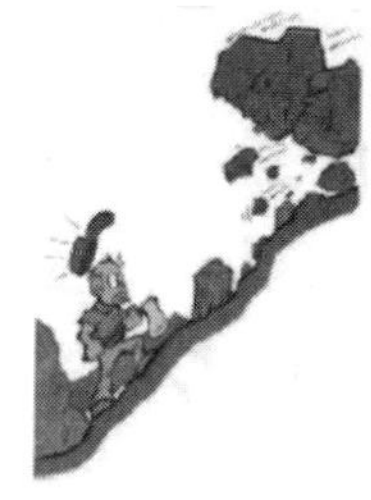

Porém, saiba que bananeiras em morros são sinais de perigo, pois concentram água no solo e tem raízes rasas, facilitando deslizamentos. Procure técnicos da Prefeitura para ajudar a substituí-las.

Faça obras em patamares. Aterros e cortes nas encostas deixam o terreno instável. Para construir sem risco junto às encostas ou morros, procure a orientação da Prefeitura.

Canos de água com vazamentos, esgoto a céu aberto, valas obstruídas e acúmulo de lixo nas encostas também provocam deslizamentos.

Trincas ou barrigas no chão e nas paredes de sua casa indicam problemas de estrutura, e até risco iminente de queda. Na dúvida, ligue 199.

Fique atento para sinais de perigo, como árvores, postes ou muros inclinados, rachaduras ou trincas no terreno e água mais barrenta que o normal. **Na dúvida, acione a Defesa Civil através do telefone 199.**

Public education material used in Brazil (Governo do Estado de São Paulo, 2000).

urban. Consistent with a worldwide wave of decentralization efforts, New Zealand made significant changes to their legal framework of city governance. Beginning in the 1970s, the government privatized a number of government functions. These reforms increased governmental transparency and accountability. Many of them affected emergency management.

The *Local Government Act* of 1974 and its 1996 amendment have increased the responsibilities of local governments. These acts authorized local governments to produce and implement their own plans. In particular, local financial and environmental management plans enabled communities to better control their quality of life. Sustainability concerns drove the Resource Management Act of 1991, which promoted performance-based environmental management by local authorities. Because the act refers to natural hazards and hazardous substances, it integrates emergency management into land-use planning and development. Other acts specifically aimed at emergency management include the Hazardous Substances and New Organisms Act of 1996, the Building Act of 1991, and the Biosecurity Act of 1993, among others. Figure 13-2 shows the relationships of factors driving changes in emergency management to the agencies, methods, and goals of change.

The most recent major disaster in New Zealand was the Napier earthquake of 1931. The long time since the last major event had allowed a certain level of complacency to arise (Britton, 2001). However, the Loma Prieta and Northridge earthquakes in the United States were instrumental in shaking this complacency. The government undertook a thorough review of emergency management legislation and agencies in the mid-1990s. One of the main reasons for the review

Figure 13-2

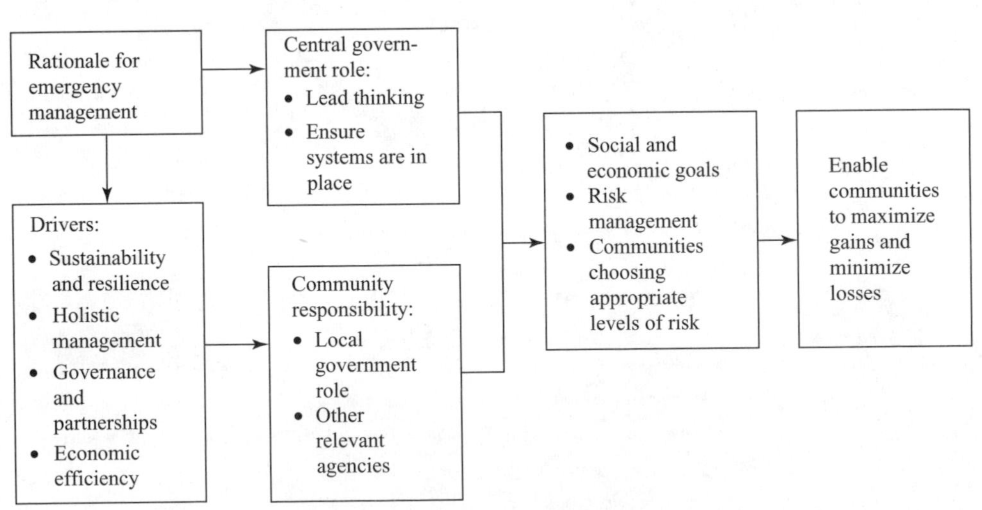

Emergency management drivers (adapted from Britton 2001).

was the belief that emergency management in New Zealand must shift away from a deterministic and reactionary orientation. It must move toward the use of hazard assessment and risk identification to promote the incorporation of hazard reduction into land use management. This effort included reports, conferences, and workshops that together produced some fundamental changes.

The review process identified several problems with the country's emergency management system. One problem was high public expectations of aid following disasters. Another problem was insufficient central and local government capacity. In order to address these problems, New Zealand's approach needed to change. The emergency management sector needed to learn from other countries' experiences, adapt to change, and better coordinate resources. To accomplish these objectives, the review process recommended replacing the existing response-focused approach. Instead, the country needed to adopt a comprehensive approach that incorporated all hazards and all phases of emergency management.

As a result of the review process, a new Ministry of Emergency Management (MEM) was created in 1999. It has three operational units, the first of which is the Sector Development and Education Unit. This unit was given responsibility for developing and translating these concepts into working models and practices. It also is responsible for evaluating emergency management effectiveness. The second unit is the Policy Unit, which is responsible for framing strategic policies. The Policy Unit is also responsible for integrating the MEM's work with that of other governmental agencies. This unit took the lead responsibility in developing a bill that provides a better legislative foundation for national emergency management. The third unit is Sector Support. This unit works with local governments and emergency services agencies to ensure their effectiveness.

In 2000, legislation developed under the MEM was introduced into Parliament to reconfigure New Zealand's emergency management practices, processes, and structures. The integration with other community goals such as growth, development, and sustainability is a major principle of the new legislation. The goal is to "improve and promote community resilience and continuity through comprehensive, integrated, and risk-based emergency management" (Britton and Clarke, 2000, p. 147).

Emergency management was a "core function" of central and local governments in 1997. The new legislation established a system of Emergency Management Groups (EMGs) throughout the country. Each EMG is composed of several local emergency management authorities. There are fewer than 20 EMGs that incorporate 86 local and regional authorities (Britton and Clark, 2000). The designers believed that this would increase local emergency management effectiveness by increasing access to regional resources. In addition, the EMGs will allow for regional coordination in mitigation, preparedness, and response.

Through its reform process, New Zealand has sought a balance between centralization and localization of emergency management. Policy, strategic, and support functions remain at the national level. However, local and regional authorities have the authority to carry out functions in the ways that are best for their situations. The current statutes emphasize sustainable management of the physical and social environment. They use performance standards to avoid, mitigate, and remedy adverse effects of social and economic activities.

13.2.3 Reconstruction and Recovery: India

India is a large country whose population is highly vulnerable to disaster impacts caused by poverty and crowded, substandard living conditions. The country suffers from frequent droughts, floods, earthquakes, tropical cyclones, and, recently, tsunamis. In spite of its drought hazard, famine has not been as serious a problem. This appears to be due, in part, to the famine relief systems begun under British colonial rule. It is also due to the existence of a free press that can publicize government failures, and a transportation system that allows for rapid large migrations (Sen, 1981). India has a federal political system that delegates the main responsibility for disaster relief and reconstruction to state governments. The national government supports their efforts with financial and logistical support. It also arranges for assistance from outside agencies such as the World Bank and the Asian Development Bank.

As in many poor countries, emergency management has been primarily reactive. India has concentrated on relief after major disasters. There are few resources available for emergency management activities before disasters strike. This makes it very important to include hazard disaster preparedness and hazard mitigation as a part of reconstruction (Vatsa and Joseph, 2003). External agencies have the power to require the inclusion of these activities into any reconstruction efforts that they fund. However, they do not always exercise this influence.

In September 1993 a 6.4 M earthquake struck the southeastern part of Maharashtra state. The earthquake caused about 8,000 deaths and 16,000 injuries. It inflicted extensive damage to 1,500 villages and completely destroyed 70 of them. The state of Maharashtra is the financial center of India and has a progressive, decentralized and competent government. Even so, outside resources were needed to help deal with the disaster's effects. The outcome was a $358 million reconstruction project funded primarily by the World Bank.

The Maharashtra Emergency Earthquake Reconstruction Program (MEERP) was initially focused on reconstruction and rehabilitation. It did not include the development of a disaster management plan. One component of the program, the Technical Assistance, Training and Equipment component, was given the task of supporting disaster management. However, no specific goals or timetables were set. The first step towards developing a disaster management plan was a workshop held in May 1995. This was almost a year after the MEERP began. The government of Maharashtra was to prepare disaster management plans for the state and four districts. These were to be used as demonstration projects. In addition, the Maharashtra government was advised to formulate 15 functional committees to develop the program. For various reasons, no immediate steps were taken. When activity began in January 1996, a Disaster Management Council was created that had the responsibility for plan development. The number of districts involved was increased to six and the number of committees was reduced to five. Each committee was assigned responsibility for specific hazards, but no committee was given responsibility for droughts. It was felt that the state already had adequate drought management plans.

The five hazard committees were useful in establishing a broad framework for assessment and response. The state government developed the actual plan with the help of national and international consultants who provided substantive and political expertise. In the end, the newly created Center for Disaster Management in the state training institute developed plans for all 31 of the state's districts rather than just the original six. In addition, the Center developed a separate plan for the city of Mumbai. This is because of the city's size and administrative complexity.

The external funding and support facilitated the rapid development of technical and administrative institutions for hazard management. It also raised the prestige of those involved in developing disaster plans. However, the state government must maintain its support for this newly developed emergency management capacity to make sure the effort has a long-term payoff. In the short term, disaster management planning gave the state a unique level of expertise. In later disasters, other Indian states were able to draw upon this expertise. Moreover, the project increased awareness at the national level of the need for disaster management throughout the country. However, it was a top-down effort. This raises questions about the long-term commitment of local governments to plan implementation.

The Maharashtra disaster management planning project contrasts with another effort, the Patanka New Life Project. A consortium of government agencies and NGOs from India, Nepal, and Japan funded this project after the 2001 Gujarat earthquake. The Patanka Project was designed as a model for community rehabilitation. It had a strong focus on teaching new skills that could mitigate the effects of future events. The project was designed with input at each stage from the citizens of Patanka village. Its main goals were to rehabilitate lives of residents by providing safer houses, better infrastructure, and "greater livelihood security." It also used a shake table demonstration to build local capacity for earthquake safe construction (Shaw, Gupta, and Sharma, 2003).

The Patanka project used a three-stage process that was developed and implemented over a two-year period. Stage I established the principles for the project, with the project team taking the lead responsibility for this activity (see Table 13-1). Stage II involved implementing the Community Action and Implementation plans developed during the consultations in Stage I. Stage II had three steps: needs assessment, capacity building, and implementation. The project team worked closely with local leaders to win community trust and develop strong local leadership. It also promoted sustainable development activities that would continue after the project itself ended. In order to build local hazard mitigation capacity, local masons were trained in building techniques that would allow them to continue using traditional, affordable materials while building more earthquake resistant homes. This training allowed them to continue their traditional livelihoods. In turn, the continuity of employment contributed to the Stage III goals of ensuring the sustainability of local social systems. It also enabled the community to take care of its own development needs and become resilient.

Table 13-1: Checklist for Sustainable Community Recovery

Stage I	*Stage II*		*Stage III*
Establish principles	Needs assessment	Capacity building	Local institutional strengthening
Rehabilitation linked to development	Dialogue	Training of masons, labor	Integration with government development schemes
Rehabilitation to be participatory	Training and demonstration	Building community confidence in disaster resistant practices	Creating assets for security
To follow minimum established standards	Community feedback	Strengthening institutional structures at community level	Ensuring means for continuous capacity building process
Rehabilitation aimed at reducing vulnerability	Damage assessments	Social mobilization	Providing new opportunities for growth
Promote empowerment	Identifying suitable options	Social calendar	
To be flexible	Preparation of local plans	Joint action	

Table 13-1: ***(Continued)***

Cooperation among stakeholders	Community preferences	Prepare sector specific action plans
Improve quality of life	Mechanism for joint action with the community	One-on-one dialogue
Strategic planning	Identifying areas of capacity building	Flexible approach
Mission	Meeting with community, involving government	Guidance and supervision of ongoing construction
Aims and objectives	Adapting government guidelines	Role clarification and transparency
Establish a team	Identifying confidence building measures	Establishing infrastructure for local storage of raw materials
	Making the first move to forge trust with the community	Establishing systems for monitoring and evaluation of construction work

(Shaw, Gupta, and Sarma, 2003)

The success of this type of intervention depends on close communication and cooperation at every stage between the project team and the community. The community must own the project for it to achieve its goals. Although foreign donors provided financial, technical and logistic support, the project was developed from the beginning with a view to local needs and perceptions. This ensured its thorough integration within the community.

13.2.4 Mitigation: Land-Use Planning in Colombia

Colombia is a unitary republic with 32 political subdivisions called departments. It is vulnerable to several hazards including volcanoes, earthquakes, floods, and landslides. Figure 13-3 shows earthquake damage that occurred in 1999. Under its centralized form of government, Colombia's national government wields many powers that are delegated to the states in the United States. In the late 1980s, Colombia began a major political restructuring. In 1988, Law 46 required the "efficient and opportune" management of resources necessary to provide for disaster

Figure 13-3

Earthquake damage in Armenia, Colombia, 1999 (World Vision International).

prevention and response. Decree Law 919 of 1989 organized the National System of Disaster Prevention and Response. It established the structures needed to implement the law. In 1997, Law 93 set forth a National Plan addressing the prevention, response, reconstruction, and "development" phases of emergency management. It addressed economic and legal issues, including education and community participation. It addressed the integration of information technology and communication systems across national, regional, and local levels. The National Plan was finally approved in 1997 (Martínez et al., 1997).

The National Plan has three goals: disaster reduction and prevention, effective disaster response, and rapid recuperation of affected areas. The first of these goals includes development planning at the sectoral and regional level. It also includes land-use planning at the municipal level. The law recommends many programs including integrated environmental policy and disaster prevention. It recommends that information on threats, risks and vulnerabilities should be incorporated into national, regional, and local environmental profiles. It also requires environmental management, and local urban development plans to incorporate this information. Basic sanitation infrastructure and other preventive measures for biological and industrial hazards should be put in place. In addition, watershed management and wildfire reduction plans should also be established.

To achieve these objectives, the Territorial Development Law includes provisions for natural hazard risk assessment, land-use planning, and urban development. The law provides detailed directions on the elements to be included in land-use plans. First, urban planning actions are to include the identification and placement of infrastructure for treating hazardous waste. It identifies areas that are unsuitable for human occupation due to hazard exposure. It also identifies the areas in need of recuperation and management in order to prevent disasters.

In addition, districts must consider ways to conserve the environment. These measures include mapping zones that have high exposure to natural hazards and developing management strategies for such zones. Both urban and rural land use plans must identify hazard risk areas and their designation as protected areas. Governments can take property and resettle people living in risk areas. Governments can also fine people who occupy risk areas illegally. The law makes executives of municipalities and districts responsible for keeping risk areas free of human settlements. Local governments must develop, implement, and enforce their own plans. Their plans must contain all the required elements. However, these plans must follow the guidance of the national government's law.

Colombia's Territorial Planning Law is remarkable because its requirements are more stringent than those that many states in the United States place on their local governments. This law is also interesting in the way it incorporates hazard mitigation directly into the fabric of local land-use planning.

13.2.5 Addressing Hazmat Through Land-Use Planning

The passage of the 1986 Emergency Planning and Community Right to Know Act (EPCRA) was an important event in the management of chemical hazards in the United States. Since that time, citizens have had a legal right to information about chemical hazards in their communities. Since the 9/11 terrorist attacks, access to chemical hazard information has been systematically eroded. These events have also shown the weakness of a policy relying on citizens to manage chemical hazards in their communities. The European Union has taken a slightly different approach that is defined by its Seveso Directives. The Seveso Directives incorporate many of the community Right-to-Know provisions in SARA Title III. However, they go further by requiring active dissemination of information to the public. They also use land-use planning to manage chemical hazards (Eijndhoven et al., 1994; Parker, 1999).

The Directives are named after an accident that occurred in Seveso, Italy, during 1976. A chemical plant released a tetrachlorodibenzoparadioxin (TCDD, or dioxin) cloud into the surrounding environment. There were no immediate deaths, but this extremely hazardous substance was deposited over ten square miles. The release resulted in the evacuation of 600 families and the treatment of some 2,000 people for dioxin poisoning. This incident resulted in the adoption of the first Seveso Directive by the European Council in 1982. Seveso I was amended in response to the Bhopal, India, accident in 1984 and the Sandoz warehouse accident in Basel, Switzerland, in 1986. The directive was reviewed in 1996 and an expanded version, called Seveso II, was adopted in December 1996.

The primary goal of Seveso II, which became effective in February 1999, is to prevent hazmat accidents. The secondary goal is to limit the health, safety, and environmental consequences of any accidents that do occur. These goals are achieved through restrictions on hazmat storage and processing. Seveso II focuses on "lower tier" and "upper tier" facilities that hold more than the minimum quantity of hazmat. Each type of facility is covered by requirements appropriate to the quantities it stores. Seveso II does not cover radiological materials or hazmat transportation (including pipelines). The chemical industry has already adopted voluntary programs for the prevention of transport accidents. They also cooperate with local authorities on emergency preparedness and response.

Seveso II requires the operator of each hazmat facility to develop an on-site emergency response plan and supply it to local authorities. The local authorities must then use the on-site plans to develop their own EOPs. On-site emergency plans must be developed in consultation with plant personnel. Off-site EOPs must be developed in consultation with the public. Information is to be shared with the public by actively distributing literature about actions to be taken in case of an accident. Accidents must be reported to the Community Documentation Center on Industrial Risks at the Major-Accident Hazards Bureau of the European Union (EU). (mahbsrv.jrc.it/). This Web site has guidance documents for safety management systems, safety reports, inspections, public information, and land-use planning. These documents do not have legal status. However, they do represent the views of all the EU member states.

Seveso II requires the integration of hazard mitigation into the land-use planning process. By doing this, it establishes a legal framework in member states for maintaining appropriate distances of hazmat facilities from residential, public-use, or environmentally sensitive areas. Member states must place controls on the siting of new hazmat facilities and modify existing hazmat facilities. In addition, they must control new urban development in the vicinity of existing hazmat facilities.

Member states have made varying degrees of progress in their implementation of the Seveso II Directive. Many northern European countries have already developed land-use planning procedures for hazmat facilities, but southern European countries have lagged behind. There are three basic approaches that have been adopted for Seveso II compliance. The first approach develops tables of "appropriate distances" based on experience with compatibility of specific types of hazmat and particular land uses. The second approach is a "consequence based" procedure that assesses the possible consequences of accidents. Third, there is a "risk-based" approach that incorporates the probability of an accident's occurrence. It considers the possible consequences in calculating vulnerable zones. At the present, it is not clear if the use of different approaches to calculate the necessary zones leads to very different results. Moreover, these three approaches are not mutually exclusive. A country could decide to adopt all three approaches and apply each one to a different category of hazmat facilities, depending on considerations such as cost-effectiveness (Christou, Amendola, and Smeder, 1999).

Currently, the first approach is used in Germany and Sweden. In Germany, the limited zones require no risk to humans or the environment from hazardous facilities. The consequence-based approach is used in France and the French-speaking area of Belgium. It is the basis for calculating vulnerable zones for emergency planning in the United States. The risk-based approach is used in the Netherlands, the United Kingdom, and the Flemish-speaking region of Belgium. It is also used in such non-European Union countries as Australia and Switzerland. Table 13-2 shows the zoning scheme that the Health and Safety Executive of the United Kingdom developed in compliance with Seveso II.

Table 13-2: United Kingdom Health and Safety Executive Siting Policy within Consultation Zones

Category of development	*Inner zone individual risk exceeds* 10^{-5}	*Middle zone individual risk exceeds* 10^{-6}	*Outer zone individual risk exceeds* 0.3×10^{-6}
Highly vulnerable or very large public facilities (schools, hospitals, nursing homes, sports stadiums)	Advise against development	Specific assessment necessary (advise against development if >25 people)	Specific assessment necessary
Residential (housing, hotel, holiday accommodation)	Advise against development (>25 people)	Specific assessment necessary (advise against development if >75 people)	Allow development
Public attractions (substantial retail, community, and leisure facilities)	Specific assessment necessary (advise against development if >100 people)	Specific assessment necessary (advise against development if >300 people)	Allow development
Low-density (small factories, open playing fields)	Allow development	Allow development	Allow development

(Christou et al., 1999)

13.2.6 Civil Society: The Chi-Chi Earthquake in Taiwan

Taiwan's democracy has recently developed from the ground up. This process began in the 1980s with the legalization of opposition political parties. It continued through the election of the first opposition president in the 2000 elections (Rigger, 1999). In early fall of 1999, a 7.6 magnitude earthquake hit the north-central part of the island. The Chi-Chi earthquake caused over 2,400 deaths and more than 11,000 injuries requiring medical attention. It destroyed or damaged thousands of buildings. Economic losses were estimated at 14 billion, which was 3.3% of Gross Domestic Product.

Taiwan had a sophisticated seismic network that gave central government officials immediate information on the earthquake's location, magnitude, and shaking intensities. This activated the national emergency management system, so representatives of essential agencies assembled at the Central Disaster EOC in Taipei. It was difficult for this group to be effective because the EOC staff only knew the size of the earthquake. They did not have detailed information from the disaster sites because damage to roads and communications links caused long delays in their receipt of information about actual effects and local needs. In turn, the lack of local information delayed the central government's response. The delayed response of the central government was a severe problem for local authorities because Taiwan's government is highly centralized. Consequently, local government lacked the resources and experience to respond until the central government could overcome its lack of emergency assessment capability.

Nonetheless, the response at the disaster site was massive and, eventually, effective. There were many totally and partially collapsed buildings. This resulted in a big need for urban search and rescue (USAR). Fortunately, Taiwan has many volunteer mountaineering groups that specialize in wilderness search and rescue. These groups, such as the International Association of Search and Rescue of the Republic of China (IASAR/ROC), activated their members immediately. They worked together with local firefighters in the mountainous areas that were hit hardest by the earthquake. IASAR/ROC teams also worked with foreign USAR teams that followed. Most of the more than 5,000 live rescues were performed by local volunteers and firefighters before the foreign teams could even arrive. As often happens after disasters, logistical, geopolitical, and bureaucratic problems delayed the outsiders' arrival. This incident reinforces the importance of local response capabilities. It also reinforces the need for local and national governments to plan for the integration of local volunteers in the crucial first hours after a major disaster.

The IASAR/ROC was founded in 1981 and now has about 10,000 members. They are organized into local teams of 50 to 90 people. The paid staff is small, about eight people at the central office in Taoyuan and one in each division office around the island. Members pay annual dues of about US$ 60 and receive subsidies from local businesses and religious groups to buy equipment and pay for training. They have experience with mountain rescues and also have participated in rescue efforts with a team from the United States when a high-rise in Taipei collapsed the year before the earthquake. This effort had given them exposure to the USAR techniques that were needed to respond to the Chi-Chi earthquake (Prater and Wu, 2002).

The educational infrastructure of the affected area was severely damaged. Government resources were insufficient to repair the nearly 800 primary and secondary schools that were damaged and destroyed. In Nantou County, the location of the epicenter, 75% of the schools closed. Here, as in other countries, NGOs stepped in to fill the gaps. The Taiwan affiliate of the International Red Cross committed to rebuild 14 primary and middle schools at a cost of US$ 15 million.

The activities of the Buddhist Compassion Relief Tz' Chi Foundation provides a particularly vivid illustration of the important role NGOs play after disasters. This group, which was founded by a Taiwanese Buddhist nun, has a long history of responding to disasters worldwide. Tz' Chi members were in the process of assisting in the relief and reconstruction effort after Turkey's Kocaeli earthquake when their own country was hit. Tz' Chi members immediately set up vegetarian soup kitchens to supplement the nonvegetarian meals served by other groups. They also began collecting equipment, clothes, and money for the victims. The organization provided tents and more substantial temporary housing. However, their most extensive recovery effort was the reconstruction of 53 schools. Their effort was unique because it delivered culturally appropriate and environmentally sensitive architectural designs for each school. Instead of using a single design for all schools, architects met with the community members to assess local needs. Once they had this information, they developed designs that incorporated local cultural motifs and environmental features. All designs were earthquake-resistant and incorporated natural ventilation and lighting to keep students comfortable while using a minimum amount of energy. Figure 13-4 shows the reinforcement being installed before pouring concrete for one of the new schools.

The government's social services were also stretched to the limit and beyond after the Chi-Chi earthquake. The Presbyterian Church in Taiwan developed a program to supplement Nantou County's six professional social workers. They did this by helping minority communities with their unique recovery needs (Prater and Wu, 2002). There are few minorities in Taiwan, but there are several thousand members of aboriginal tribes. These tribes live in isolated villages in the mountainous areas of central Taiwan that were heavily damaged by the earthquake. Community recuperation centers were created to provide counseling and day care. They also helped with navigating the bureaucracy to acquire recovery assistance and to design economic development projects. In many cases,

Figure 13-4

Rebuilding begins in Taiwan (Buddhist Compassion Relief Tz' Chi Foundation).

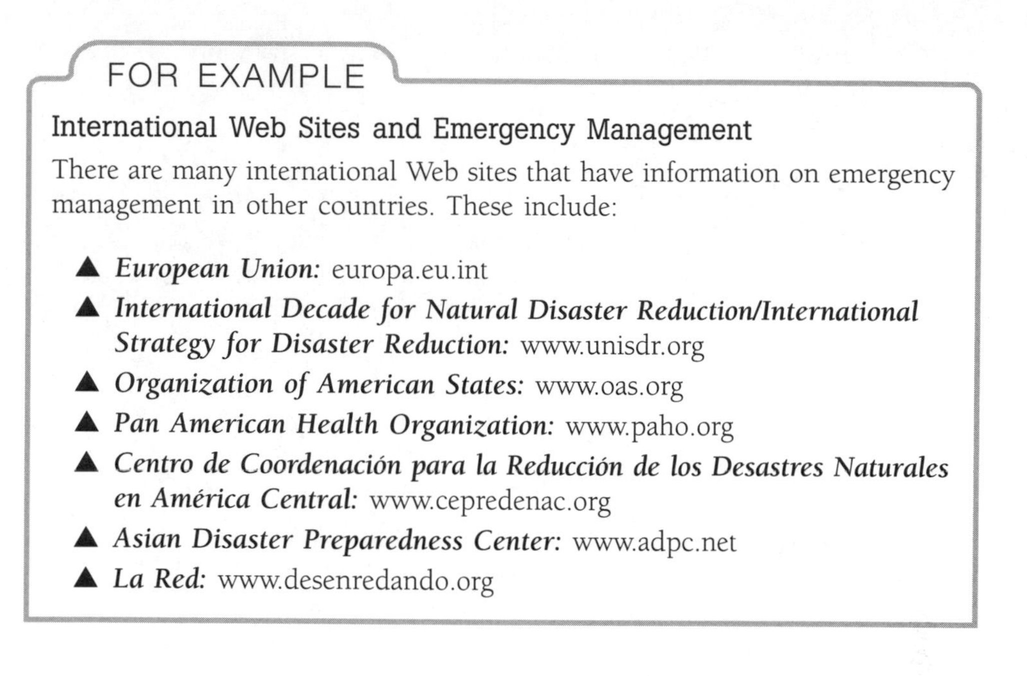

FOR EXAMPLE

International Web Sites and Emergency Management

There are many international Web sites that have information on emergency management in other countries. These include:

▲ *European Union:* europa.eu.int
▲ *International Decade for Natural Disaster Reduction/International Strategy for Disaster Reduction:* www.unisdr.org
▲ *Organization of American States:* www.oas.org
▲ *Pan American Health Organization:* www.paho.org
▲ *Centro de Coordenación para la Reducción de los Desastres Naturales en América Central:* www.cepredenac.org
▲ *Asian Disaster Preparedness Center:* www.adpc.net
▲ *La Red:* www.desenredando.org

local workers were hired and given the necessary training to staff the centers. These volunteers provided significant assistance in promoting the full recovery of households in earthquake impact zone.

SELF-CHECK

- Explain why famine has not been as serious a problem in India in recent years.
- Outline the stages used in the Patanka project.
- Identify the three goals of The National Plan of Colombia.
- Identify the goals of Seveso II.

SUMMARY

Globalization has significant effects on the world. Today, people are much more aware of natural disasters than they ever have in the past. And, countries are able to collaborate more easily to provide assistance. Responding to natural disasters often times brings the world together, as evidenced by the great tsunami in December of 2004. Perhaps the world has more to learn from these disasters than just the disasters' effects.

KEY TERMS

Civil society A society that includes all groups that are independent of the government, including religious groups, civic clubs, political parties, and other groups with specific interests.

La Red A network of Latin American social scientists that publishes scholarly work on disasters in the region. Their work highlights the issues of social vulnerability and sustainable development.

Pan American Health Organization The regional office of the World Health Organization. It emphasizes retrofitting hospitals and strengthening public health programs.

ASSESS YOUR UNDERSTANDING

Go to www.wiley.com/college/lindell to assess your knowledge of international emergency management.
Measure your learning by comparing pre-test and post-test results.

Summary Questions

1. True civil society groups are grass roots organizations that emerge independently of government. True or False?
2. The Natural Hazards Project is a division of the United Nations. True or False?
3. In New Zealand, which act integrates emergency management into land-use planning and development?
   - (a) the Local Government Act of 1974
   - (b) the Resource Management Act of 1991
   - (c) the Hazardous Substances and New Organisms Act of 1996
   - (d) the Building Act of 1991
4. Countries with high levels of exposure have disaster cultures. True or False?
5. In poorer countries
   - (a) emergency management is top priority.
   - (b) disasters are more likely to occur.
   - (c) emergency management is low on the priority list.
   - (d) there is always rapid urbanization.
6. Too much emphasis on large disasters leads to
   - (a) decentralization.
   - (b) quick, effective responses.
   - (c) over-centralization.
   - (d) an empowerment of local governments to handle emergencies.

Review Questions

1. What are the specialized assets mentioned that affect emergency management?
2. How is the United Nations devoted to promoting improved emergency management practices?

3. What are some factors that cause landslides to be the most common cause of death from natural disasters in Brazil?
4. What does the Buddhist Compassion Relief Tz' Chi Foundation do to help victims after disasters?

Applying This Chapter

1. In recent years, a large number of natural disasters have struck the continent of Asia. What would you do to better prepare that continent for disasters?
2. What can Americans learn from the disasters world-wide that should be implemented into our emergency response preparedness?
3. What characteristics of the United States hinder emergency management and emergency response efforts? What characteristics of the United States are beneficial to you as an emergency manager?

YOU TRY IT

International Emergency Response Preparedness

A New Orleans group of emergency managers is studying how other countries handle hurricane disaster emergency response peparedness procedures. The group believes that the U.S. and New Orleans emergency response personnel can learn from how Asian countries handled the Tsunami. The group decides to research what happened. Write a report that describes what the New Orleans emergency managers learned and how it can impact how they handle future disasters in New Orleans.

Case Studies

You have been asked for specific examples of emergency responses to three different types of disasters that have occurred in other countries. These examples are to be used to support a plan you and a group of emergency managers are writing for the city of New Orleans. Develop three case studies of disasters and responses to those disasters that have occurred in other countries. Then, explain how you would use these to support your ideas for an emergency disaster response plan for New Orleans.

14

PROFESSIONAL ACCOUNTABILITY

Being a Professional

Starting Point

Go to www.wiley.com/college/lindell to evaluate your knowledge of professional accountability.
Determine where you need to concentrate your effort.

What You'll Learn in This Chapter

- ▲ Role of an emergency manager versus the role of an emergency responder
- ▲ The requirements and characteristics of a profession
- ▲ The maturity of emergency management as a profession
- ▲ Certification programs in the field of emergency management
- ▲ Growth and availability of academic programs in emergency management
- ▲ Liability issues that emergency managers face

After Studying This Chapter, You'll Be Able To

- ▲ Distinguish between emergency managers and emergency responders
- ▲ Analyze the definition of emergency managers
- ▲ Examine certification programs available to emergency managers
- ▲ Analyze educational opportunities
- ▲ Prepare a job description for an emergency manager
- ▲ Examine liability issues emergency managers face

Goals and Outcomes

- ▲ Evaluate the emergency management profession
- ▲ Assess ethical issues that emergency managers face
- ▲ Build the skills of an emergency manager
- ▲ Evaluate academic opportunities
- ▲ Evaluate professional development opportunities
- ▲ Limit personal legal liability

INTRODUCTION

Emergency management has changed over the past fifty years. The job duties have changed. The job title has changed. Other job titles that came before emergency manager include civil defense director, disaster planner, and emergency planner. It has only been in the past two decades that government departments changed from emergency services to emergency management agencies. There is still debate as to whether emergency management is a profession or an occupation. In this chapter, we examine this debate. We also look at the certifications and academic development programs available to emergency managers. Finally, we look at liabilities of the job and how you can protect yourself.

14.1 Distinguishing Emergency Management

The International Association of Emergency Managers defines an **emergency manager** as "one who possesses the knowledge, skills and abilities to effectively manage a comprehensive [emergency] management program" (Ditch, 2003, p. 12). This includes knowledge about a wide range of hazards. In addition, you must know how to manage your community's vulnerability. To be categorized as a profession, there must be agreement on the main features of the field and the job duties. This sounds simple, but it is difficult. For years, researchers have been trying to define a disaster. They have had limited success (Perry and Quarantelli, 2004). Emergency management can be seen as a field of applied practice, a field of public policy, and a field of academic research. Yet there are important differences for those who pursue each of these fields.

There are different roles within emergency management. Two of these roles are emergency managers and emergency responders. **Emergency responders** directly respond to the disaster. They attack the threat to reduce the potential or actual losses of a disaster. Brunacini (2001) tells fire fighters that their job, when appropriate, is to put their bodies between citizens and a threat. Emergency responders include fire fighters, police officers, and emergency medical technicians. They usually respond to events that:

- Occur frequently.
- Affect only a few people in a small geographical area.
- Cause limited economic loss.
- Require small teams of responders within a few government agencies.

They are also the "hands-on" responders in community-wide disasters. In such situations, there may also be other personnel who are included within the category of emergency responders. For example, highway department personnel help manage large-scale evacuations and public works personnel frequently remove debris during recovery. In a biological incident, hospital physicians, nurses, and technicians are emergency responders because they are directly exposed to the hazard agent.

By contrast, emergency managers develop expertise on a wide range of threats. They must also know the full range of hazard management strategies. They rarely appear at an incident scene to personally deliver services. Instead, they coordinate the many classes of emergency responders during all phases. However, emergency management duties are often performed by people who initially trained as emergency responders within fire or police departments (Drabek, 1987). Such intensive training is valuable experience. Nonetheless, the scope of their later emergency management duties is broader than the scope of their earlier emergency responder duties.

Another distinction is between public and private sector emergency managers. Public sector emergency managers work for any level of government—federal, state, or local. Private-sector emergency managers might work for chemical facilities, nuclear power plants, or railroads. Private sector emergency managers have many of the same duties as public sector emergency managers, but there are important differences as well. First, private-sector emergency managers generally work for a single business, site, or industry—although some work as consultants for public or private sector organizations. Second, private-sector emergency managers are responsible for the facility's employees, but not the public. Private sector organizations might be held liable in courts for personal and property damages caused by their actions. However, they have limited responsibility to engage in emergency management (Lindell, in press).

By contrast, public-sector emergency managers address the needs of government, government employees, citizens, and other private-sector organizations within their communities (Perry and Lindell, 1987). Since the 9/11 terrorist attacks, the government has been concerned that private organizations develop good emergency plans (Perry and Lindell, 2003). It is the obligation of the government to protect its citizens. The government may be held legally liable for failures to recognize and plan for threats. The government could be held liable to citizens and other governments. This does not mean that one setting is better than the other. Rather, private and public employees operate with different resources, duties, and accountability.

The final distinction is among local, state, and federal emergency managers. For a city or town, emergency management rarely exists as a separate department. It is often located within a fire or police department. Sometimes it is overseen by a county organization. Local emergency management functions vary in their presence as well as in their degree of success (U.S. General Accountability Office, 2003). Local managers are closest to the disaster impact and the people affected. At the same time, local managers are subject to federal and state mandates, but have the fewest resources. Municipal emergency managers must often rely on other agencies, outside experts, the media, and private sector organizations to accomplish their objectives. At the county level, emergency management constraints are similar.

State and federal emergency managers have positions that are quite different from those of local managers. Each state has an emergency management agency that must work with other departments that perform emergency management tasks. State emergency management agencies conduct state-wide hazard vulnerability analyses and provide technical guidance to LEMAs. They also provide financial support to LEMAs and evaluate LEMA performance. However, most state emergency management agencies face financial constraints that are similar to those of local emergency management agencies. They have much to do and not enough people and money to do it.

FEMA works with other agencies such as the Environmental Protection Agency (EPA), Coast Guard, and Department of Transportation. Together, they develop programs and provide technical and financial assistance to LEMAs (see Figure 14-1). This support role has long been a feature of federal emergency management policy that has continued with the National Incident Management System (NIMS) that was adopted in 2004. NIMS emphasizes that federal and state resources "flow downward" into structures created by local managers. At federal and state levels, the emergency manager's job emphasizes program management. The job also

Figure 14-1

Michael Brown, former head of FEMA discusses federal efforts during Hurricane Katrina with Lousiana Governor Kathleen Blanco and other officials.

FOR EXAMPLE

The Importance of Emergency Management Experience

During Hurricane Katrina, many Americans saw the head of FEMA, Michael Brown, on television for the first time. Many Americans were astonished to learn that, prior to 2001, Brown had no experience in emergency management. Brown had overseen successful relief efforts for several hurricanes during the previous year but his performance during Hurricane Katrina left much to be desired. His lack of emergency management experience was not a problem during the smaller disasters, but a major catastrophe proved his undoing.

emphasizes working with organizations and coordinating their efforts. As Drabek (1990) indicates, critical skills also include:

- Agenda control.
- Constituency support building.
- Financial analysis.
- Coalition building skills.
- Entrepreneurial skills.

When you add these skills to the need for expert knowledge about all the different hazard agents and government emergency management programs, the job description might seem to require superhuman abilities. However, it is important to recognize the difference between required and desirable job qualifications. It is desirable to be an expert on all of the different aspects of emergency management. However, it is only necessary to have a basic knowledge of different hazard agents and government emergency management programs. The principal requirements are your willingness and ability to work with multidisciplinary teams.

SELF-CHECK

- Define **emergency managers.**
- Define **emergency responders.**
- Describe the differences between public and private sector emergency managers.
- Describe the differences among local, state, and federal emergency managers.

14.2 The Requirements and Characteristics of a Profession

Emergency managers are clearly a *group* whose members share duties that differentiate them from others regardless of the differences in their work settings. What makes emergency management a *profession* is that it meets certain requirements. A **profession** requires an advanced education and training. A profession defines and applies a body of knowledge. It also sets minimum standards of relevant knowledge as a requirement for membership. In addition, a profession has methods for developing new knowledge and a system for teaching the body of knowledge to newcomers. Finally, a profession holds its members accountable to their peers for behavior that is relevant to the profession. A profession accomplishes all these functions through a professional society. This organization promotes public recognition of the profession as an organized group with specialized expertise.

This definition might seem to imply a degree of homogeneity that is at odds with the diversity of work done by managers. After all, there are important differences in the work done by various managers. One way to accommodate such differences into the vision of a profession is to adopt the approach of Trank and Rynes (2003). These authors, who focused on the field of business administration, suggest that any profession can be viewed as composed of a variety of occupations. Each occupation is distinct to some degree. However, all can be grouped in terms of shared knowledge and goals. Consistent with this view, Evetts (2003, p. 397) argues "professions are the structural, occupational and institutional arrangements for dealing with work." In both of these views, a profession is a class or category of activity. Within a profession, one finds a variety of occupations.

Within this concept of a profession, there is agreement on essential features of all professions. The first is that professions have membership rules to exclude unqualified people (Trank and Rynes, 2003). These membership rules are usually education and training requirements. In the case of mature professions that have developed standards by consensus, education often takes the form of acquiring a degree from an accredited college or university. An example is the Master of Urban Planning degree. An external professional accrediting board oversees the content of the degree programs. This board reviews the staffing and content of the degree program. The board provides accreditation to programs that meet its standards.

There is an important distinction between education and training. Education consists of broad principles that can be applied in a wide variety of situations. Training has a narrower aim. It helps to develop skill in performing specific tasks for specific situations. In less mature professions, where degree programs and specific accrediting bodies have not evolved, training becomes the "marker" by which practitioners can be identified. In this context, training is seen as multidimensional. You may require specific training in a variety of skills to adequately claim professional status. Also, training is based on current practices. Training materials change as the problems to be solved change. In addition, training programs usually demand refresher training to ensure that critical skills are maintained.

Training is closely associated with certification. Broadly speaking, a **certification** is an assurance that an individual has mastered the knowledge and the methods used to solve specific problems. Certification might follow classroom-based training or other types of educational activities. However, certification requires you to demonstrate your knowledge, frequently by taking a written exam. In addition, certifications often require performance tests that require the applicant to demonstrate a specific skill. During a certification exam, you are evaluated by a certified professional, because the legitimacy of any certification depends upon the authority of the association or organization that grants it.

Education and training also have a significant impact on the body of knowledge (see Figure 14-2). The second characteristic of a profession is that it has an "evolving and agreed-upon body of knowledge" and world-view (Hays and Reeves, 1984, p. 137). The body of knowledge might be science-based. However, this is not the case for all professions (e.g., religious professions). The important point is that the knowledge is systematic. There are consensus-based rules for generating,

Figure 14-2

Ongoing training opportunities for emergency managers contributes to the body of knowledge.

evaluating, and using that knowledge. The body of existing knowledge and the rules for developing new knowledge "constitutes the foundation from which professionals innovate and extend the knowledge base" (Trank and Rynes, 2000, p. 191).

Finally, the third defining feature of professions is that they have ethical standards. Professions socialize members to act in terms of professional norms. These norms may differ from the views of either the public or the management of organizations in which the professional is employed. As Friedson (2001, p. 122) indicates, the ideology of a profession provides members with "a larger and putatively higher goal that may reach beyond that of those they are supposed to serve." This attitude defines the professional identity or culture that supports the use of discretion in identifying problems and solving them. Professional ideologies have been embodied in ethical codes. These statements reflect the values embraced by members of the profession. Ethical codes encourage compliance as proof of professionalism. They also describe the punishments for those who fail to comply.

These are the three features of professions: membership certification, organized body of knowledge, and ethical standards. They provide a framework within which to discuss emergency management as a profession. They also capture the notion of accountability in that professionals are required to define education and training. This community also creates, changes, and applies the body of knowledge as well as educates, trains, and socializes newcomers to the profession. The community also enforces its ethical standards.

SELF-CHECK

- **Define profession and certification.**
- **Name three qualities of a profession.**

14.3 Emergency Management as a Profession

Most would agree that emergency management is a profession. However, they might disagree about the extent to which it is a *mature* profession. Perhaps at this point it is most useful to look at emergency management as a developing profession. After all, the concept of emergency management has been rapidly changing. Emergency management meant civil defense or wartime attack preparations as recently as the 1980s. During the 1990s, the vision of the field became more firmly represented as the management of natural and technological hazards. Most recently, the practice has expanded to include terrorist threats. Over this same period, the emphasis of practice has changed from mostly reactive to significantly more proactive. Beginning with the administration of FEMA Director James Lee Witt, hazard mitigation came to be regarded as important. In part, this was a response to the devastating effects of Hurricane Andrew. Mitigation began to be recognized as just as important as preparedness, response, and recovery. In addition, there have been increases in the threat environment and the tools available for dealing with those threats. This very fluid situation has slowed the development of consensus on the definition of the field. It has also slowed defining the body of practitioners.

The vision of practitioners has changed over the past fifty years. Perry (1985) has pointed out that, if emergency managers were distinguished from those delivering police and fire services, the lead emergency management role was often embodied in the Civil Defense Director. People in this role tended not to have training with the exception of experience in the military. They were not college educated. They were not well known in the local government. Thus, "the vision was one of a largely invisible person, presumably attached in some way to defense authorities (whoever they were), charged for the most part with civil defense duties (whatever they were)" (Perry, 1985, p. 135). This vision has given way to a career-oriented, college-educated professional who has acquired knowledge from the physical and social sciences (Blanchard, 2004). Emergency managers are now seen as people who must possess communication skills. They must have organizational skills. They must also grasp the technical fundamentals of a range of threats. Drabek interviewed emergency managers in 2003. He found the most frequent advice offered to new recruits is to work on interpersonal networking and communication skills.

Emergency managers need many skills, including:

- ▲ Comprehensive knowledge about the full range of natural and technological threats.
- ▲ Knowledge about integrated systems for managing community vulnerability.
- ▲ Knowledge about hazard vulnerability analysis, hazard mitigation, and recovery and response.
- ▲ Communication skills.
- ▲ Organizational skills.
- ▲ Strategic planning and management skills.
- ▲ Political management skills.
- ▲ Human resources management skills.

It is this knowledge that distinguishes emergency managers from emergency responders. Accordingly, emergency managers must integrate contributions from many different technical disciplines. They don't need to be competent in all technical skill areas. However they must understand how these different disciplines fit into the mosaic of emergency management. They are generalists who know where to find and how to request the services of specialists.

14.3.1 Professional Development

The next step is for emergency mangers to continue to develop professionally. There are many opinions regarding "what we need" as a profession. There are no definite answers. Indeed, the question of professionalization has received much attention in recent years. The following discussion addresses three widely agreed upon ways to grow professionally.

The first activity is to continue to define emergency management as a profession and a distinctive professional identity. This includes enhancing our ideology and ethics as emergency managers. Emergency management draws upon many disciplines. It must serve as an umbrella for many technical fields. However, it is also necessary to identify the unique features. This can be accomplished by distinguishing emergency managers from emergency responders, environmental planners, and others. This should be done without placing any field above another.

Emergency management is related to many professional associations. Emergency managers should participate in one or more of these groups. Some groups have a narrow focus. Other groups cover a wider subject matter. The International Association of Emergency Managers (IAEM) is one group. IAEM offers continuing education and a professional certification program. Similarly, the National Emergency Management Association is for state emergency management directors. The Disaster Preparedness and Emergency Response Association is an international organization with members from both public and private sectors. It offers training and a variety of school-based educational programs. Participation in any of these groups provides opportunities for learning and networking with other members of the profession. Participation also allows gaining a sense of self as an emergency manager.

It is important for you to receive additional education. Continuing education and professional development programs are also important to the profession. You can receive continuing education credits through a variety of colleges and universities. Finally, the FEMA Emergency Management Institute offers professional development opportunities. There is a *Professional Development Series* and an *Advanced Professional Series*, each of which offers not only specialized training, but also a certificate for completion of a full course of study.

14.3.2 Professional Ethics

Professional ethics has not received as much attention as other issues. The IAEM emphasizes ethics among its members and has adopted a formal ethical code. The IAEM code includes three parts. The first part focuses upon the need to respect people, laws, regulations, and fiscal resources. The second part emphasizes gaining trust, acting fairly, and being effective stewards of resources. Finally, the IAEM code asserts members should embrace professionalism founded on education, safety, and protection of life and property. One interesting feature of this code is the apparent concern for acting within the regulations and resources of the organizations served. Moore (1995) argues that such stands place the administrator in the role of "faithful servant" of the managers. As they consolidate their position as "experts," some professional groups move into a more assertive role in management. They focus ethical aspects of service on more discipline-specific principles and rules. By adopting and publishing an ethical code, the IAEM is leading the discussion of ethics in emergency management. Because emergency management is a diverse field, practitioners are subject to a variety of ethical codes and certification by different associations. For example, emergency managers who are Certified Environmental Professionals are subject to the Academy of Board Certified Environmental Professionals Code of Ethics and Standards of Practice for Environmental Professionals. Similarly, the Business Continuity Institute maintains a code of ethics for those who accept its membership or certification. Emergency managers employed by government are subject to the formal ethical codes of their jurisdictions. These overlapping sets of ethical guidelines rarely present problems. This values expressed in most statements of ethical standards apply to any profession, not just emergency management. Moreover, it is common for people to have multiple sets of ethical guidelines. It remains important, however, that emergency managers continue to attend to the issue of ethics. This practice stresses to the public that emergency management is a profession.

14.3.3 Body of Knowledge

A second activity that is crucial for the profession is to build and grow an identifiable body of knowledge for practitioners. Emergency management is an interdisciplinary profession, so you always draw upon many bodies of knowledge in the physical and social sciences. However, there is a specific portion of each body of knowledge that is most directly relevant to emergency management. What you need to know about hazards is how they can affect your community. Similarly, what you need to know about hazard adjustments is how effective they are in protecting your community and what resources your community must allocate to implement them. This is the organized body of knowledge that emergency management needs to develop.

The National Research Council has argued that "education and research, in addition to the role of technology . . . play vital roles in the advancement of emergency management" (2003, p.5). To date, however, there has been only a modest amount of research evaluating emergency management programs and policies (Lindell and Perry, 2001). Emergency managers are practitioners, not researchers. No one expects that, in addition to their other duties, emergency managers want to conduct their own research. However, there is a tradition of hazards and disasters research in the academic community. It is these researchers that can conduct the studies emergency management needs to provide a sound scientific basis for its programs and practices. The practicing emergency manager should:

- Endorse the need for well-designed research.
- Participate in research.
- Identify areas for needing research.
- Examine the research findings.
- Use the resulting knowledge.

The quality of the research depends on the degree to which practitioners and researchers work together (Mileti, 1999). There has long been an awkward relationship between practitioners and researchers. This is not unique to emergency management. Similar relationships exist in other professions such as medical, public health, business administration, urban planning, and public administration. The reward systems for researchers and practitioners are quite different (Fischer, 1998). These differing reward systems lead each group to have distinct goals and interests. Nonetheless, it is crucial for researchers and practitioners to seek opportunities for collaboration wherever possible.

There are many benefits from the growth of an emergency management body of knowledge. Practitioners are increasingly involved in higher education. Degree programs for training emergency managers are emerging to supplement the long-term programs for training hazards and disaster researchers. Both of these factors will help to bridge the gap between practitioners and researchers.

It is difficult to overstate the importance of an organized body of knowledge in establishing emergency management as a profession. An established body of tested knowledge is critical. It supports the claim that emergency managers can make credible contributions to community decisions about managing hazards. An accepted body of professional knowledge forms a basis for practice. The extent to which managers can use the body of professional knowledge provides a standard for evaluating their performance. Moreover, it forms the basis for the design and execution of training and education. In asserting their credibility to the public, the body of knowledge is what sets emergency managers apart from other professions.

The third critical feature for advancing the profession involves asserting control over the body of professional knowledge and its dissemination. This goal is being accomplished through training, degrees, and certifications. Every teacher's

FOR EXAMPLE

Emergency Management Is Interdisciplinary

Federal and state programs define a substantial amount of the specific body of knowledge for emergency management. For example, emergency managers need to understand how federal agencies operation within the National Response Plan and the National Incident Management System. However, this body of knowledge focuses on incident management. Emergency managers also need to be familiar with a variety of academic disciplines so they understand what emergency assessment, hazard operations, and population protection actions they can take. For example, emergency managers need to be able to understand vulcanologists when they describe the behavior of an active volcano. In addition, they need to be able to understand social scientists when they describe how people respond to warnings about different types of hazards. Ultimately, it is the emergency manager—not the vulcanologist or social scientist—that must construct and disseminate warnings to residents threatened.

choices of what material to teach and every evaluator's choices of what skills to test define—at least implicitly—what material is relevant to emergency management. In addition, organizations are attempting to explicitly define the emergency management body of knowledge. The FEMA Higher Education Project is contributing to this effort by examining how to develop an accreditation system for degree programs (Walker, 1998). Until an accrediting body is established, the Higher Education Project is defining the body of knowledge in three ways:

1. Studying emergency management practice and developing lists of essential competencies (Blanchard, 2003).
2. Disseminating course outlines for existing courses and developing new ones. These materials are made widely available through the FEMA Higher Education Web site. The content of these courses has been developed by highly regarded emergency managers and researchers and reviewed by experts. The resulting courses have become a basis for an expanding curriculum.
3. Compiling and updating a directory of college level programs, including those that offer undergraduate and graduate degrees, certificates, academic minors, and diplomas. These materials direct prospective students to the schools that are most suitable for them.

SELF-CHECK

- Name the three ways that the Higher Education Project is defining the emergency management body of knowledge until an accrediting body is established.
- Describe how practicing emergency managers should be involved with research.
- Name two groups that emergency managers can belong to in an effort to further their professional development.

14.4 Certification Programs in Emergency Management

Training courses consolidate the emergency management body of knowledge and teach it to practitioners. The next step is to establish a program that includes multiple courses into a certificate. When academic degrees are unavailable, certificates are especially important. Certificates ensure that you have received the necessary technical training. They also ensure that an expert has verified your knowledge of the training material. Professionally relevant certificates are available in many areas. Each of these programs has its own audience and level of credibility. Many of these certificates are even older than the term emergency management. Some test broader skills whereas others test narrower skills.

Business continuity planning offers a wide range of certificates. The two main sponsors are the Business Continuity Institute (BCI) and the Disaster Recovery Institute International (DRII) (Mallet, 2002). BCI offers a series of progressive certifications. DRII offers three graded certifications based on experience and skills. A number of universities and colleges also offer certifications. These can be found on the FEMA Higher Education Web site.

There are also many certificates available in the areas of security. For example, if you are a security professional, you can receive a certification in physical or information security. You can also get a certificate in areas such as homeland security or terrorism. The ASIS International Foundation has offered certificates since 1977. For example, ASIS has a program for professional investigators. There are 47 certification or diploma programs offered by colleges in security (Blanchard, 2004). There are also 21 programs in areas such as homeland security. There are nine international disaster management and humanitarian assistance programs. In addition, there are dozens of new programs under development.

With the new areas of studies, you must be cautious in pursuing training. There are many new programs of unproven quality. Some programs are meaningful and others are little more than academic diploma mills (Kaplan, 2004). For example, some programs provide certification based on experience immediately after a fee is paid (Kaplan, 2004). You can be more confident about programs that:

▲ Have been established for a long period of time.
▲ Have an independent board of examiners.
▲ Use established training and education programs.
▲ Have the endorsement of relevant professional associations.

You can also seek guidance from the National Fire Protection Association Standard 1000 Fire Service Professional Qualifications, Accreditation and Certification Systems. This standard identifies requirements for accrediting bodies and certifying entities.

The critical certification is the Certified Emergency Manager (CEM), which is offered through the IAEM. The CEM program was established with the goal of increasing and maintaining standards of knowledge, skills, and abilities. In 2002, 560 individuals had received the CEM. It is renewed on a five-year cycle based on continuing education and service (Ditch, 2003). More than 60 percent of current CEMs have held certification for more than five years. The CEM program is overseen by a Certification Commission composed of emergency managers from a variety of areas (government, allied fields, military services, and private industry), FEMA, and several professional associations.

The process of becoming a Certified Emergency Manager involves four phases (Ditch, 2003):

1. Completing an application.
2. Fulfilling credential requirements.
3. Passing the examination.
4. Obtaining recertification after five years.

The credentialing requirements address education, training, and experience. You need to have a bachelor's degree to advance to the CEM. You are required to have a minimum of three years in emergency management. However, it is possible to substitute years of experience for years of college education. You can substitute at the rate of two years of experience for one year of education. You must also have managed a disaster event or have had a significant role in managing a full-scale disaster exercise. You must also submit three professional references. You must have completed 100 hours of emergency management training and another 100 hours of general management training. There is a 25 hour limit placed on training in any single area. The collective training experience should cover all four phases of emergency management—mitigation, preparedness, response, and recovery. You must also demonstrate "contributions to the profession" in at least six areas. These include

FOR EXAMPLE

Certification Opportunity

The Academy of Board Certified Environmental Professionals, has offered the Certified Environmental Professional Program since 1979. This is a very broad certification that offers five functional areas—environmental assessment, documentation, operations, planning, and research and education. This is a challenging certification. The minimum requirements include a bachelor's degree and nine years of professional environmental experience.

teaching, publishing, serving on boards, course development, membership in associations, giving speeches, state certifications, or assuming leadership roles. Finally, all applicants must complete a technical essay on comprehensive emergency management. After meeting the requirements, the next step is to pass an examination. Following certification, CEMs must recertify on a five-year calendar.

The CEM is the only certification that assures competence in comprehensive emergency management and integrated emergency management systems. Other certifications are useful, but they only measure knowledge and skills in specific areas related to emergency management. Because of its comprehensive nature, the CEM is widely accepted by those outside the profession as proof of expertise. Many employers now ask that job applicants possess the CEM. If you do not have the CEM, many employers will ask you to earn it within five years of accepting employment. Other associations recognize the CEM as either a criterion for membership or proof of expertise. These groups include the American College of Contingency Planners, the American Society of Professional Emergency Planners, and the U.S. Department of Defense.

SELF-CHECK

- Explain how to assess the worth of certification programs.
- Describe the CEM certification.

14.5 Academic Programs in Emergency Management

Certifications often include educational requirements. However, they do not replace academic degrees just as academic degrees do not replace certifications. The educational requirement establishes that you have acquired a broad base of knowledge and skills. Job-specific training and experience builds on this knowledge. The growth of academic degree programs to support a profession represents maturing of that profession. There are two aspects to this development. First, professional degree programs help you acquire principles and procedures from different theoretically-organized disciplines. You can then use this knowledge by applying it to solve problems. Emergency management degree programs use material from the physical and social sciences. They also use material from other professions such as engineering, planning, and public health. In turn, the programs must organize this information into a coherent body of knowledge that addresses the problems that confront emergency management. There must be a depth to education that is difficult to acquire in training or jobs. Education emphasizes principles, models, and theories. Training focuses on specific tasks and the appropriate methods for performing those tasks. It is this knowledge of broad principles that helps define a professional—the ability to improvise solutions to new problems that were not explicitly addressed in planning and training (Drabek, 2003). It is impossible to identify in advance every problem that will confront you. It is impossible to devise specific procedures to address every potential problem. Knowledge of basic principles will help you. For example, knowing how people will react to warnings helps you to design warning procedures for a variety of different hazards and community situations.

The second positive aspect of developing degree programs is that they bring together practitioners and researchers. Practitioners often use the body of knowledge. However, they have little time to conduct, refine, or extend it. It is usually academics who conduct research. There have long been a few university-based centers that focus on emergency management. There is the Disaster Research Center that was founded in 1963 at Ohio State University and is now based at the University of Delaware. Tierney, Lindell, and Perry (2001) identified 28 such university-based centers. However, the academic connection with emergency management needs to be more extensive. This is especially true when supporting an entire profession, and a diverse one at that. Degree programs allow faculty to disseminate, refine, and extend knowledge. This allows faculty to interact with students and practitioners in the teaching process. Faculty can then integrate that experience into research designs.

The path to developing academic support for a profession is long and multifaceted. There must first be sufficient demand for training and for education to capture the attention of educational institutions. The job market must be such that a large number of people are seeking available positions. They need a means of demonstrating their expertise. The process of obtaining credentials for professions with small staffing levels rests with training and certifications. This is because these types of programs can be sustained on relatively small volume. In the past decade, the number of emergency management jobs has begun to increase. The levels are now high enough to provide a market for academic programs. Typically, a few institutions recognize the needs of an unserved niche in the labor market. Then they begin the process of serving that niche. In the case of emergency management, the University of North Texas was the first to initiate an emergency management degree program. In the two decades since then, over a hundred other programs have responded to the need for educating future emergency managers.

Market recognition is normally a gradual process. However, emergency management has been both hastened and promoted by the FEMA Higher Education Project. Through its Web site and by advocacy, the Higher Education Project has made it known that there is a wide market for emergency managers. The Higher Education Project has tracked and publicized the available degree programs. It has also made public the challenges they have faced. This information speaks not just to faculty. It tells practitioners of available educational opportunities. It also informs the public and other professionals that there is a firm intellectual grounding for emergency management.

This visibility does not allow emergency management to escape the gradual nature of the evolution of degree programs however. Darlington (2000) studied a sample of 1,886 schools. She found that 11.6% offered at least one course with emergency management content. She also reported that less than 1% of colleges offered a bachelor's degree in emergency management and 1.5% had an emergency management related postgraduate degree program. These gains may seem small. However, when you consider that there were no programs prior to 1983, you can see how far educational opportunities have progressed.

To serve the profession, degree programs must achieve some level of standardization. There must be an assurance that graduates of degree programs know the body of professional knowledge. Both the FEMA Higher Education Project and the IAEM have taken active roles in shaping the vision of emergency management knowledge, skills, and abilities. The Higher Education Project has created a forum for exchange as well as partnerships with government agencies, associations, and private institutions. The FEMA staff have also created a learning resource center and posted sample syllabi available for a wide range of classes. They have developed full college courses with instructor

FOR EXAMPLE

Academic Programs

To view a list of academic programs in emergency management, visit the list on the FEMA Higher Education Project Web site at www.training.fema.gov/EMIWeb/edu/collegelist.

guides, readings, exercises, field trips, and student notes. Perhaps most critically, the Higher Education Project has developed and given suggestions for program curricula.

At some point, these efforts to standardize programs will mature into a system of accreditation. An independent accrediting body will be established. This body will establish standards and develop a systematic process reviewing for programs. Typically, this body will be national or international in scope. In addition to developing standards and conducting evaluations, it can foster information exchange. Program accreditation links the body of professional knowledge with academic institutions.

SELF-CHECK

- **Discuss why it's important to develop degree programs for emergency management.**
- **Describe how the FEMA Higher Education Project and the IAEM have taken active roles in shaping the vision of emergency management knowledge, skills, and abilities.**

14.6 Issues of Legal Liability

Legal liability applies more to organizations and government agencies than to individuals. Thus, an employer may hold an individual responsible for professional behavior. However, these obligations and sanctions rest with the employing organization. An association may hold its members responsible for ethical practices. However, the sanctions depend on the level of control the association can exert over an individual member. The most severe sanction a professional association can impose is expulsion. For the most part, professional associations do not impose technical competence obligations directly upon their members. Accrediting bodies can withdraw accreditation of those who fail to meet standards. This usually takes place during the accreditation rather than in the context of professional practice. That is, accreditation is not granted (or renewed) rather than being explicitly revoked. As yet, we do not have anything for the emergency management profession that compares to the concept of medical malpractice for physicians.

A detailed discussion of legal liability is far beyond the scope or purpose of this chapter. However, it is appropriate to examine two aspects of legal liability that are relevant to the practice of emergency management. At the outset, it is important to note that emergency management statutes vary widely among the states, as do the emergency powers that are available to address disasters. To understand state laws, you must consult the appropriate statutes. At the federal level, there is a maze of laws that defines liability and immunity. One of these is the Stafford Act, which provides aid to states and under which the president may declare an emergency or major disaster. The Defense Against Weapons of Mass Destruction Act is important as well. This act assigns rights, duties, and resources relative to WMD. Other important acts include the Comprehensive Environmental Response Act, the Compensation and Liability Act, the Clean Water Act, and the Homeland Security Act of 2002. The most thorough discussion of federal law and legal liability was produced by the Defense Threat Reduction Agency in its 2004 *Domestic DWM Incident Management Legal Deskbook*. However, the best advice for someone who has questions about legal liability is to directly question your jurisdiction's legal authorities.

There are two areas of legal concern that commonly arise. Both have to do with damage to people or property. The first concern is a claim that government officials in responding to an emergency caused damage to persons or property. The second concern is a claim that a failure of the government to plan for or respond to a disaster resulted in damage to persons or property. Legal decisions pertaining to each of these situations are often based on the same statutes and obligations. The complexity of the discussion is reduced when each is considered separately.

With regard to emergency response, federal liability is addressed in at least three statutes. The Federal Tort Claims Act introduces the notion of liability in three ways. First, this act waives sovereign immunity of the federal government when employees are negligent in their duties (28 U.S. Code, section 2671). Sovereign immunity, which is derived from English common law, means a citizen cannot file a civil suit against the government. The government retains sovereign immunity for governmental functions, but not for proprietary functions (Pine, 1991). Governmental functions are actions that an ordinary citizen would not be able to undertake. These include regulatory and public safety actions. Proprietary functions are activities, such as operating a bus line or parking lot, that ordinary citizens could take. However, the act permits civil suits, in accordance with the law where the negligence took place, for negligent action by a government employee involved in a proprietary function. Three exceptions to immunity are important for emergency managers. First, claims for damages may not be brought in connection with the imposition of quarantine. Second, one may not bring suit if federal agencies or employees can demonstrate that they exercised "due care" in carrying out a statute. Finally, the "discretionary function exception" provides immunity for federal agencies and employees when the claim is based in the "exercise or performance or failure to exercise or perform" a discretionary action (28 U.S. Code, section 2860, subsection a). A discretionary action, which involves the establishment of a policy, is different from an operational action that implements that policy (Pine, 1991). The Stafford Act provides for government immunity in response. Specifically

> *The Federal Government shall not be liable for any claim based upon the exercise or performance of or the failure to perform a discretionary function or duty on the part of a Federal agency or an employee of the Federal Government in carrying out the provisions of this Act (42 U.S. Code, section 5148).*

Consequently, the Stafford Act offers fairly broad immunity for agencies and employees that must improvise during emergency response and recovery. Finally, there is some liability protection under the Homeland Security Act of 2002 that has been less well tested in the courts. The act (Public Law 107-296, section 302, subsection c) addresses the power of the Secretary of the Department of Health and Human Services to declare a public health emergency and require medical antidotes for the public. There are limited remedies offered to those who experience death or injury from such countermeasures. Most of the provision offers immunity to those who manufacture, distribute, or administer medical countermeasures under an official emergency declaration.

Each state recognizes the rights of individuals and businesses to be protected by tort law. This is the portion of civil law that addresses a person's right to seek compensation when harmed by another. Like the federal government, all states have some form of statutory immunity for emergency management activities. Clearly, there are circumstances when elected officials and employees are not liable for injury or property damage stemming from the impacts of disasters or from the impacts of officials' emergency response actions. These protections stem from state tort claims acts or from specific emergency management statutes, particularly those pertaining to emergency powers.

Immunity is recognized unless it can be shown that some form of negligence exists. Negligence can exist when someone takes an action that unintentionally harms persons or property. Negligence can also exist when someone fails to take "reasonable and prudent" actions when they have a duty to do so (Pine, 1991). For the most part, actions brought against individual emergency responders or governments have been decided in favor of the responder or government. To mount a successful defense, a jurisdiction needs to document that it has a technically sound emergency plan. It then needs to show that actions were guided by that plan. Then there will be little risk of successful lawsuits against either the individual emergency responders or the jurisdiction employing them. Indeed, in most states, where these two conditions exist, immunity is almost always extended during a disaster.

The failure to plan for or respond to an event is also addressed under tort law. Many of the same statutes and regulations discussed in connection with response apply to this situation as well. There are specific circumstances under which public officials and governments are not legally accountable for damage caused by disaster agents. The issues determining accountability are complex and involve intergovernmental responsibilities. They also rest on whether a disaster declaration is in force and on state and federal statutes. However, two plaintiff strategies have met with some success in the courts. The first strategy is to demonstrate that the government failed to plan effectively for a disaster whose impact subsequently produced losses.

The second strategy is to demonstrate that a government failed to perform an effective hazard vulnerability analysis (HVA). To be successful, this strategy must also show the inadequate HVA led the government to fail to plan for a disaster that produced losses. Cases following this strategy tend to be successful most often when there is a mandate that a disaster plan be developed and implemented. Although such mandates can come from any level of government, federal mandates are binding on all levels of government. The failure to develop a plan, despite a mandate to do so, places governments and emergency managers at a serious disadvantage in court. However, the definition of a mandate is subject to interpretation by the courts. For example, it is clear that the National Flood Insurance Program requires floodplain management. However, this guidance might also be interpreted to indicate that emergency response plans for floods are also mandated. Even when a mandate does not exist, officials are not immune from litigation. A plaintiff might argue convincingly that a mandate should have existed and hence a plan should have been in force. In such a case, the jurisdictions and its officials might be held accountable. These arguments are usually made on the grounds that a threat existed and local officials should have recognized and planned for it.

A critical consideration in these cases is the determination of what is an effective plan. This issue has arisen both in cases where plans were mandated and those where they were not mandated. Thus, the mere presence of a plan is not in itself a reasonable basis for arguing that a jurisdiction has acted responsibly and is immune from claims. Usually a plan is judged to be effective if it is consistent with government (usually federal) guidelines for such planning or if it can be demonstrated to follow professional standards. The notion of compliance with "generally accepted standards" is difficult to define. Federal agencies have published planning guidance for a wide range of threats. If you develop a plan, you have two choices. You can follow federal guidance and document compliance. Alternatively, you can

FOR EXAMPLE

Legal Immunity for Drills and Exercises

Four states—Alaska, Kansas, South Carolina, and Utah—extend discretionary immunity when emergency responders are operating within an existing plan to conduct disaster drills and exercises. Thus, legal protection for responding to officially declared disasters is high.

establish and document the rationale used to develop a plan that is not explicitly covered by available federal or professional guidance. Meeting the standards adopted in National Fire Protection Association Standard 1600, FEMA's State Capability Assessment for Readiness (CAR) program, and the Emergency Management Accreditation Program is an excellent way to demonstrate the professional competence of the planning process and the plans it produces.

SELF-CHECK

- **Explain the two areas of legal concern that commonly arise and what they have to do with damaging people or property.**
- **Explain when negligence can exist.**

SUMMARY

Emergency management is a challenging career that is maturing as a profession. Whereas there were once limited opportunities for growth in the field, there are now abundant opportunities for professional growth. There are educational, training, and certification opportunities. There are professional associations you can belong to that will help you in your career. During this chapter, you also examined the ethical and legal responsibilities that accompany the emergency management profession.

KEY TERMS

Certification An assurance that an individual has mastered the knowledge and the methods used to solve specific problems.

Emergency manager A person who manages a comprehensive program for hazards and disasters. An emergency manager is responsible for aspects of the program involving mitigation, preparation, response, and recovery.

Emergency responders People who directly respond to a disaster. They attack the threat to reduce the potential or actual losses of a disaster.

Profession An occupation that requires an advanced education and training.

ASSESS YOUR UNDERSTANDING

Go to www.wiley.com/college/lindell to evaluate your knowledge of professional accountability.
Measure your learning by comparing pre-test and post-test results.

Summary Questions

1. Emergency responders usually respond directly to events that occur frequently, affect few people in a geographical area, cause limited economic loss, and require small teams, whereas emergency managers must develop expertise on a wide range of threats. True or False?
2. What are the three features of emergency manager professions?
   (a) membership certification
   (b) organized body of knowledge
   (c) ethical standards
   (d) all of the above
3. Which of the following are requirements of the CEM?
   (a) bachelor's degree
   (b) 3 years of emergency management experience
   (c) 3 professional references
   (d) all of the above
4. The IAEM offers the CEM certification. True or False?
5. It is possible to identify in advance every problem that will confront you. True or False?
6. Legal liability applies more to organizations and government agencies than to individual emergency managers as individuals. True or False?
7. An example of a diploma mill is an organization that
   (a) grants a degree immediately after a fee has been paid.
   (b) has been established for many years.
   (c) graduates more than 200 people per year.
   (d) uses an independent board of examiners.
8. The definition of what constitutes a mandate is
   (a) always clearly written in legislation.
   (b) subject to the interpretation of the courts.
   (c) subject to interpretation by an individual emergency manager.
   (d) subject to interpretation by local political officials.

Review Questions

1. What type of events do emergency responders respond to?
2. What is the distinction between education and training?
3. What are the skills emergency managers need?
4. What are some of the professional associations related to emergency management?
5. What are the four phases of becoming a Certified Emergency Manager?
6. What is involved in developing academic support for a profession?
7. How are the Stafford Act and the Defense Against Weapons of Mass Destruction Act examples of laws that define liability and immunity?

Applying This Chapter

1. You are writing a job description for the director of FEMA. What experience and credentials do you want the director of FEMA to have and why?
2. You are writing a job description for the emergency manager of Chattanooga, TN. It is a mid-size city and is close to the Sequoyah nuclear plant. What experience and credentials do you want the Chattanooga emergency manager to have and why?
3. You are presenting information on academic programs to a group of emergency managers. You are asked by one of the audience members why it took so many years for universities and colleges to offer academic programs for emergency managers. What do you tell the audience?
4. You have been asked to speak to prospective emergency managers. What skills do you describe as important to emergency managers?
5. You are presenting information on academic programs to a group of emergency managers. You are asked by one of the audience members when hazard mitigation started to become recognized as important. What do you tell the audience?
6. You are writing a report about the legal responsibilities of an emergency manager. What do you include in the report?

YOU TRY IT

Liability

During Hurricane Katrina, Mayor Ray Nagin urged people who could not evacuate to seek shelter in the Superdome. There was a lack of supplies available, the power went out, and clean water and food was not available. Some people with preexisting health conditions died in the Superdome. Should the city be held responsible?

Certificate Programs

You want to become certified in homeland security. How do you find a certificate program? How can you determine if it is a quality program and not the equivalent of a diploma mill?

List of Competencies

You are working with the FEMA Higher Education Project to define a list of areas where emergency managers should be competent. What list of competencies would you create and why?

15

FUTURE DIRECTIONS IN EMERGENCY MANAGEMENT

Challenges and Opportunities

Starting Point

Go to www.wiley.com/college/lindell to evaluate your knowledge of future directions in emergency management.
Determine where you need to concentrate your effort.

What You'll Learn in This Chapter

- ▲ Global challenges that will affect emergency management
- ▲ Global and national opportunities for emergency managers
- ▲ National challenges that emergency managers will face in the coming years
- ▲ Professional challenges of working with new professions and professionals in other academic disciplines
- ▲ Professional opportunities including involvement with hazard mitigation and recovery planning

After Studying This Chapter, You'll Be Able To

- ▲ Examine global challenges that will affect emergency management
- ▲ Demonstrate an increased scientific understanding of the hazards and societal responses and revolutionary technologies
- ▲ Examine ways to reduce disaster losses
- ▲ Examine the challenges that emergency managers have in working with disaster researchers
- ▲ Appraise the benefits of higher education in emergency management

Goals and Outcomes

- ▲ Assess how global challenges will affect your community
- ▲ Assess national trends that will affect disasters
- ▲ Model a plan after the principles of Project Impact to reduce disaster losses
- ▲ Evaluate ways to better communicate with disaster researchers
- ▲ Propose ideas on how to better educate all emergency personnel

INTRODUCTION

This is an exciting time to be involved in emergency management. There are many challenges and opportunities facing you. To face these challenges, you must be prepared. To take advantage of the opportunities, you must understand them. This chapter presents the challenges and opportunities that lie ahead in emergency management. As you read this chapter, keep in mind that facing these challenges and opportunities means you need to utilize all of the skills and the strategies discussed in this book. Whether those skills involve planning, communicating, or using your creative talents, face the challenges with an understanding of them and with the willingness to use your skills.

15.1 Global Challenges

There are many global challenges facing you. These include global climate change, increasing population and population density, increasing resource scarcities, and rising income inequality. To meet these challenges, you must first understand them.

15.1.1 Global Climate Change

You may have heard that environmental scientists disagree about global climate change. The disagreements are only about the rate of change and how bad the global consequences will be. Scientists agree that climate change is a fact and that the consequences are likely to be serious. During the twentieth century we saw (Intergovernmental Panel on Climate Change, 2001) the following:

- ▲ Increase in the global average temperature.
- ▲ Increase in the sea level.
- ▲ Decrease in snow cover.
- ▲ Changes in precipitation patterns.

These changes will result in a global average temperature increase of 1.4 to 5.8°C (2.7–10.4°F) and a sea level rise of 0.1 to 0.9 m (4–35 in) over the next 40 years.

These changes would be challenging even if the rate of change was constant and was spread out over a long period of time. The Sea level rise would gradually inundate many coastal communities around the world, but there would be lots of time to adjust to the impact. However, the rate of change seems to be increasing. Accelerating change is a significant problem because it reduces the amount of time we have to adapt to new conditions. Nature does not have enough time to adapt to new conditions either. Moreover, the effects of climate change at the local level are difficult to predict because variation around the average conditions can be quite large. These variations can have unforeseen impacts on households, businesses, and government agencies.

Climate change is especially important for emergency managers because it will increase the number of extreme events. Climate change will increase the number of severe storms and floods (Mileti, 1999). One of the most serious effects of climate change will be an increase in drought conditions across the plains of Africa, North America, and South America. Such droughts could cause severe disruption of agriculturally based economies in those areas. Some of the effects of climate change are already becoming conflicts across sub-Saharan Africa. These conflicts are driven, in part, by competition for increasingly scarce water resources and the shrinking availability of land that can be used for growing crops. Even in countries where droughts do not cause famines and political unrest, they will increase the incidence of massive wildfires in parched forests and grasslands.

15.1.2 Increasing Population and Population Density

The world population will increase as much as 50% in the next 50 years. Most of this increase will occur in the developing countries of Asia and Africa (Organization for Economic Cooperation and Development [OECD], 2003). The population boom in these countries will create increasing demands for food, water, and energy. If these needs cannot be met, the countries will experience great political instability. We know now that politically unstable nations are breeding grounds for terrorists. Afghanistan and Somalia are recent examples.

The effect of population growth is magnified by increasing population concentration in major cities. Movement to cities also increases the wealth, in the form of buildings and infrastructure, that is located there. The resulting urban concentration of population and wealth creates the potential for a catastrophe. A catastrophe is an extreme event that might have cost hundreds of lives and millions of dollars in a small city but cost thousands of lives and billions of dollars in a major city. The cost of a major earthquake in Tokyo could be $1 to 3 trillion. This is 25 to 75% of Japan's Gross Domestic Product. In most countries, even a loss of 5% is catastrophic (Cherveriat, 2000). Closer to home, Hurricane Andrew cost $20 billion when it struck just south of Miami. If it had made landfall just a few miles farther north, it could have cost $100 billion. Hurricane Katrina was the deadliest American disaster in over 80 years. It also was the costliest natural disaster in American history and is expected to cost $200 billion. The earthquake-prone cities of the West Coast have yet to experience "the Big One." The impact of a powerful earthquake could be far worse than the 57 lives and $25 billion lost in the Northridge earthquake.

These major cities are also increasingly connected to outside geographic areas through social, economic, and political transactions. Disrupted flows of information, materials, and money can affect the economic system in the entire country. The immense size of the United States tends to limit the geographic, social, and economic impact of these disasters. Nonetheless, the economic losses from these megadisasters affect areas far away from the location of the physical destruction. For example, Hurricane Katrina's disruption of the oil industry in Louisiana increased the price of gas for residents of Oregon.

15.1.3 Increasing Resource Scarcities

The world's supply of available water will shrink as increasing population raises demand. There is over 12,500 km^3 of fresh water available for human use worldwide. Half of this supply is currently being used. At the current rate of increase in consumption, the fresh water consumed will reach 90% by 2030. Availability will not be uniform across the world. Two-thirds of the world's population will experience chronic shortages of safe drinking water by that time (see Figure 15-1). Increasing reliance on polluted water will:

- ▲ Increase the spread of infectious diseases.
- ▲ Force people to move to areas where they can find water.
- ▲ Encourage international political instability.

Figure 15-1

Two-thirds of the world's population will experience a chronic shortage of safe drinking water by the year 2030.

There will be similar problems in energy and food. Energy consumption is expected to increase by 350% in developing countries over the next 50 years (Brown, Gardner, and Halweil, 1998). In the same countries, the use of cereal grains will increase by 50% in 2010 from the 1990 levels. This could impact resources and climate. If the needs of the developing countries are not met, them:

- ▲ The populations could become vulnerable to disasters.
- ▲ The people could try to farm unsustainable land.
- ▲ The population could deplete available fresh water resources.
- ▲ There would be social, economic, and political instability.
- ▲ There would be mass migrations following disasters.

15.1.4 Rising Income Inequity

Social, economic, and political systems are changing in ways that increase the vulnerability of large numbers of people. Income and wealth are becoming increasingly concentrated in the hands of the wealthiest. This is apparent between countries as well as within countries. In 1900, the richest five countries had an average per capita Gross Domestic Product (GDP in 1990 dollars) of $4,000. The poorest five had GDPs about 1% of that value (OECD, 2003). A century later, the richest five countries had an average per capita GDP of about $20,000. The poorest five had made little progress. Rich countries control most of the world's economic resources. Poor countries have become increasingly burdened with levels of debt that are impossible for them to repay.

15.1.5 Increasing Risk Aversion

Residents of developed countries are demanding that many technologies become safer (OECD 2003). Disaster victims angrily demand to have power restored in

FOR EXAMPLE

A Record Number of Storms

The 2005 hurricane season set a record for duration. Hurricanes continued to form long after the usual end of the season on November 1. In addition, the 2005 season set a record for the number of named storms. For the first time in its history, the National Hurricane Center ran out of names before it ran out of named storms. Meteorologists named the last few storms of the season with letters of the Greek alphabet.

hours when they might have waited silently for weeks in the past. Unfortunately, few are willing to pay higher taxes, which is needed to support an enhanced emergency management capability. In the case of electric power, increased rates are needed to have power lines placed underground where they are safe from wind damage. As an emergency manager, you must be prepared to make do with current levels of resources, respond quickly during and after disasters, and be patient with some people's unrealistic expectations.

SELF-CHECK

- Define **megadisaster.**
- Explain the effects of the population growth.
- Explain what could happen if the needs of developing countries are not met.
- Explain what is meant by rising income inequity.

15.2 Global and National Opportunities

As an emergency manager, you also have many exciting opportunities. These include increased scientific understanding of the hazards and societal responses as well as revolutionary technologies.

15.2.1 Increased Scientific Understanding of Hazards and Societal Responses

Policy makers need to understand the causes of hazards to manage them effectively. People are becoming aware that the state of our environment is a major factor in the occurrence of disasters. There is recognition that the economy is part of the ecology, not vice versa (Davidson, 2001). All economic benefits are based on the ability of the soil, air, and water to perform their life-sustaining functions. Quality of air, soil, and water is unevenly distributed around the world. It is deteriorating in most places. It is expensive to recover the life sustaining functions of the environment once it is lost. A nation that cannot feed its population is more vulnerable to disasters. It also is less able to recover from those disasters without international assistance. The increased scientific understanding of these principles has created opportunities for more effective hazard management. However, it has also created conflict between those who want to use these principles as a guide to smart growth and those who wish to pursue economic growth at any cost.

Natural disturbances can be healthy events that lead to a better use of land (Abramowitz, 2001). For example, seasonal flooding improves the health of rivers. It also maintains delta lands that are habitat for protein-rich stocks of marine life. Scientists have found that excessive flood control measures are producing a loss of coastal land in Louisiana by confining the Mississippi River, speeding the stream flow, and causing silt to be transported into the Gulf of Mexico. Naturally occurring wildfires maintain the health of prairie land and some forest habitats. The wildfires eliminate diseased trees and reduce the amount of fuel (trees, brush, leaves, and grass) available for burning. Scientists now understand that suppressing small to medium-sized fires increases the amount of fuel and the potential for catastrophic firestorms. Accommodating nature improves our patterns of settlement and patterns of resource use. Many resources have been exploited for short-term economic purposes with little regard to the long-term effects on the environment. Businesses around the world are beginning to understand their impact on the environment. Businesses are adjusting their product lines and distribution systems to protect the environment.

15.2.2 Revolutionary Technologies

People can now do many things online, from buying goods to voting. Information can be stored in databases that can be interconnected and linked. Communities are using enterprise software systems that store, manipulate, and retrieve all community data within a single system. Integration of information and access to data can make your job easier. However, any information on any computer system can be destroyed or damaged by cyber terrorists.

Emergency managers first used emergency management information technology to develop decision support systems for the emergency response (Marston, 1986). We can now conduct hazard vulnerability analyses with computers. With the help of powerful computing hardware and software, we can identify areas at risk and project damage (Dash, 1997). In addition, there are many rapidly developing technologies we will be able to use. These include:

- ▲ Remote sensing systems.
- ▲ Global Positioning System (GPS).
- ▲ Cellular and satellite communication.

Hazard Vulnerability Analysis

Emergency managers need information on:

- ▲ How often different types of disasters occur in their communities.
- ▲ The cost of these disasters.
- ▲ The cost-effectiveness of emergency management in reducing disaster losses.

All communities have tight budgets and every department wants increased funding. Major disasters, which might never occur, have a difficult time competing with daily demands such as street repairs. You will be more effective if you can show that it is less expensive to mitigate hazards than rebuild after disaster. For example, strengthening the levees in New Orleans prior to Hurricane Katrina would have cost tens of millions of dollars. Repairing the damage after the levees failed will cost tens of billions. The data you need, together with computer programs such as HAZUS, is becoming increasingly available to local emergency managers. Information can be found at state emergency management Web sites as well as sites of federal agencies such as:

- FEMA.
- National Oceanographic and Atmospheric Administration (NOAA).
- U.S. Geological Survey (USGS).

Other sources for information are as follows:

- Flood insurance rate maps (FIRMs) are available in hardcopy and also on the Internet. Using these, you can identify areas in the 100-year and 500-year floodplains. However, FIRMs must be updated because floodplain boundaries often change as upstream areas are developed. The high cost of the necessary hydrological studies slows the process.
- The SLOSH computer model helps identify coastal areas that are exposed to storm surge.
- HAZUS has been upgraded to address wind and floods in addition to its original earthquake component.
- Hazard analysis data for some toxic chemicals is available. However, the 9/11 terrorist attacks has led to the restriction of information because of concerns the information could allow terrorists to identify high value targets. This information has been increasingly limited to the staff of chemical facilities, LEPCs, SERCs, and LEMAs.

The greatest advance in technology for emergency management has been the increased availability of geographical information systems (GISs). We can use GISs for database management, mapping, and spatial analysis. We can use GIS for many important applications. Some universities strongly encourage their students to take GIS courses. However, even if you have not acquired GIS skills, you can work with land-use planners who do (Lindell et al., 2002).

Hazard Mitigation

GISs also support hazard mitigation by providing an effective method of storing and retrieving property data. With GISs, you can develop different versions of land-use plans that can be compared to determine which of them is best. You can change the information displayed at different points in presentations and avoid overwhelming audiences with details. Also, making maps available on a Web site provides greater information access for citizens.

GIS data can also be used to help determine what percentage of a community's buildings can withstand hazards. Structural assessment by trained building inspectors is the easiest and most accurate way, but it can only be done during initial construction. After construction, the building's structure is covered by siding. Without being able to see the core of the building, it is difficult to determine how strong it is. It is also difficult and expensive to determine a building's resistance to air infiltration after it has been built. You can use GISs to map each building's year of construction. This tells you which building code a structure was built under and allows you to make judgments about its hazard resistance.

Emergency Preparedness

The primary advances in technology for this area are computer software and the internet. Plans and procedures have long been stored and updated on computers. We are now able to prepare PowerPoint presentations to facilitate training. Digital photography and video also allows you to quickly distribute visual aids at modest expense. With software such as GISs and CAMEO/ALOHA, we can create databases. These databases can be used during emergency preparedness and response. For example, these databases can be used during training and exercises to find emergency resources ranging from bulldozers to hazmat response teams.

There have been numerous developments in evacuation modeling. Different programs have been developed, but they are difficult to use. Typically, evacuation-modeling programs use different scenarios to calculate how long it would take a community to evacuate. If the situation in an actual emergency is different from any of the scenarios used in analyses, then the evacuation time estimates will be inaccurate. In many cases, emergency managers do not know the model's assumptions, so they cannot make adjustments when they need to. Evacuation modeling programs such as OREMS (Oak Ridge National Laboratory, 2003) and EMDSS (Lindell and Prater, in press) are being developed for emergency managers to use.

Emergency Response

We have improved forecast and warning systems for a variety of different hazards. Technology has improved so we can now detect many hazards and predict their impacts. We can predict meteorological (hurricanes and tornadoes) and hydrological (floods and tsunamis) hazards with much greater accuracy than even a few decades ago. We have also seen improvements in the prediction of volcanic eruptions but not in the predication of earthquakes (Sorensen, 2000).

Emergency managers also have more ways they can record and communicate information (Tierney 1995).

Sensing and recording devices include:

- ▲ Hazmat detection systems.
- ▲ Satellite and aerial remote sensing.
- ▲ GPS.
- ▲ Portable weather stations/scanners.
- ▲ Digital cameras.

Communications devices include

- ▲ Cell and satellite telephones.
- ▲ Pagers.
- ▲ Fax machines.
- ▲ Personal computers that are connected through systems such as satellite dishes.
- ▲ Local and wide area network connections.
- ▲ Radio.

There will continue to be significant advances in the communication technology used to warn residents about hazards. As late as 1991, we had only sirens to warn large populations of imminent hazards. We now have telephone-based community alerting systems. We also have the NOAA **Weather Radio** system which provides tone-activated notification of emergencies over most of the United States. Its coverage has been extended over the years. In addition, the old Emergency Broadcast System has been replaced by the new **Emergency Alert System.** This new system provides a greater range of capabilities in the digital age. The Partnership for Public Warning promoted use of cell phones and other digital technologies to notify emergency managers and responders of hazards as well as warning risk area residents.

Disaster Recovery

You can use many of these same tools during disaster recovery. With cell phones, GPS devices, and powerful laptop and notebook computers, emergency responders can quickly assess damage and send the information back to you at the EOC. With computers, you can also quickly access databases. From these databases, you can locate critical facilities, hazardous facilities, infrastructure, and historic buildings. In addition, damage assessors can enter their findings directly into those databases. This saves time by bypassing paper forms. Assessors can also use these wireless capabilities to transmit each day's disaster assessments back to the EOC.

Summary of Technological Advances

Advances in technology will aid you in your job. The main reasons for implementing new technology are (Drabek, 1991b):

FOR EXAMPLE

FEMA Tracking Supplies

An internal review of FEMA's computer system showed that it was overwhelmed by the 2004 hurricane season. FEMA's computer system could not track supplies, although it is now testing a GPS that will track supplies. For example, FEMA's system could not track the delivery of ice and water to Florida, resulting in millions of dollars worth of ice left unused at response centers and $1.6 million in leftover water returned to storage.

- ▲ Increased office efficiency (e.g., word processing capability).
- ▲ Networking potential.
- ▲ Budget management.
- ▲ Resource management.
- ▲ Public warning/evacuation applications (e.g., flash flood warnings).
- ▲ Automated emergency notification for staff.
- ▲ Decision support systems (e.g., hurricane or hazmat plume tracking).

However, it is not always easy to transition to new technology. The challenges in using new technology include:

- ▲ Staff shortages.
- ▲ Computer/software incompatibility.
- ▲ Software inadequacies.
- ▲ Lack of training materials.
- ▲ Expense.

SELF-CHECK

- Explain some uses of emergency management information technology.
- Name Web sites that are helpful to mitigate hazards.
- Define **Emergency Alert System.**
- List three reasons for adopting new information technology.

15.3 National Challenges

There are many challenges that you and other emergency managers will face in the coming years.

15.3.1 Increasing Urbanization and Hazard Exposure

The United States population is projected to increase over the next 50 years. Much of the increase will occur in hazard-prone areas (Schwab et al., 1998). Despite this increasing rate of exposure to environmental hazards, the United States has not had a significant increase in the annual loss of life to these events. Indeed, casualties have declined for many hazards because of improved detection, forecast, and warning systems (Sorensen, 2000). What has increased is property loss (Mileti, 1999). Property losses will continue to increase. However, the death toll in Hurricane Katrina may be an indication that we have reached a limit in our ability to reduce casualties though improved forecast and warning systems.

15.3.2 Interdependencies in Infrastructure

With just-in-time manufacturing, many companies now have minimal inventory. This has increased profits for investors and lowered prices for consumers. In some cases, this has reduced companies' vulnerability. Because companies have less inventory, they have less to lose when a disaster does strike. In other cases, however, undamaged companies have become more vulnerable. This is because companies must be able to continually restock their inventories or they will have to shut down. Natural disasters and accidental technological disasters are likely to have relatively small effects on the national economy. However, deliberate attacks on electronic government and commerce could be much more damaging.

We have also increased short-term efficiency by reducing the amount of resources needed to produce each unit of goods and services. This benefits consumers who receive reduced prices. However, short-term efficiency eliminates the reserve resources that are needed to cope with disruptions. Disruptions prevent customers from obtaining needed products and services. Therefore, producers need to increase diversity and redundancy in their distribution channels to assure customers' needs are met.

15.3.3 Continued Emphasis on Growth

Public policy in most communities is significantly affected by growth coalitions of real estate, construction, and other commercial interest groups that benefit from "public subsidies to and private investments in infrastructure, civic capital, construction, and related activities that help to attract people, employers and jobs to a local area" (Buttel, 1997, p. 47). Such interests often develop hazard-prone land and sell it to others. They do not have to live with the long-term consequences of their decisions. Attempts to stop this by opposing all growth are generally not feasible. What is needed is smart growth that develops less exposed locations. If that is not possible (e.g., in the case of tornadoes where there is no local variation in hazard exposure), proponents of smart growth use hazard resistant building practices. As an emergency manager, you need to form coalitions with other local government agencies, businesses, and community groups to promote smart growth that minimizes disaster losses.

15.3.4 Rising Costs of Disaster Recovery

Federal disaster relief programs pay for many of the losses from environmental hazards. Those who live in disaster prone areas benefit from this system. However, the cost is spread among all taxpayers. This would be a fair system if those who received the most disaster relief also paid the highest taxes, but there is no evidence this is the case. Thus, property owners in disaster prone areas are being subsidized by the rest of the taxpayers. A fairer and more efficient system would be to expand the present system of flood insurance to cover all hazards and to charge premiums in proportion to policyholders' loss potential (Kunreuther and Roth, 1998). This solution was recognized over 30 years ago (White and Haas, 1975) but has made limited progress. During the 1990s, FEMA was able to improve federal flood insurance by promoting the Community Rating System and reducing repetitive losses. Nonetheless, further progress in flood insurance has been limited. Moreover, insurance for earthquakes and hurricanes has become increasingly problematic. The cost of disaster recovery seems likely to continue to be paid by all taxpayers through the process of disaster declarations.

15.3.5 Increasing Population Diversity

Recent years have seen an increase in the cultural and language diversity of the American population. Large numbers of Hispanics continue to immigrate here. In some jurisdictions, such as Los Angeles County, there are more than 100 major languages or dialects in daily use. Emergency information must be translated into all of these languages. In addition, the percentage of the population greater than 65 years of age is rapidly increasing. This means many more risk area residents will have physical or mental limitations. Some of these will be in nursing homes where they can be readily warned and evacuated. Others who live independently will require additional assistance when disasters strike (Tierney et al., 2001).

There is increasing inequality in household incomes throughout the United States just as there is throughout the world. Over the past decade, the incomes of the top 20% of households have increased, whereas those of the bottom 20% of households have decreased. Some rural counties have income levels substantially below their urban neighbors. This has negative implications for communities whose households have incomes below the national average. With a low income base, there are fewer tax monies collected. This leads to small budgets for LEMAs and the other public safety departments. This gap between the richest and poorest jurisdictions will continue to fuel the gap between those that do and those that do not have enough money and training to adopt advanced emergency management technologies. Some communities can adopt new technology and others will not have the money to. As we saw in New Orleans during Hurricane Katrina, the poorest are the most vulnerable to disasters and are not well equipped to put measures in place to reduce hazard damage.

15.3.6 Terrorist Threats

Most terrorist threats involve familiar explosive and flammable materials (see Figure 5-2). Toxic chemicals have been used much less frequently. Radiological and biological agents remain a potential threat. These threats initiate the familiar emergency response functions:

- Emergency assessment.
- Hazard operations.

Figure 15-2

Wreckage from the World Trade Center in September of 2001. Terrorist threats will continue to pose challenges to emergency managers.

▲ Population protection.
▲ Incident management.

The methods of emergency assessment and hazard operations for terrorist attacks will differ from those for natural hazards. However, terrorists are likely to use the same types of hazmat that can be released accidentally from fixed-site facilities. Consequently, your plans for responding to terrorist threats can use many of the procedures you expect to use for technological accidents.

However, with terrorist threats come new challenges. Exotic chemicals such as sarin gas, "dirty bombs", and biohazards present different problems from technological accidents. For example, a biohazard is likely to spread throughout the population much more than chemical contamination. Moreover, response to terrorist attack will require coordination with agencies you might not have worked with before. Consequently, preparedness for terrorist threats requires modifications to your EOP, not an entirely different EOP.

To prepare for terrorist threats:

▲ Link preparedness for each hazard agent into your existing emergency management network.
▲ Anticipate the impact on each risk area population segment.
▲ Assess the capabilities of population segment for self-protection.
▲ Develop clear lines of authority and mechanisms for interagency coordination.
▲ Assign response functions according to agency abilities and resources.
▲ Identify potential sources of extra-community assistance.
▲ Promote emergency resource acquisition at household, organization, community, and supra-community levels.
▲ Provide training, drills, and exercises on hazard-specific methods of emergency assessment, expedient hazard mitigation, and population protection.

An intelligent adversary also creates some important information security problems. An intelligent adversary can take advantage of predictable population protective responses to inflict even greater casualties in secondary attacks during mass evacuations.

15.3.7 Priority of Emergency Management

For many years, emergency management has had a low priority on the government agenda. This changed dramatically in the aftermath of the 9/11 terrorist attacks. This is similar to the increased priority of emergency management after the 1979 Three Mile Island nuclear power plant accident and the 1984 chemical plant accident in Bhopal, India. The history of previous events suggests interest in any hazard agent is highest when it conveys what Slovic (1987) calls **signal value.** Signal value is an indication that a previously unnoticed threat warrants attention. Despite the recent attention to terrorism, it is certainly not a new phenomenon. In fact, terrorism was the proximate cause of World War I. The assassination of Austrian Archduke Ferdinand and his wife in Sarajevo in 1914 led to an outbreak of war all over Europe. Terrorism has been used for decades in Ireland and Great Britain. Terrorism has been institutionalized in the relationship between the Israeli state and the Palestinians.

Indeed, terrorism was not even new in the United States. Only six years before the 9/11 attacks, terrorists destroyed the Murrah Federal Building and killed 168 people in Oklahoma City. What was new about 9/11 was that it was a very dramatic and very deadly strike by foreign nationals on U.S. soil. The political implications of the attack and its aftermath are extraordinary and will not be fully understood for some time to come. The challenge for you is to hold on to and apply what you already know, while adapting as necessary to any changes that are truly required. An additional challenge will be to cope with the restrictions on mobility and access to information that have been implemented in the aftermath of the 9/11 attacks.

Past history suggests that concern about terrorism will likely decrease in coming years. Many events have attracted attention and government action in their immediate aftermath. As time passes, the media, the public, and government revert to an indifferent attitude toward these hazards. It remains to be seen how long terrorism will dominate the news and the spending priorities of government. Even if the political salience of emergency management does not drop in coming years, it is likely that funding will. The increase in the national debt and budget restrictions at state and local levels will create fierce competition among government agencies. This will be one more reason for you to build coalitions with other agencies, NGOs, and private sector organizations.

15.3.8 Legal Liability

Legal liability is a major issue because each state has different rules regarding liability in an emergency response (Drabek, 1991a). You must be aware of the areas in which your actions might conflict with individuals' rights (Anderson and Mattingly, 1991). You should get advice from your community's legal counsel about specific situations, but some general rules apply to all situations. First, you should be transparent and reasonable in any decision making process. Second, you should be fair in implementing and enforcing any actions resulting from your decisions. That is, you should have reasonable grounds for taking an action (i.e., a threat to public safety) and notify those who will be affected by the action. You should provide those who will be affected with an opportunity to be heard. You should take only those actions that will clearly reduce threats to public safety.

15.3.9 Intergovernmental Tensions

One area that exemplified the tensions among federal, state, and local governments during the late 1980s was preparedness for nuclear attack (Anderson and Mattingly, 1991; Drabek, 1991a). That issue disappeared with the collapse of the Soviet Union. Nonetheless, the fact that intergovernmental tensions have continued is testimony to their structural nature. There are fundamental conflicts among levels of government that are natural within a federal system of government. Further, there are significant differences in technical expertise, which is usually greater at higher levels of government. In addition, there are differences in the availability of site-specific data, which is usually greater at lower levels of government. There are also differences in financial resources, which are usually greater at higher levels of government. Finally, there are differences in direct accountability to those at risk, which is usually greater at lower levels of government.

15.3.10 Conflicting Values

There is a conflict between the goals of economic development and private property rights, on the one hand, and public safety and welfare, on the other hand (May and Deyle, 1998). The balance between these two sets of goals is managed by a system of case law, legislation, and executive orders and regulations. Each of these originates in a different branch of government—judicial, legislative, and executive, respectively. Each of these three branches of government exists at three different levels (federal, state, and local). Given the complexity of the system, it is no wonder there is a "patchwork" of requirements. Indeed, flood hazards are managed by 12 federal agencies, all 50 states, 3,000 conservation districts, and 20,000 local governments in flood-prone areas (Federal Interagency Floodplain Management Task Force, 1992). The federal government alone has more than 50 hazard management laws and executive orders that have conflicting requirements.

There is limited control over land use at the federal level and, indeed, at the state or county levels in some states. Moreover, the federal land-use provisions that do exist are weakly enforced. Federal programs also indirectly encourage development of hazardous areas. The federal government provides funding for roads and sewers in hazardous areas, subsidized flood insurance, grants, subsidized loans, and tax write-offs to disaster victims. The predictable consequence is more people and property at risk. Moreover, individual communities often find themselves competing with each other to attract economic development. They compete by offering more favorable terms to commercial and industrial developers who seek the most profitable financial outcomes for themselves. Many times this involves construction in hazard-prone areas. Communities need to have agreements with each other that they will restrict building in hazardous

FOR EXAMPLE

Using What Is Available

The 9/11 terrorists used our airplanes and transportation system against us. They did not import any chemicals or weapons into this country; they used what was already here. Timothy McVeigh, the Oklahoma City bomber, also used commonly available items (fertilizer and fuel oil) to make his bomb that killed 168 people. Since these events, the federal government and many industries have made an effort to identify and eliminate opportunities for other terrorists to use similar strategies (National Academy of Sciences, 2006).

areas. These regulations can be supported by financial and technical assistance from higher levels of government. Such approaches can avoid minimal compliance with standardized programs.

15.3.11 Applying the Principles of Project Impact

There was a shift in the federal emphasis within emergency management from response and recovery to preimpact action during the 1990s. Project Impact, in particular, played a major role in fostering public-private partnerships to identify hazard-prone areas and promote mitigation actions by government, businesses, and households. Unfortunately federal funding for this initiative has been eliminated. Nonetheless, some of the local programs continue to exist and some are funded at the local level (Prater, 2001). A challenge for you will be to see how you can use the principles of Project Impact to reduce disaster losses in your community.

SELF-CHECK

- Explain how property owners in disaster prone areas are being subsidized by the rest of the taxpayers.
- Explain how a diverse population makes disaster recovery challenging.
- Define **signal value**.
- Explain why there is a conflict between the goals of economic development and private property rights.

15.4 Professional Challenges

There are two important professional challenges confronting you in the coming years.

15.4.1 Linkage of Emergency Management with New Professions

For years, emergency managers have worked with the fire service, law enforcement, and emergency medical services. Frequent contact among these agencies has generally promoted effective interagency performance in emergency preparedness and response. Because of Project Impact, many emergency managers began to work with land-use planners. Similarly, increased concern about biological threats has increased the interaction of emergency managers with public health departments. Nonetheless, the contacts with these new agencies appear to have remained limited. Contact and cooperation could revert to their previous low levels in the aftermath of the termination of Project Impact. If years pass with no biological attacks, then contact with public health departments could revert to low levels as well.

15.4.2 Linkage of Emergency Management Practitioners and Academic Disciplines

Emergency managers and disaster researchers try to work together, but cooperation is not easy. Each group claims the other fails to understand its problems or meet its needs. This gap between academics and emergency managers is not unique to emergency management. This complaint arises in all professions, including architecture, business, construction, engineering, public administration, public health, and urban and regional planning. There is no easy way to bridge the gap. Academics and emergency managers are employed by organizations that have very different cultures. They also have very different job responsibilities. Emergency managers face problems that demand immediate solutions. By contrast, professors have many duties in addition to the service activities they perform for emergency management agencies. At teaching universities, professors must teach many courses and must make their time available to many undergraduate students. At research institutions, professors have fewer courses to teach but spend many hours supervising graduate students and conducting research.

This situation seems to be changing for the better. Professors in emergency management and related fields have greater contact with emergency managers than they did in the past. There are increasing numbers of research projects designed to solve practical problems rather than address purely theoretical issues. There are more reports written for practitioners rather than for other academics. The increasing number of emergency management programs will enhance this trend. These programs will produce even more professors who want to try to establish good relationships with emergency managers.

FOR EXAMPLE

The Bird Flu

One potential health crisis, the Avian flu also known as the "bird flu" will force emergency managers to work with others. Not only will emergency managers need to work with public health officials, they will also need to work with veterinarians and other experts in animal health as the disease spreads from animals to humans.

SELF-CHECK

- List two important professional challenges confronting emergency managers.
- Explain the duties that professors must perform in addition to the service activities they perform for emergency management agencies.

15.5 Professional Opportunities

There are several important professional challenges confronting you in the coming years.

15.5.1 Increased Professionalization of Emergency Management

In the past, emergency managers were public safety officers from police and fire departments or retired military personnel. They did not have degrees in emergency management. However, recent years have seen an increase in graduates holding post-secondary degrees in emergency management. As late as 1995, there were only three emergency management programs and two certificate programs. By June 2004, there were 115 programs listed with the FEMA Higher Education Program. Nineteen of these were associate degrees and eleven were bachelor degrees, requiring as many as ten courses. Another 50 were certificate programs or minors that generally require half as many courses. In addition, there were 35 graduate (28 master and seven Ph.D.) programs. These programs are currently found in 39 states. An additional eight states have programs proposed for initiation in the near future. The advantage of hiring someone with a degree in emergency management is that the person begins with a base of knowledge that can be enriched by on-the-job training.

There is much less information available about other aspects of the staffing process. The human resource management process for a given job consists of (Cascio, 1998):

- ▲ The labor market.
- ▲ The applicant pool.
- ▲ The selection process.
- ▲ Job demands and job context.
- ▲ Education and training.
- ▲ Performance evaluation.
- ▲ Tangible compensation and intangible rewards.
- ▲ Job tenure and turnover.

Information is needed about each stage of the emergency management staffing process. This information will identify patterns in the profession as well as differences by region, state, and jurisdiction size.

15.5.2 Involvement in Hazard Mitigation

The need for emergency managers to become involved in hazard mitigation has been recognized for many years (Anderson and Mattingly, 1991). As noted earlier, Project Impact created public/private partnerships to promote hazard mitigation before being discontinued. A major question for the future is whether local emergency managers will use the lessons of Project Impact and apply them without the support of federal funding.

15.5.3 Involvement in Preimpact Disaster Recovery Planning

There has been increasing recognition that the policies, plans, and procedures needed to facilitate a rapid disaster recovery must be developed before—not after—disaster strikes. Preimpact recovery planning accomplishes two objectives. The first objective is to accelerate housing recovery. The second objective is to achieve hazard mitigation activities at the same time as disaster recovery. Preimpact recovery planning also provides additional opportunities to work with land-use planners and building construction officials. The challenge for the future will be for you to work with these other professions to develop preimpact recovery plans.

15.5.4 Expansion of the Professional Domain

The challenges of working with other professions and academic disciplines are matched by exciting opportunities. You must be ready to respond to a wide variety of hazards. This responsibility can mean dealing with natural events as rare as tsunamis or as common as heat waves. Similarly, your responsibility for technological events can be as unusual as nuclear power plant accidents or as routine as major fires.

Your knowledge and skills in helping communities prepare for a wide range of unexpected events makes you an invaluable consultant to senior administrators. Indeed, to the degree that you are successful in promoting hazard mitigation, they will see a steady decrease in the amount of time they spend in emergency response. Thus, your role as *emergency* managers could expand to fill a broader function as *environmental hazard* managers.

15.5.5 Regional Collaboration

Regional collaboration is an important solution for many types of hazards. Many local jurisdictions first established their own hazmat response teams and later found out how expensive it is to staff, train, and equip these organizations. The same problems arise in coping with large-scale incidents such as hurricanes and wildfires. Regional collaboration is an extension of the Mutual Aid Pacts that ensure assistance from neighboring jurisdictions in emergencies (Lindell and Perry, 2001). Collaboration allows communities to combine to pay for services that neither one can afford by itself. Typical candidates will be functions and equipment that have high cost and infrequent use. A regional hazmat response team would take longer to assemble and respond but would be significantly less expensive than one for each jurisdiction. Longer response times would be a significant issue for emergency response teams but not for teams performing hazard vulnerability analysis, hazard mitigation, emergency preparedness, or disaster recovery.

Collaboration among jurisdictions within a region or between levels of government requires some degree of organizational standardization. This is because assisting organizations are of no use if they have incompatible structures, training, and equipment. Many have endorsed the Incident Command System (ICS) or Incident Management System (IMS). Unfortunately, most jurisdictions modified these systems for their own use. The more ICS/IMS has been adapted to local conditions, the more versions there are, and the less standardized they become. If everyone has a different version, it will not have many advantages for joint operations.

The National Incident Management System (NIMS) might yet prove successful in promoting a higher level of standardization. This is not to say that complete standardization is likely to be achieved; only that a higher level might result. The degree of standardization will depend on the compatibility of the new system with the structures of thousands of cities and counties. Its success depends on the willingness of federal agencies, especially the Department of Homeland Security, to engage in a dialogue with representatives from a wide range of organizations. In addition, the federal government must mandate the new program or provide incentives to implement it. One powerful incentive the

FOR EXAMPLE

FEMA and Education

One of FEMA's goals is to encourage and support emergency management-related education. FEMA's Emergency Management Institute (EMI), in Emmitsburg, Maryland, which focuses on skills-based training for existing emergency management personnel, has undertaken several projects which promote college-based emergency management education for future emergency managers. In 1995, EMI devoted a full-time staff officer to the task of working with academics to develop and promote emergency management-related college courses.

federal government has used previously is to make funds contingent on local adoption of federal programs. The success of this approach depends on the ability to monitor compliance. It is easy to obtain letters from local jurisdictions saying they have adopted a program. It is difficult to determine if they have implemented the program without on-site audits. The U.S. Nuclear Regulatory Commission (NRC) conducts such audits at one hundred nuclear power plants throughout the country. However, it is not feasible for any agency to conduct audits in more than ten thousand state and local jurisdictions. Consequently, mandatory national standardization will be a greater challenge than voluntary regional standardization.

SELF-CHECK

- Explain why information is needed about each stage of the emergency management staffing process.
- Explain how professional opportunities have changed in emergency management.
- Explain how your role as an emergency manager can expand to fill a broader function as an environmental hazard manager.
- Explain why collaboration among jurisdictions within a region or between levels of government requires some degree of organizational standardization.

SUMMARY

Many things are changing in emergency management. As this chapter describes, there are many challenges and opportunities facing you, both locally and nationally. In addition, it is essential that communication improves so collaboration with other professionals can better prepare you for emergency management. There are global challenges you need to face, and better education and preparation will help you tackle these difficult problem.

KEY TERMS

Emergency Alert System A new alert system that replaced the Emergency Broadcast System and provides a greater range of capabilities in the digital age.

Signal value An indication that a previously unnoticed threat warrants attention.

Weather Radio A system that provides tone-activated notification of emergencies throughout most of the United States.

ASSESS YOUR UNDERSTANDING

Go to www.wiley.com/college/lindell to assess your knowledge of future directions of emergency management.
Measure your learning by comparing pre-test and post-test results.

Summary Questions

1. The world population will increase as much as _____ in the next 50 years.
   - (a) 25%
   - (b) 50%
   - (c) 75%
   - (d) 100%
2. Technology has improved greatly, and we can now accurately detect all hazards. True or False?
3. Which of the following is *not* an emergency response function after a threat?
   - (a) emergency assessment
   - (b) hazard operations
   - (c) population protection
   - (d) recovery management
4. Increased concern about biological threats has increased the interaction of emergency managers with public health departments. True or False?
5. The human resource management process for a given job consists of which of the following?
   - (a) the applicant pool
   - (b) performance evaluation
   - (c) the selection process
   - (d) all of the above
6. Hurricane Katrina was a megadisaster. True or False?
7. The problem with Project Impact was
   - (a) it forced emergency managers to work with people from other disciplines.
   - (b) it forced emergency managers to work with researchers.
   - (c) federal funding for the project was discontinued.
   - (d) the scope was too large.

8. Emergency management became a priority of the federal government after:
   (a) Hurricane Andrew.
   (b) Hurricane Katrina.
   (c) the terrorist attacks on September 11, 2001.
   (d) the 1989 San Francisco Earthquake.

Review Questions

1. Why is climate change especially important for emergency managers?
2. How can natural disturbances be healthy events that lead to a better use of land?
3. How do you prepare for a terrorist threat?
4. Why is cooperation not easy between emergency managers and disaster researchers?
5. What are the two objectives that preimpact recovery planning accomplishes? Explain each objective.
6. Why is standardization important?

Applying This Chapter

1. What are some examples of technology that aid in emergency management?
2. If wildfires burned hundreds of acres in California, destroying many homes, how would you explain the environmental benefits to the homeowners?
3. In preparing for a terrorist chemical attack in your city, what chemicals would you anticipate being used and why?
4. You are asked to make a presentation about the new opportunities in emergency management education. What would you say in your presentation?
5. You are rolling out a new software and database system in your office that will coordinate and track relief and response efforts. What are some of the challenges in transitioning to this new system?
6. In preparing for a biological attack, who would you want to work with and why?

YOU TRY IT

Clean Water

Think about your community and what would happen if there was a shortage of safe drinking water. Write a paper on the problems this would pose and what you would do to mitigate these problems.

Working with Others

If you were recruiting new employees at a job fair, what qualities in the applicant would you look for? Write a job description.

Your Education

Now that you have completed an emergency management course, what lessons can you apply to your job? How does this knowledge give you an advantage over someone who does not have any formal education in emergency management?

BIBLIOGRAPHY

Abkowitz, M., and E. Meyer. 1996. Technological advancements in hazardous materials evacuation planning. *Transportation Research Record* 1522:116–121.

Abramowitz, J. N. 2001a. *Unnatural disasters.* Worldwatch Paper 158. Washington, DC: Worldwatch Institute.

Abramowitz, J. N. 2001b. Averting disaster. In *State of the world 2001,* ed. Worldwatch Institute, 123–142. New York. W. W. Norton.

Adams, W. C., S. D. Burns, and P.G. Handwerk. 1994. *Nationwide LEPC survey*. Washington, DC: George Washington University Department of Public Administration.

Aguirre, B., D. Wenger, and G. Vigo. 1998. A test of the emergent norm theory of collective behavior. *Sociological Forum* 13:301–320.

Alam, S. A., and K. G. Goulias. 1999. Dynamic emergency evacuation management system using geographic information system and spatiotemporal models of behavior. *Transportation Research Record* 1660:92–99.

Aldrich, D. C., D. M. Ericson, and J. D. Johnson. 1978. *Public protection strategies for potential nuclear reactor accidents: Sheltering concepts with existing public and private structures.* Albuquerque, NM: Sandia National Laboratories.

Aldrich, D. C., D. M. Ericson, and J. D. Johnson. 1982. *Technical guidance for siting criteria development.* Albuquerque, NM: Sandia National Laboratories.

Alesch, D. J., and J. N. Holly. 1998. Small business failure, survival, and recovery: Lessons from the January 1994 Northridge earthquake. *NEHRP conference and workshop on research on the Northridge, California, earthquake of January 17, 1994.* Richmond, CA: Consortium of Universities for Research in Earthquake Engineering.

Alesch, D. J., J. N. Holly, E. Mittler, and R. Nagy. 2001. When small businesses and not-for-profit organizations collide with environmental disasters. Paper presented at the 1st annual IIASA-DPRI Meeting on Integrated Disaster Risk Management, Laxenburg, Austria.

Alesch, D. J., C. Taylor, S. Ghanty, and R. A. Nagy. 1993. Earthquake risk reduction and small business. In *1993 National Earthquake Conference monograph 5: Socioeconomic impacts,* ed. Committee on Socioeconomic Impacts, 133–160. Memphis, TN: Central United States Earthquake Consortium.

Alexander, D. 1993. *Natural disasters.* New York: Chapman and Hall.

Alfaro, J. 1998. The role of federal disaster relief assistance to local communities for historic preservation. In *Disaster management programs for historic sites,* ed. D. H. R. Spennemann and D. W. Look, 39–42. San Francisco: Association for Preservation Technology. 1976. Participation of Blacks, Puerto Ricans and Whites in voluntary associations. *Social Forces* 56:1053–1071.

American Institute of Architects. 1992. *Buildings at risk: Seismic design basics for practicing architects.* Washington, DC: American Institute of Architects.

American Institute of Architects. 1995. *Buildings at risk: Wind design basics for practicing architects.* Washington, DC: American Institute of Architects.

Anderson, J. E. 1994. *Public policymaking: An introduction.* Boston: Houghton Mifflin.

Anderson, W. A. 1969a. Disaster warning and communication in two communities. *Journal of Communication* 19:92–104.

Anderson, W. A. 1969b. *Local civil defense in natural disaster.* Columbus: Ohio State University Disaster Research Center.

Anderson, W. A., and S. Mattingly. 1991. Future directions. In *Emergency management: Principles and practice for local government,* ed. T. S. Drabek and G. J. Hoetmer, 311–335. Washington, DC: International City/County Management Association.

Anno, G. H., and M. A. Dore. 1978a. *Protective action evaluation: The effectiveness of sheltering as a protective action against nuclear accidents involving gaseous releases.* Pt. 1. Washington, DC: Environmental Protection Agency.

Anno, G. H., and M. A. Dore. 1978b. *Protective action evaluation: The effectiveness of sheltering as a protective action against nuclear accidents involving gaseous releases.* Pt. 2. Washington, DC: Environmental Protection Agency.

Anthony, D. F. 1994. Managing the disaster. *Fire Engineering* 147 (8): 22–40.

Arizona Division of Emergency Management. 2003. *Model local hazard mitigation plan.* Phoenix, AZ: Arizona Division of Emergency Management. http://www.dem.state.az.us/operations/mitigation/ pdm.htm.

Arlikatti, S., M. K. Lindell, C. S. Prater, and Y. Zhang. Forthcoming. Risk area accuracy and hurricane evacuation expectations of coastal residents. *Environment and Behavior.*

Army Corps of Engineers National Floodproofing Committee. 1993. *Floodproofing: How to evaluate your options.* Washington, DC: Army Corps of Engineers National Floodproofing Committee.

Ashford, N. A., J. V. Gobbell, J. Lachman, M. Matthiesen, A. Minzner, and R. Stone. 1993. *The encouragement of technological change for preventing chemical accidents: Moving firms from secondary prevention and mitigation to primary prevention.* Cambridge, MA: Massachusetts Institute of Technology Center for Technology, Policy, and Industrial Development.

Auf der Heide, E. 1994. *Disaster response: Principles of preparation and coordination.* St. Louis: Mosby.

Bachrach, P., and M. Baratz. 1962. Decisions and non-decisions: An analytical framework. *American Political Science Review* 57:632–642.

Baker, E. J. 1979. Predicting response to hurricane warnings. *Mass Emergencies* 4:9–24.

Baker, E. J. 1991. Hurricane evacuation behavior. *International Journal of Mass Emergencies and Disasters* 9:287–310.

Baker, E. J. 1993. Empirical studies of public response to tornado and hurricane warnings in the United States. In *Prediction and perception of natural hazards,* ed. J. Nemec, J. Nigg, and F. Siccardi, 65–74. London: Kluwer Academic Publishers.

Banerjee, M. M., and D. F. Gillespie. 1994. Strategy and organizational disaster preparedness. *Disasters* 11:421–436.

Barber, B. 1995. *Jihad vs. McWorld: How globalization and tribalism are reshaping the world.* New York: Times Books.

Barlow, H. D. 1993. Safety officer accounts of earthquake preparedness at riverside industrial sites. *International Journal of Mass Emergencies and Disasters* 11:421–436.

Baron, J. 2000. *Thinking and deciding.* New York: Cambridge University Press.

Barrett, B., B. Ran, and R. Pillai. 2000. Developing a dynamic traffic management modeling framework for hurricane evacuation. *Transportation Research Record* 1733:115–121.

Barry, D. 2001. The search: A few moments of hope in a mountain of rubble. *New York Times,* September 13.

Barton, A. 1969. *Communities in disaster.* New York: Doubleday.

Bartosh, D. 2003. *Incident command in the era of terrorism.* Washington, DC: Police Executive Research Forum.

Bates, F. L., and W. G. Peacock. 1993. *Living conditions, disasters and development: An approach to cross-cultural comparisons.* Athens, GA: Univ. of Georgia Press.

Bates, F. L., and C. Pelanda. 1994. An ecological approach to disasters. In *Disasters, collective behavior, and social organization,* ed. R. R. Dynes and K. J. Tierney, 149–159. Newark, DE: Univ. of Delaware Press.

Baumgartner, F. R., and B. D. Jones. 1993. *Agendas and instability in American politics.* Chicago: Univ. of Chicago Press.

Beatley, T. 1998. The vision of sustainable communities. In *Cooperating with nature: Confronting natural hazards with land use planning for sustainable communities,* ed. R. J. Burby, 233–262. Washington, DC: Joseph Henry Press.

Berke, P. R. 1995. Natural-hazard reduction and sustainable development: A global assessment. *Journal of Planning Literature* 9:370–382.

Berke, P. R., J. Kartez, and D. E. Wenger. 1993. Recovery after disaster: Achieving sustainable development, mitigation and equity. *Disasters* 17:93–109.

Berke, P., T. Larsen, and C. Ruch. 1984. A computer system for hurricane hazard assessment. *Computers in Environmental and Urban Systems* 9:259–269.

Bianchi, S., and R. Farley. 1979. Racial differences in family living arrangements and economic well-being. *Journal of Marriage and the Family* 41:537–551.

Birkland, T. A. 1997. *After disaster: Agenda setting, public policy and focusing events.* Washington, DC: Georgetown Univ. Press.

Blaikie, P., T. Cannon, I. Davis, and B. Wisner. 1994. *At risk: Natural hazards, people's vulnerability and disasters.* London: Routledge.

Blair, J. P., and D. Bingham. 2000. Economic analysis. In *The practice of local government planning,* 3rd ed., ed. C. J. Hoch, L. C. Dalton, and F. S. So, 119–137. Washington, DC: International City/County Management Association.

Blakely, E. J. 2000. Economic development. In *The practice of local government planning,* 3rd ed., ed. C. J. Hoch, L. C. Dalton, and F. S. So, 283–305. Washington, DC: International City/County Management Association.

Blanchard, B. W. 1986. *American civil defense 1945–1984.* Emmitsburg, MD: FEMA Emergency Management Institute.

Blanchard, B. W. 2003. *Outlines of competencies to develop successful 21st century hazard or disaster or emergency or risk managers.* Emmitsburg, MD: FEMA Emergency Management Institute.

Blanchard, B. W. 2004. *FEMA higher education project.* Emmitsburg, MD: FEMA Higher Education Workshop. http://training.fema.gov/EMIWeb/edu.

Blanchard-Boehm, R. D. 1998. Understanding public response to increased risk from natural hazards. *International Journal of Mass Emergencies and Disasters* 16:247–278.

Boileau, A., B. Catrtarinussi, G. Delli Zotti, C. Pelanda, R. Strassoldo, and B. Tellia. 1979. *Friuli: La provca del terremoto.* Milan: Franco Angeli.

Bolin, R. C. 1982. *Long-term family recovery from disaster.* Boulder: Univ. of Colorado Institute of Behavioral Science.

Bolin, R. C. 1985. Disaster characteristics and psychosocial impacts. In *Disasters and mental health: Selected contemporary perspectives,* ed. B. J. Sowder, 3–28. Rockville, MD: National Institute of Mental Health.

Bolin, R. C. 1993. *Household and community recovery after earthquakes.* Boulder: Univ. of Colorado Institute of Behavioral Science.

Bolin, R.,C., and P. Bolton. 1986. *Race, religion, and ethnicity in disaster recovery.* Boulder: Univ. of Colorado Institute of Behavioral Science.

Bolin, R. C., and D. Klenow. 1983. Older people in disaster. *Journal of Aging* 26:29–45.

Bolin, R. C., and L. Stanford. 1991. Shelters, housing and recovery: A comparison of U.S. disasters. *Disasters* 45:25–34.

Bolin, R. C., and L. Stanford. 1998. Community-based approaches to unmet recovery needs. *Disasters* 22:21–38.

Bolin, R. C., and P. A. Trainer. 1978. Modes of family recovery following disaster: A cross-national study. In *Disasters: Theory and research,* ed. E. L. Quarantelli, 233–247. Beverly Hills, CA: Sage.

Bourque, L., L. Russell, and J. Goltz. 1993. Human behavior during and immediately after the earthquake. In *The Loma Prieta, California, earthquake of October 17, 1989,* ed. P. Bolton, 3–22. Reston, VA: U.S. Geological Survey.

Britton, N. R. 2001. A new emergency management for the new millennium? *Australian Journal of Emergency Management* 16:44–54.

Britton, N. R., and G. R. Clark. 2000. From response to resilience: EM reform in New Zealand. *Natural Hazards Review* 1:145–150.

Brown, L. R., G. Gardner, and B. Halweil. 1998. *Beyond Malthus: Sixteen dimensions of the population problem.* Washington, DC: Worldwatch Institute.

Brunacini, A. V. 1985. *Fire command.* Quincy, MA: National Fire Protection Association.

Brunacini, A. V. 2001. *Fire command.* 2nd ed. Quincy, MA: National Fire Protection Association.

Bryant, E. A. 1997. *Natural hazards.* Cambridge: Cambridge Univ. Press.

Buck, G. 1998. *Preparing for terrorism.* Albany, NY: Del Mar.

Buckle, P., G. Mars, and S. Smale. 2000. New approaches to assessing vulnerability and resilience. *Australian Journal of Emergency Management* 15:8–14.

Bullard, R. D. 1996. Environmental justice for all. In *Unequal protection: Environmental justice and communities of color,* ed. R. D. Bullard, 3–22. San Francisco: Sierra Club Books.

Burby, R. J., ed. 1998. *Cooperating with nature: Confronting natural hazards with land use planning for sustainable communities.* Washington, DC: Joseph Henry Press.

Burby, R. J., and S. P. French. 1985. *Flood plain land use management: A national assessment.* Boulder, CO: Westview Press.

Burnstein, E., and A. Vinokur. 1977. Persuasive argumentation and social comparison as determinants of attitude polarization. *Journal of Experimental Social Psychology* 13:315–332.

Burson, Z. G., and A. E. Profio. 1977. Structure shielding in reactor accidents. *Health Physics Journal* 33:287–299.

Burton, I., R. Kates, and G. F. White. 1993. *The environment as hazard.* 2nd ed. New York: Guildford Press.

Bush, G. W. 2002. *The Department of Homeland Security.* Washington, DC: The White House.

Buttel, F. 1997. Social institutions and environmental change. In *The international handbook of environmental sociology,* ed. M. Redclift and G. Woodgate, 40–54. Cheltenham, UK: Edward Elgar.

Caiden, G. E. 1982. *Public administration.* 2nd ed. Pacific Palisades, CA: Palisades Publishers.

Cannon, T., J. Twigg, and J. Rowell. 2003. *Social vulnerability, sustainable livelihoods and disasters.* London: Univ. College Benefield Hazard Research Centre.

Canter, D., J. Breaux, and J. Sime. 1980. Domestic, multiple occupancy and hospital fires. In *Fires and human behavior,* ed. D. Canter, 117–136. New York: Wiley.

Caplow, T., H. Bahr, and B. Chadwick. 1981. *Analysis of the readiness of local communities for integrated emergency management planning.* Washington, DC: Federal Emergency Management Agency.

Carlson, G. P. 1983. *Incident command system.* Stillwater, OK: Oklahoma State Univ. Fire Protection Publications.

Carr, L. 1932. Disaster and the sequence-pattern concept of social change. *American Journal of Sociology* 38:209–215.

Cascio, W. F. 1998. *Applied psychology in human resource management.* 5th ed. Upper Saddle River, NJ: Prentice Hall.

Chapanis, A. 1970. Human factors in systems engineering. In *Systems psychology,* ed. K. B. DeGreene, 51–78. New York: McGraw-Hill.

Chapin, F. S., and E. J. Kaiser. 1985. *Urban land use planning.* 3rd ed. Urbana: Univ. of Illinois Press.

Charvériat, C. 2000. *Natural disasters in Latin America and the Caribbean: An overview of risk.* Working paper no. 434. Washington, DC: Inter-American Development Bank.

Chavis, D. M., and A. Wandersman. 1990. Sense of community in the urban environment: A catalyst for participation and community development. *American Journal of Community Psychology* 18:55–82.

Chin, J. 2000. *Control of communicable diseases manual.* 17th ed. Washington, DC: American Public Health Association.

Ching, F. D. K., and C. Adams. 1991. *Building construction illustrated.* 2nd ed. New York: Van Nostrand Reinhold.

Christen, H. 2004. NIMS: The National Incident Management System. *Firehouse* (July): 96–104.

Christensen, L., and C. Ruch. 1978. Assessment of brochures and radio and television presentations on hurricane awareness. *Mass Emergencies* 3:209–216.

Christou, M. D., A. Amendola, and M. Smeder. 1999. The control of major accident hazards: The land-use planning issue. *Journal of Hazardous Materials* 65:151–178.

Christou, M. D., and S. Porter. 1999. *Guidance on land use planning as required by council directive 96/82/EC (Seveso II).* Ispra, Italy: Major-Accident Hazards Bureau, Joint Research Centre, European Union.

Church, R. L., and T. J. Cova. 2000. Mapping evacuation risk on transportation networks using a spatial optimization model. *Transportation Research Part C: Emerging Technologies* 8:321–336.

Churchill E. R. 1997. Effective media relations. In *The public health consequences of disasters,* ed. E. K. Noji, 122–132. London: Kluwer Academic.

Cliver, E. B. 1998. Assessing the character and systems of historic buildings. In *Disaster management programs for historic sites,* ed. D. H. R. Spennemann and D. W. Look, 49–50. San Francisco: Association for Preservation Technology.

Cohen, S., and R. Kapsis. 1978. Participation of Blacks, Puerto Ricans, and Whites in voluntary associations. *Social Forces* 56:1053–1071.

Cohen, B. L., and I. Lee. 1979. A catalog of risks. *Health Physics* 36:707–722.

Coleman, J. S. 1964. *Introduction to mathematical sociology.* New York: Free Press.

Coleman, R., and J. Granito. 1988. *Managing fire services.* 2nd ed. Washington, DC: International City Management Association.

Comerio, M. C. 1998. *Disaster hits home: New policy for urban housing recovery.* Berkeley, CA: Univ. of California Press.

Committee on Assessing the Costs of Natural Disasters. 1989. *The impacts of natural disasters: A framework for loss estimation.* Washington, DC: National Academy Press.

Committee on Risk Perception and Communication. 1989. *Improving risk communication.* Washington, DC: National Academy Press.

Congresso de Colombia. 1997. Ley No. 388, de Desarollo Territorial.

Conklin, C., and J. Edwards. 2000. Selection of protective action guides for nuclear incidents. *Journal of Hazardous Materials* 75:131–144.

Connor, D. 2005. *Outreach assessment: How to implement an effective tsunami preparedness outreach program.* State of Oregon Department of Geology and Mineral Industries Open File Report OFR 0–05–10. Portland, OR: Nature of the Northwest Information Center.

Cook, T. D., D. T. Campbell, and L. Peracchio. 1990. Quasi experimentation. In *Handbook of industrial and organizational psychology,* 2nd ed, vol. 3, ed. M. D. Dunnette and L. M. Hough, 491–576. Palo Alto, CA: Consulting Psychologists Press.

Cooke, D. 1995. L.A. earthquake puts city disaster planning to test. *Disaster Recovery Journal* 7:10–14.

Cova, T. J., and J. P. Johnson. 2002. Microsimulation of neighborhood evacuations in the urban-wildland interface. *Environment and Planning* 34:763–784.

Covello, V. T. 1987. Case studies of risk communication: Introduction. In *Risk communication,* ed. J. C. Davies, V. T. Covello, and F. W. Allen, 63–65. Washington, DC: Conservation Foundation.

Covello, V. T. 1991. Risk comparisons and risk communication: Issues and problems in comparing health and environmental risks. In *Communicating risks to the public: International perspectives,* ed. R. E. Kasperson and P. J. M. Stallen, 79–124. London: Kluwer Academic.

Covello, V. T., D. B. McCallum, and M. T. Pavlova. 1989. *Effective risk communication.* New York: Plenum.

Cross, J. A. 1980. Residents' concerns about hurricane hazard within the lower Florida Keys. In *Hurricanes and coastal storms,* ed. E. J. Baker, 61–66. Tallahassee: Florida State Univ.

Crouch, E., and R. Wilson. 1982. *Risk/benefit analysis.* Cambridge, MA: Ballinger.

Cutter, S. 1996. Vulnerability to environmental hazards. *Progress in Human Geography* 20:529–539.

Cutter, S. L. 2001. *American hazardscapes: The regionalization of hazards and disasters.* Washington, DC: Joseph Henry.

Cutter, S. J., B. J. Boruff, and W. L. Shirley. 2003. Social vulnerability to environmental hazards. *Social Science Quarterly* 84:242–261.

Dacy, D. C., and H. Kunreuther. 1969. *The economics of natural disaster.* New York: Free Press.

Dahlhamer, J. M., and M. J. D'Sousa, eds. 1997. Determinants of business-disaster preparedness in two U.S. metropolitan areas. *International Journal of Mass Emergencies and Disasters* 15:265–281.

Dahlhamer, J. M., and L. M. Reshaur. 1996. *Businesses and the 1994 Northridge earthquake: An analysis of pre- and post-disaster preparedness.* Newark, DE: Univ. of Delaware Disaster Research Center.

Dahlhamer, J. M., and K. J. Tierney. 1998. Rebounding from disruptive events: Business recovery following the Northridge earthquake. *Sociological Spectrum* 18:121–141.

Daines, G. E. 1991. Planning, training, and exercising. In *Emergency management: Principles and practice for local government,* ed. T. E. Drabek and G. J. Hoetmer, 161–200. Washington, DC: International City Management Association.

Daly, H. E. 1996. *Beyond growth: The economics of sustainable development.* Boston: Beacon Press.

Danzig, E., P. Thayer, and L. Galanter. 1958. *The effects of a threatening rumor on a disaster-stricken community.* Washington, DC: National Academy of Sciences.

Darlington, J. 2000. *The profession of emergency management: Educational opportunities and gaps.* Macomb, IL: Western Illinois Univ.

Dash, N. 1997. The use of geographic information systems in disaster research. *International Journal of Mass Emergencies and Disasters* 15:135–146.

Dash, N., W. G. Peacock, and B. H. Morrow. 1997. And the poor get poorer: A neglected Black community. In *Hurricane Andrew: Ethnicity, gender and the sociology of disaster,* ed. W. G. Peacock, B. H. Morrow, and H. Gladwin, 206–225. London: Routledge.

Davidson, E. A. 2001. *You can't eat GNP: Economics as if the ecology mattered.* Cambridge, MA: Perseus Publishing.

Defense Threat Reduction Agency. 2004. *Domestic WMD incident management legal deskbook.* Washington, DC: Defense Threat Reduction Agency.

DeGreene, K. B. 1970. Systems analysis techniques. In *Systems psychology,* ed. K. B. DeGreene, 79–130. New York: McGraw-Hill.

Diggory, J. 1956. Some consequences of proximity to a disease threat. *Sociometry* 19:7–53.

Disaster Research Group. 1961. *A review of disaster studies.* Washington, DC: National Academy Press.

Ditch, R. 2003. *Professionalism in emergency management.* Falls Church, VA: International Association of Emergency Managers.

Dolowitz, D. P., and D. Marsh. 2000. Learning from abroad: The role of policy transfer in contemporary policy-making. *Governance: An International Journal of Policy and Administration* 13:5–24.

Donaldson, W. 1998. The first ten days: Emergency response and protection strategies for the preservation of historic structures. In *Disaster management programs for historic sites,* ed. D. H. R. Spennemann and D. W. Look, 25–29. San Francisco: Association for Preservation Technology.

Dow, K., and S. L. Cutter. 1998. Crying wolf: Repeat responses to hurricane evacuation orders. *Coastal Management* 26:237–252.

Dow, K., and S. L. Cutter. 2002. Emerging hurricane evacuation issues: Hurricane Floyd and South Carolina. *Natural Hazards Review* 3:12–18.

Drabek, T. E. 1969. Social processes in disaster. *Social Problems* 16:336–347.

Drabek, T. E. 1985. Managing the emergency response. *Public Administration Review* 45:85–92.

Drabek, T. E. 1986. *Human system responses to disaster: An inventory of sociological findings.* New York: Springer-Verlag.

Drabek, T. E. 1987. *The professional emergency manager.* Monograph no. 44. Boulder: Univ. of Colorado Program on Environment and Behavior.

Drabek, T. E. 1990. *Emergency management: Strategies for maintaining organizational integrity.* New York: Springer-Verlag.

Drabek, T. E. 1991a. Introduction. In *Emergency management: Principles and practice for local government,* ed. T. E. Drabek and G. J. Hoetmer, xvii–xxxiv. Washington, DC: International City Management Association.

Drabek, T. E. 1991b. The evolution of emergency management. In *Emergency management: Principles and practice for local government,* ed. T. E. Drabek and G. J. Hoetmer, 3–29. Washington, DC: International City Management Association.

Drabek, T. E. 1991c. Anticipating organizational evacuations: Disaster planning by managers of tourist-oriented private firms. *International Journal of Mass Emergencies and Disasters* 9:219–245.

Drabek, T. E. 1991d. *Microcomputers in emergency management: Implementation of computer technology.* Boulder: Univ. of Colorado Institute of Behavioral Science.

Drabek, T. E. 1994a. *Disaster evacuation and the tourist industry.* Boulder: Univ. of Colorado Institute of Behavioral Science.

Drabek, T. E. 1994b. Risk perceptions of tourist business managers. *The Environmental Professional* 16:327–341.

Drabek, T. E. 1995. Disaster responses within the tourist industry. *International Journal of Mass Emergencies and Disasters* 13:7–23.

Drabek, T. E. 1996. *Disaster evacuation behavior: Tourists and other transients.* Monograph no. 58. Boulder: Univ. of Colorado Institute of Behavioral Science.

Drabek, T. E. 1997. *The social dimensions of disaster.* FEMA Higher Education Project college course. Emmitsburg, MD: FEMA Emergency Management Institute. http://training.fema.gov/emiweb/edu/completeCourses.asp.

Drabek, T. E. 1999. Understanding disaster warning responses. *Social Science Journal* 36:515–523.

Drabek, T. E. 2003. *Strategies for coordinating disaster responses.* Monograph no. 61. Boulder: Univ. of Colorado Program on Environment and Behavior.

Drabek, T. E., and K. Boggs. 1968. Families in disaster: Reactions and relatives. *Journal of Marriage and the Family* 30:443–451.

Drabek, T. E., and W. Key. 1976. The impact of disaster on primary group linkages. *Mass Emergencies* 1:89–106.

Drabek, T. E., A. H. Mushkatel, and T. S. Kilijanek. 1983. *Earthquake mitigation policy: The experience of two states.* Boulder: Univ. of Colorado Institute of Behavioral Science.

Drabek, T. E., and J. Stephenson. 1971. When disaster strikes. *Journal of Applied Social Psychology* 1:187–203.

Durkin, M. 1984. The economic recovery of small businesses after earthquakes: The Coalinga experience. Paper presented at the International Conference on Natural Hazards Mitigation Research and Practice, New Delhi, India.

Dynes, R. 1970. *Organized behavior in disaster.* Lexington, MA: Heath-Lexington Books.

Dynes, R. R. 1983. Problems in emergency planning. *Energy* 8:633–660.

Dynes, R. R. 1994. Community emergency planning: False assumptions and inappropriate analogies. *International Journal of Mass Emergencies and Disasters* 12:141–158.

Dynes, R. R., and E. L. Quarantelli. 1975. *The role of civil defense in disaster planning.* Columbus: Ohio State Univ. Disaster Research Center.

Dynes, R., and E. L. Quarantelli. 1976. The family and community context of individual reactions to disaster. In *Emergency and disaster management,* ed. H. Parad, L. Resnik, and L. Parad, 231–244. Bowie, MD: Charles Press.

Dynes, R., E. L. Quarantelli, and G. Kreps. 1972. *A perspective on disaster planning.* Columbus: Ohio State Univ. Disaster Research Center.

Dynes, R., R. Russell, E. L. Quarantelli, and D. E. Wenger. 1990. *Individual and organizational response to the 1985 earthquake in Mexico City, Mexico.* Book and monograph series no. 24. Newark, DE: Univ. of Delaware Disaster Research Center.

Eagly, A. H., and S. Chaiken. 1993. *The psychology of attitudes.* Ft. Worth, TX: Harcourt Brace.

Earthquake Engineering Research Institute. 1996. *Construction quality, education and seismic safety.* Oakland, CA: Earthquake Engineering Research Institute.

Ebert, C. H. V. 1988. *Disasters: Violence of nature and threats by man.* Dubuque, IA: Kendall/Hunt.

Edelman, P., E. Herz, and L. Bickman. 1980. A model of behavior in fires applied to a nursing home fire. In *Fires and fire behavior,* ed. D. Canter. Chichester, UK: Wiley.

Edwards, R. 1994. *Fire Chem I: The basics of II.T.M.* 5th ed. St. Augustine, FL: S.A.F.E. Films.

Eijndhoven, J. C. M., R. A. P. M. Weterings, C. W. Worrell, J. de Boer, J. van der Pligt, and P. J. M. Stallen. 1994. Risk communication in the Netherlands: The monitored introduction of the EC "Post-Seveso" directive. *Risk Analysis* 14:87–96.

Elazar, D. J. 1994. *The American mosaic: The impact of space, time, and culture on American politics.* Boulder, CO: Westview Press.

Elliott, D., E. Swartz, and B. Herbane. 1999. *Business continuity management.* London: Routledge.

Ellis, S. M., Sr., and W. L. Waugh, Jr. 2001. Emergency managers for the new millenium. In *Handbook of crisis and emergency management,* ed. A. Farazmand, 693–702. New York: Marcel Dekker.

Emergency Management Accreditation Program. 2004a. *Candidate's guide to accreditation.* Lexington, KY: Emergency Management Accreditation Program.

Emergency Management Accreditation Program. 2004b. *EMAP standard.* Lexington, KY: Emergency Management Accreditation Program.

Enarson, E., C. Childers, B. H. Morrow, D. Thomas, and B. Wisner. 2003. *A social vulnerability approach to disasters.* Emmitsburg, MD: Federal Emergency Management Agency Emergency Management Institute. http://training.fema.gov/emiweb/edu/completeCourses.asp.

Environmental Systems Research Institute. 2000. About GIS. http://www.esri.com/library/gis/abtgis/what_gis.html.

Ericsson, K. A., and H. A. Simon. 1993. *Protocol analysis: Verbal reports as data.* 2nd ed. Cambridge, MA: MIT Press.

Erikson, K. T. 1976. *Everything in its path.* New York: Simon and Schuster.

Escobedo, L. G., T. L. Chorba, and P. L. Remmington. 1992. The influence of safety belt laws on self-reported safety belt use in the United States. *Accident Analysis and Prevention* 24:643–653.

European Commission. 2003. Chemical accident prevention, preparedness and response. http://europa.eu.int/comm/environment/seveso.

Evetts, J. 2003. The sociological analysis of professionalism. *International Sociology* 18:395–415.

Expert Review Committee. 1987. *The National Earthquake Hazard Reduction Program.* Washington, DC: FEMA Office of Earthquakes and Natural Hazards.

Federal Emergency Management Agency. 1983. *Emergency program manager: An orientation to the position.* Washington, DC: Federal Emergency Management Agency.

Federal Emergency Management Agency. 1984. *Emergency operating center (EOC) handbook: CPG 1–20.* Washington, DC: Federal Emergency Management Agency.

Federal Emergency Management Agency. 1986. *Making mitigation work: A handbook for state officials.* Washington, DC: Federal Emergency Management Agency.

Federal Emergency Management Agency. 1987. *The California FIRESCOPE program.* Emmitsburg, MD: FEMA Emergency Management Institute.

Federal Emergency Management Agency. 1993. *The emergency program manager.* Washington, DC: Federal Emergency Management Agency.

Federal Emergency Management Agency. 1995a. *Introduction to emergency management: Student manual 230.* Emmitsburg, MD: FEMA Emergency Management Institute.

Federal Emergency Management Agency. 1995b. *National mitigation strategy: Partnerships for building safer communities.* Washington, DC: Federal Emergency Management Agency.

Federal Emergency Management Agency. 1995c. *Rapid assessment planning workshop in emergency management: Resource guide.* Washington, DC: Federal Emergency Management Agency.

Federal Emergency Management Agency. 1996a. *Emergency response to a criminal/terrorist incident: Participant handbook.* Washington, DC: Federal Emergency Management Agency.

Federal Emergency Management Agency. 1996b. *Guide for all-hazard emergency operations planning.* Washington, DC: Federal Emergency Management Agency.

Federal Emergency Management Agency. 1997. *Multihazard identification and risk assessment: A cornerstone of the national mitigation strategy.* Washington, DC: Federal Emergency Management Agency.

Federal Emergency Management Agency. 1998a. *Introduction to mitigation: IS-393.* Emmitsburg, MD: FEMA Emergency Management Institute.

Federal Emergency Management Agency. 1998b. *Principles of emergency management: Student manual and instructor guide, G230.* Emmitsburg, MD: FEMA Emergency Management Institute.

Federal Emergency Management Agency. 1999. *FEMA hazard mitigation grant program desk reference.* Washington, DC: Federal Emergency Management Agency. http://www.fema.gov/fima/hmgp/hmgp_ref.shtm

Federal Emergency Management Agency. 2002. *Getting started: Building support for mitigation planning, FEMA 386–1.* Washington, DC: Federal Emergency Management Agency.

Federal Emergency Management Agency. 2003a. *Developing and managing volunteers, IS 244.* Emmitsburg, MD: FEMA Emergency Management Institute.

Federal Emergency Management Agency. 2003b. *Exercise design course, IS 139.* Emmitsburg, MD: FEMA Emergency Management Institute.

Federal Emergency Management Agency. 2004. *National Incident Management System (NIMS): An introduction.* Washington, DC: Federal Emergency Management Agency.

Federal Emergency Management Agency. n. d., a. *Taking shelter from the storm: Building a safe room inside your house.* Washington, DC: Federal Emergency Management Agency.

Federal Emergency Management Agency. n. d., b. *Emergency management guide for business and industry.* Washington, DC: Federal Emergency Management Agency.

Federal Emergency Management Agency/United States Fire Administration. 1999. *Developing effective standard operating procedures for fire and EMS departments.* Washington, DC: Federal Emergency Management Agency/United States Fire Administration.

Federal Emergency Management Agency/United States Fire Administration. 2000. *Hazardous materials response technology assessment.* Washington, DC: Federal Emergency Management Agency/United States Fire Administration.

Federal Emergency Management Agency, U.S. Department of Transportation, and U.S. Environmental Protection Agency. n. d., a. *Handbook of chemical hazard analysis procedures.* Washington, DC: Federal Emergency Management Agency, U.S. Department of Transportation, and U.S. Environmental Protection Agency.

Federal Interagency Floodplain Management Task Force. 1992. *Floodplain management in the United States: An assessment report.* Vol. 1 summary report. Washington, DC: Federal Emergency Management Agency.

Fink, S. 1986. *Crisis management: Planning for the inevitable.* New York: AMACOM.

Fischer, H. W. 1998. *Response to disaster: Fact versus fiction and its perpetuation.* 2nd ed. Lanham, MD: Univ. Press of America.

Fiske, S. T., and S. E. Taylor. 1991. *Social cognition.* 2nd ed. New York: McGraw-Hill.

Flavin, C. 1994. Storm warnings: Climate change hits the insurance industry. *World Watch* 7 (6): 10–20.

Florin, P., and A. Wandersman. 1990. An introduction to citizen participation, voluntary organizations, and community development: Insights for empowerment through research. *American Journal of Community Psychology* 18:41–54.

Ford, J., and A. Schmidt. 2000. Emergency preparedness training: Strategies for enhancing real-world performance. *Journal of Hazardous Materials* 75:195–215.

Foster, H. D. 1980. *Disaster planning: The preservation of life and property.* New York: Springer-Verlag.

Fraiser, J. 1999. A review of the substantive provisions of the Mississippi Governmental Immunity Act. *Mississippi Law Journal* 68 (703): 774–75.

Freedy, J. R., D. L. Shaw, M. P. Jarrell, and C. R. Masters. 1992. Towards an understanding of the psychological impact of natural disasters: An application of the conservation of resources stress model. *Journal of Traumatic Stress* 5:441–454.

Freidson, E. 2001. *Professionalism.* Chicago: Univ. of Chicago Press.

French, J. R. P., and B. H. Raven. 1959. The bases of social power. In *Studies in social power,* ed. D. Cartwright, 150–167. Ann Arbor, MI: Institute for Social Research.

French, S. P. 1986. The evolution of decision support systems for earthquake hazard mitigation. In *Terminal disasters: Computer applications in emergency management,* ed. S.A. Marston, 57–68. Boulder: Univ. of Colorado Institute of Behavioral Science.

Friesma, H. P., J. Caporaso, G. Goldstein, R. Linberry, and R. McCleary. 1979. *Aftermath: Communities after natural disasters.* Beverly Hills, CA: Sage.

Fritz, C. E. 1957. Disasters compared in six American communities. *Human Organization* 16:6–9.

Fritz, C. E. 1961. Disaster. In *Contemporary social problems,* ed. R. K. Merton and R. A. Nisbet, 651–694. New York: Harcourt, Brace, and World.

Fritz, C. E. 1968. Disasters. In *International encyclopedia of the social sciences,* ed. D. Sills, 202–207. New York: MacMillan and Free Press.

Fritz, C. E. and E. Marks. 1954. The NORC studies of human behavior in disaster. *Journal of Social Issues* 10:26–41.

Fritz, C. E., and J. Mathewson. 1957. *Convergence behavior in disasters.* Washington, DC: National Academy Press.

Frosdick, S. 1997. The techniques of risk analysis are insufficient in themselves. *Disaster Prevention and Management* 6:165–77.

Gabor, T. 1981. Mutual aid systems in the United States for chemical emergencies. *Journal of Hazardous Materials* 4:343–356.

Geis, D. E. 1996. *Creating sustainable and disaster resistant communities.* Aspen, CO: Aspen Global Change Institute.

Geis, D. E. 2000. By design: The disaster resistant and quality-of-life community. *Natural Hazards Review* 1:151–160.

Gerrity, E. T., and B. W. Flynn. 1997. Mental health consequences of disasters. In *The public health consequences of disasters,* ed. E. K. Noji, 101–121. New York: Oxford Univ. Press.

Gillespie, D., R. Colignon, M. Banerjee, S. Murty, and M. Rogge. 1993. *Partnerships for community preparedness.* Boulder: Univ. of Colorado Institute of Behavioral Science.

Gillespie, D. F., and R. W. Perry. 1976. An integrated systems and emergent norm approach to mass emergencies. *Mass Emergencies* 1:303–312.

Gillespie, D. F., and C. L. Streeter. 1987. Conceptualizing and measuring disaster preparedness. *International Journal of Mass Emergencies and Disasters* 5:155–176.

Gilovich, T., D. Griffin, and D. Kahneman. 2002. *Heuristics and biases: The psychology of intuitive judgment.* New York: Cambridge Univ. Press.

Girard, C., and W. G. Peacock. 1997. Ethnicity and segregation: Post-hurricane relocation. In *Hurricane Andrew: Ethnicity, gender and the sociology of disasters,* ed. W. G. Peacock, B. H. Morrow, and H. Gladwin, 191–205. New York: Routledge.

Gist, R., B. Lubin, and B. G. Redburn. 1999. Psychosocial, community, and ecological approaches on disaster response. In *Response to disaster: Psychosocial, community, and ecological approaches,* ed. R. Gist and B. Lubin, 1–20. Philadelphia: Brunner/Mazel.

Gist, R., and S. B. Stolz. 1982. Mental health promotion and the media: Community response to the Kansas City hotel disaster. *American Psychologist* 37:1136–1139.

Gladwin, C. H., H. Gladwin, and W. G. Peacock. 2001. Modeling hurricane evacuation decisions with ethnographic methods. *International Journal of Mass Emergencies and Disasters* 19:117–143.

Glass, J. 1979. Citizen participation in planning. *Journal of the American Planning Association* 45:180–189.

Glass, R., R. Craven, D. Bregman, B. Stoll, N. Horowitz, P. Kerndt, and J. Winkle. 1980. Injuries from the Wichita Falls tornado. *Science* 207:734–738.

Gleser, G., B. Green, and C. Winget. 1981. *Prolonged psychosocial effects of disaster.* New York: Academic Press.

Glickman, T. S., and A. M. Ujihara. 1989. Conclusions. In *Proceedings of the conference on in-place protection during chemical emergencies,* ed. T. S. Glickman and A. M. Ujihara, 2–4. Washington, DC: Resources for the Future.

Godschalk, D. R., T. Beatley, P. Berke, D. J. Brower, E. J. Kaiser, C C. Bohl, and R. M. Goebel. 1999. *Natural hazard mitigation: Recasting disaster policy and planning.* Washington, DC: Island Press.

Godschalk, D. R., E. J. Kaiser, and P. R. Berke. 1998. In *Cooperating with nature: Confronting natural hazards with land-use planning for sustainable communities,* ed. R. J. Burby, 85–118. Washington, DC: Joseph Henry Press.

Goetsch, D. L. 1996. *Occupational safety and health: In the age of high technology for technologists, engineers, and managers.* 2nd ed. Englewood Cliffs, NJ: Prentice-Hall.

Goggin, M. L., A. O. Bowman, J. P. Lester, and L. J. O'Toole. 1990. *Implementation theory and practice: Toward a third generation.* Glenview, IL: Scott Foresman.

Goldstein, I. L. 1993. *Training in organizations: Needs assessment, development and evaluation.* 3rd ed. Pacific Grove, CA: Brooks/Cole.

Golec, J. A., and P. J. Gurney. 1977. The problem of needs assessment in the delivery of EMS. *Mass Emergencies* 2:169–177.

Goltz, J., L. Russell, and L. Bourque. 1992. Initial behavioral response to a rapid onset disaster. *International Journal of Mass Emergencies and Disasters* 10:43–69.

Gordon, P., H. W. Richardson, B. Davis, C. Steins, and A. Vasishth. 1995. *The business interruption effects of the Northridge earthquake.* Los Angeles: Univ. of Southern California Lusk Center Research Institute.

Governo do Estado de São Paulo, Gabinete do Governador, Casa Militar, Coodenadoria Estadual de Defesa Civil. 2000. *Manual do Cidadão, Volume 1: Como proceder nas emergencies do verão.* São Paulo, Brazil: Governo do Estado de São Paulo, Gabinete do Governador, Casa Militar, Coodenadoria Estadual de Defesa Civil.

Graham, C. B., Jr., and S. W. Hays. 1993. *Managing the public organization.* 2nd ed. Washington, DC: CQ Press.

Grannis, D. 2003. Sustaining domestic preparedness: Changes in a post 9/11 world. In *First to arrive: State and local responses to terrorism,* ed. J.N. Kayyem and R. L. Pangi, 207–220. Cambridge, MA: MIT Press.

Greene, M. R., R. W. Perry, and M. K. Lindell. 1981. The March 1980 eruptions of Mt. St. Helens: Citizen information and threat perception. *Disasters* 5:49–66.

Griffith, D. A. 1986. Hurricane emergency management applications of the SLOSH numerical storm surge prediction model. In *Terminal disasters: Computer applications in emergency management,* ed. S. A. Marston, 83–94. Boulder: Univ. of Colorado Institute of Behavioral Science.

Gruntfest, E., T. Downing, and G. F. White. 1978. Big Thompson flood exposes need for better flood reaction system. *Civil Engineering* 78:72–73.

Gudykunst, W. B. 1998. *Bridging differences: Effective intergroup communication.* Thousand Oaks, CA: Sage.

Haas, J. E., H. Cochrane, and D. Eddy. 1977. Consequences of a cyclone for a small city. *Ekistics* 44:45–51.

Haas, J. E., and T. E. Drabek. 1973. *Complex organizations: A sociological perspective.* New York: Macmillan.

Haddow, G. B., and J. A. Bullock. 2003. *Introduction to emergency management.* New York: Butterworth-Heinemann.

Hamilton, R., R. M. Taylor, and G. Rice. 1955. *The social psychological interpretation of the Udall, Kansas, tornado.* Wichita: Univ. of Wichita Press.

Hance, B., C. Chess, and P. Sandman. 1988. *Improving dialogue with communities.* New Brunswick, NJ: New Jersey Department of Environmental Protection.

Haney, T. 1986. Application of computer technology for damage/risk projections. In *Terminal disasters: Computer applications in emergency management,* ed. S. A. Marston, 95–108. Boulder: Univ. of Colorado Institute of Behavioral Science.

Harding, D. M., and D. J. Parker. 1974. Flood hazard at Shrewsbury, United Kingdom. In *Natural hazards: Local, national and global,* ed. G. F. White, 43–52. New York: Oxford Univ. Press.

Harris, M. 1975. *Significant events in United States civil defense history.* Washington, DC: Civil Defense Preparedness Agency.

Hays, S., and T. Reeves. 1984. *Personnel management in the public sector.* Boston: Allyn and Bacon.

Heady, F. 1996. *Public administration: A comparative perspective.* New York: Marcel Dekker.

Hess, C., and J. Harrald. 2004. The national response plan: Process, prospects and participation. *Natural Hazards Observer* 28 (July): 1–3.

Hobeika, A. G., and C. Kim. 1998. Comparison of traffic assignments in evacuation modeling. *IEEE Transactions on Engineering Management* 45:192–198.

Hobeika, A. G., C. Kim, and R. Beckwith. 1994. A decision support system for developing evacuation plans around nuclear power stations. *Interfaces* 24:22–35.

Hobeika, A. G., and B. Jamei. 1985. MASSVAC: A model for calculating evacuation times under natural disaster. *Proceedings of the Conference on Computer Simulation in Emergency Planning* 15(1).

Hoetmer, G. J. 2003. *Characteristics of effective emergency management organizational structures.* Fairfax, VA: Public Entity Risk Institute.

Hornsby, R. I., G. Ortloff, and M. Smith. 1978. A highway accident which involved a spill of natural uranium oxide concentrate. In *Proceedings of the 5th International Symposium on Packaging and Transportation of Radioactive Materials,* 623–630. Washington, DC: U.S. Department of Transportation.

Houts, P. S., P. D. Cleary, and T. W. Hu. 1988. *The Three Mile Island crisis: Psychological, social and economic impacts on the surrounding population.* University Park, PA: Pennsylvania State Univ. Press.

Hovland, C., I. Janis, and H. Kelley. 1953. *Communication and persuasion.* New Haven, CT: Yale Univ. Press.

Howlett, M. 1991. Policy instruments, policy styles, and policy implementation: National approaches to theories of instrument choice. *Policy Studies Journal* 19:1–21.

Huebner, R. S., K. S. McLeary, G. P. Partridge, Jr., W. M. Stayer, and J. F. White. 2000. Accidental and catastrophic releases. In *Air pollution engineering manual,* 2nd ed., ed. W. T. Davis, 839–849. New York: Wiley.

Hunt, M. 2005. *Creating a national standard for computer-aided management for prevention and emergency response (CAMPER).* College Station, TX: Texas A&M Univ. Hazard Reduction and Recovery Center.

Hwang, S. N., W. G. Sanderson, and M. K. Lindell. 2001. Analysis of state emergency management agencies' hazard analysis information on the Internet. *International Journal of Mass Emergencies and Disasters* 19:85–106.

Hyndman, D., and D. Hyndman. 2006. *Natural hazards and disasters.* Belmont, CA: Brooks/Cole.

Inglehart, R. 1997. *Modernization and postmodernization: Cultural, economic, and political change in 43 societies.* Princeton, NJ: Princeton Univ. Press.

Institute for Business and Home Safety. 1997. *Is your home protected from hurricane disaster?* Boston, MA: Institute for Business and Home Safety.

Interagency Floodplain Review Committee. 1994. *Sharing the challenge: Floodplain management into the 21st century.* Washington, DC: U.S. Government Printing Office.

Intergovernmental Panel on Climate Change. 2001. *Climate change 2001: The scientific basis.* New York: Cambridge Univ. Press.

International City Management Association. 1981. *Local government disaster protection.* Washington, DC: International City Management Association.

Jackson, E. L. 1977. Public response to earthquake hazard. *California Geology* 30:278–280.

James, L., and S. Sells. 1981. Psychological climate: Theoretical perspectives and empirical research. In *Toward a psychology of situations: An interactional perspective,* ed. D. Magnusson, 275–295. Hillsdale, NJ: Erlbaum.

Janerich, D. T., A. Stark, P. Greenwald, W. Burnett, H. Jacobson, and P. McCuster. 1981. Increased leukemia, lymphoma and spontaneous abortion in western New York following a flood disaster. *Public Health Reports* 96:350–366.

Janis, I. 1962. Psychological effects of warnings. In *Man and society in disaster,* ed. G. Baker and D. Chapman, 290–321. New York: Basic Books.

Janis, I., and L. Mann. 1977. *Decision making.* New York: Free Press.

Jelesnianski, C. P., J. Chen, and W. Shaeffer. 1992. *SLOSH: Sea, lake, and overland surges from hurricanes.* NOAA Technical Report NWS 48. Silver Spring, MD: National Oceanographic and Atmospheric Administration, National Weather Service.

Jirsa, J. O. 1993. Buildings: General issues and characteristics. In *Mitigation of damage to the built environment,* ed. Central United States Earthquake Consortium, 3–18. Memphis, TN: Central United States Earthquake Consortium.

Johnson, N. R. 1988. Fire in a crowded theatre. *International Journal of Mass Emergencies and Disasters* 6:7–26.

Johnson, N. R., W. Feinberg, and D. Johnston. 1994. Microstructure and panic. In *Disasters, collective behavior, and social organization,* ed. R. Dynes and K. Tierney, 168–189. Newark, DE: Univ. of Delaware Press.

Joint Committee on Defense Production. 1976. *Federal, state and local emergency preparedness.* 94th Congress, 2 Sess. Washington, DC: U.S. Senate.

Jones, A., and L. James. 1979. Psychological climate: Dimensions and relationships of individual and aggregated work environment perceptions. *Organizational Behavior and Human Performance* 23:201–250.

Kanaisty, K., and F. Norris. 1995. In search of altruistic community: Patterns of social support mobilization following Hurricane Hugo. *American Journal of Community Psychology* 23:447–477.

Kane, J. S., and E. E. Lawler. 1978. Methods of peer assessment. *Psychological Bulletin* 85:555–586.

Kang, J. E., M. K. Lindell, and C. S. Prater. Forthcoming. Hurricane evacuation expectations and actual behavior in Hurricane Lili. *Journal of Applied Social Psychology.*

Kaplan, J., and M. Demaria. 1995. A simple empirical model for predicting the decay of tropical cyclone winds after landfall. *Journal of Applied Meteorology* 34:2499–2512.

Kaplan, S. 2004. You're certifiable! *CSO: The Resource for Security Executives.* http://csoonline.com/read/images/100702_certificate_head.gif.

Kariotis, J. 1998. The tendency to demolish repairable structures in the name of "life safety." In *Disaster management programs for historic sites,* ed. D. H. R. Spennemann and D. W. Look, 55–59. San Francisco: Assoc. for Preservation Technology.

Kartez, J. 1992. *LEPC roles in toxic hazards reduction: Implementing Title III's unwritten goals.* College Station, TX: Texas A&M Hazard Reduction and Recovery Center.

Kartez, J. D., and M. K. Lindell. 1987. Planning for uncertainty. *Journal of the American Planning Association* 53:487–498.

Kartez, J. D., and M. K. Lindell. 1990. Adaptive planning for community disaster response. In *Cities and disaster,* ed. R. Sylves and W. Waugh, 5–31. Springfield, IL: Charles C. Thomas.

Kasperson, R. 1987. Panel discussion on "Trust and credibility: The central issue?" In *Risk communication,* ed. J. C. Davies, V. T. Covello, and F. W. Allen, 43–62. Washington, DC: Conservation Foundation.

Kasperson, R., and P. J. M. Stallen. 1991. Risk communication: The evolution of attempts. In *Communicating risks to the public: International perspectives,* ed. R. E. Kasperson and P. J. M. Stallen, 1–12. London: Kluwer Academic.

Kates R. W. 1977. Major insights: A summary and recommendations. In *Reconstruction following disaster,* ed. J. E. Haas, R. W. Kates, and M. J. Bowden, 261–293. Cambridge, MA: MIT Press.

Kates R. W., and D. Pijawka. 1977. From rubble to monument: The pace of reconstruction. In *Reconstruction following disaster,* ed. J. E. Haas, R. W. Kates, and M. J. Bowden, 1–23. Cambridge, MA: MIT Press.

Katz, D., and R. L. Kahn. 1978. *The social psychology of organizations.* 2nd ed. New York: Wiley.

Keating, J., E. Loftus, and M. Manber. 1983. Emergency evacuations during fires. In *Advances in applied social psychology,* ed. R. Kidd and M. Saks, 83–99. Hillsdale, NJ: Lawrence Erlbaum Associates.

Kidd, J. S., and H. P. Van Cott. 1972. System and human engineering analyses. In *Human engineering guide to equipment design,* ed. H. P. Van Cott and R. G. Kinkade, 1–16. Washington, DC: U.S. Government Printing Office.

Kilbourne, E. M. 1997a. Heat waves and hot environments. In *The public health consequences of disasters,* ed. E. K. Noji, 245–269. New York: Oxford Univ. Press.

Kilbourne, E. M. 1997b. Cold environments. In *The public health consequences of disasters,* ed. E. K. Noji, 270–286. New York: Oxford Univ. Press.

Killian, L. M. 1952. The significance of multi-group membership in disaster. *American Journal of Sociology* 57:309–314.

Kimmelman, A. 1998. Cultural heritage and disaster management in Tucson, Arizona. In *Disaster management programs for historic sites,* ed. D. H. R. Spennemann and D. W. Look, 31–37. San Francisco: Association for Preservation Technology.

Kingdon, J. W. 1984. *Agendas, alternatives and public policy.* Boston: Little, Brown.

Klepeis, N. E., W. C. Nelson, W. R. Ott, J. P. Robinson, A. M. Tsang, P. Switzer, J. V. Behar, S. C. Hern, and W. H. Englemann. 2001. The national human activity pattern survey (NHAPS): A resource for assessing exposure to environmental pollutants. *Journal of Exposure Analysis and Environmental Epidemiology* 11:231–252.

Klonglan G. E., Beal G. M., Bohlen J. M., and Schafer R. B. 1967. *Analysis of change in role performance of local civil defense directors.* Rural Sociology Rep. No. 65. Ames, IA: Iowa State Univ. Department of Sociology and Anthropology.

Klonglan, G. E., C. L. Mulford, and D. A. Hay. 1973. *Impact of career development program upon local coordinators: Final report.* Rural Sociology Rep. No. 65. Ames, IA: Iowa State Univ. Department of Sociology and Anthropology.

Kontratyev, K. Y., A. A. Grigoryev, and C. A. Varotsos. 2002. *Environmental disasters: Anthropogenic and natural.* New York: Springer.

Kramer, M. L., and W. M. Porch. 1990. *Meteorological aspects of emergency response.* Boston: American Meteorological Society.

Kramer, W. M., and C. W. Bahme. 1992. *The fire officer's guide to disaster control.* Saddlebrook, NJ: Fire Engineering Books.

Kreps, G. A. 1981. The worth of the NAS-NRC and DRC studies of individual and social response to disasters. In *Social science and natural hazards,* ed. J. Wright and P. Rossi, 41–86. Cambridge, MA: Abt Books.

Kreps, G. A. 1989. *Symposium on social structure and disaster.* Newark, DE: Univ. of Delaware Press.

Kreps, G. A. 1991. Organizing for emergency management. In *Emergency management: Principles and practice for local government,* ed. T. S. Drabek and G. J. Hoetmer, 30–54. Washington, DC: International City/County Management Association.

Kroll, C. A., J. D. Landis, Q. Shen, and S. Stryker. 1990. The economic impacts of the Loma Prieta earthquake: A focus on small business. *Berkeley Planning Journal* 5:39–58.

Kunreuther, H. 1998. Insurability conditions and the supply of coverage. In *Paying the price: The status and role of insurance against natural disasters in the United States,* ed. H. Kunreuther and R. J. Roth, Sr., 17–50. Washington, DC: Joseph Henry Press.

Kunreuther, H. C. 2001. Protective decisions: Fear or prudence. In *Wharton on making decisions,* ed. S. J. Hoch and H. C. Kunreuther, 259–72. New York: Wiley.

Kunreuther, H., R. Ginsberg, L. Miller, P. Sagi, P. Slovic, B. Borkan, and N. Katz. 1978. *Disaster insurance protection: Public policy lessons.* New York: Wiley.

Kunreuther, H., and R. J. Roth, Sr. 1998. *Paying the price: The status and role of insurance against natural disasters in the United States.* Washington, DC: Joseph Henry Press.

Labadie, J. 1984. Problems in local emergency management. *Environmental Management* 8:489–494.

Langness, D. 1994. The Northridge earthquake: Planning and fast action minimize devastation. *California Hospitals* 8:8–13.

Lasswell, H. 1948. The structure and function of communication in society. In *Communication of ideas,* ed. L. Bryson, 43–71. New York: Harper.

Lavell, A. 1994. Opening a policy window: The Costa Rican hospital retrofit and seismic insurance programs, 1986–1992. *International Journal of Mass Emergencies and Disasters* 12:95–115.

Lavell, A. 2002. *Iniciativas de reducción de riesgo a desastres en centroamérica y Republica Dominicana: Una revisión de recientes desarollos,* 1997–2002. Panama City, Panama: Centro de Coordenación para la Prevención de los Desastres Naturales en América Central.

Lazarus, R. S., and S. Folkman. 1984. *Stress, appraisal, and coping.* New York: Springer.

Leonard, V. A. 1973. *Police pre-disaster preparation.* Springfield, IL: Charles C. Thomas.

Lerner, K. 1991. Governmental negligence liability exposure in disaster management. *Urban Law* 23 (333): 341–45.

Lesak, D. M. 1989. Operational decision making. *Fire Engineering* 142:63–69.

Lesak, D. M. 1999. *Hazardous materials: Strategies and tactics.* Upper Saddle River, NJ: Prentice-Hall.

Lewis, D. C. 1985. Transport planning for hurricane evacuations. *ITE Journal* 55 (8): 31–35.

Lindell, M. K. 1994a. Motivational and organizational factors affecting implementation of worker safety training. In *Occupational medicine state of the art reviews: Occupational safety and health training,* ed. M. J. Colligan, 211–240. Philadelphia: Hanley and Belfus.

Lindell, M. K. 1994b. Are local emergency planning committees effective in developing community disaster preparedness? *International Journal of Mass Emergencies and Disasters* 12:159–182.

Lindell, M. K. 1994c. Perceived characteristics of environmental hazards. *International Journal of Mass Emergencies and Disasters* 12:303–326.

Lindell, M. K. 1995. Assessing emergency preparedness in support of hazardous facility risk analyses: An application at a U.S. hazardous waste incinerator. *Journal of Hazardous Materials* 40:297–319.

Lindell, M. K. 2006. Hazardous materials. In *Planning and urban design standards,* ed. American Planning Association, 168–170. New York: Wiley.

Lindell, M. K. Forthcoming. Regulation of hazardous chemicals and chemical wastes. In *Encyclopedia of chemical processing,* ed. S. Lee. New York: Marcel Dekker.

Lindell, M. K., with D. Alesch, P. A. Bolton, M. R. Greene, L. A. Larson, R. Lopes, P. J. May, J. P. Mulilis, S. Nathe, J. M. Nigg, R. Palm, P. Pate, R. W. Perry, J. Pine, S. K. Tubbesing, and D. J. Whitney. 1997. Adoption and implementation of hazard adjustments. *International Journal of Mass Emergencies and Disasters* 15:327–453.

Lindell, M. K., and V. E. Barnes. 1986. Protective response to technological emergency: Risk perception and behavioral intention. *Nuclear Safety* 27:457–467.

Lindell, M. K., P. Bolton, R. W. Perry, G. Stoetzel, J. Martin, and C. Flynn. 1985. *Planning concepts and decision criteria for sheltering and evacuation in a nuclear power plant emergency.* Washington, DC: Atomic Industrial Forum.

Lindell, M. K., and C. J. Brandt. 2000. Climate quality and climate consensus as mediators of the relationship between organizational antecedents and outcomes. *Journal of Applied Psychology* 85:331–348.

Lindell, M. K., and T. Earle. 1983. How close is close enough?: Public perceptions of the risks of industrial facilities. *Risk Analysis* 3:245–253.

Lindell, M. K., J. C. Lu, and C. S. Prater. 2005. Household evacuation decision making in response to Hurricane Lili. *Natural Hazards Review* 6:171–179.

Lindell, M. K., and M. J. Meier. 1994. Effectiveness of community planning for toxic chemical emergencies. *Journal of the American Planning Association* 60:222–234.

Lindell, M. K., and R. W. Perry. 1980. Evaluation criteria for emergency response plans in radiological transportation. *Journal of Hazardous Materials* 3:335–348.

Lindell, M. K., and R. W. Perry. 1983. Nuclear power plant emergency warning: How would the public respond? *Nuclear News* 26:49–53.

Lindell, M. K., and R. W. Perry. 1987. Warning mechanisms in emergency response systems. *International Journal of Mass Emergencies and Disasters* 5:137–153.

Lindell, M. K., and R. W. Perry. 1990. Effects of the Chernobyl accident on public perceptions of nuclear plant accident risks. *Risk Analysis* 10:393–399.

Lindell, M. K., and R. W. Perry. 1992. *Behavioral foundations of community emergency planning*. Washington, DC: Hemisphere.

Lindell, M. K., and R. W. Perry. 1996a. Identifying and managing conjoint threats: Earthquake-induced hazardous materials releases in the U.S. *Journal of Hazardous Materials* 50:31–46.

Lindell, M. K., and R. W. Perry. 1996b. Addressing gaps in environmental emergency planning: Hazardous materials releases during earthquakes. *Journal of Environmental Planning and Management* 39:531–545.

Lindell, M. K., and R. W. Perry. 1997a. *Hazardous materials releases and risk reduction following the Northridge earthquake*. College Station, TX: Texas A&M Univ. Hazard Reduction and Recovery Center.

Lindell, M. K., and R. W. Perry. 1997b. Hazardous materials releases in the Northridge earthquake. *Risk Analysis* 17:147–156.

Lindell, M. K., and R. W. Perry. 1998. Earthquake impacts and hazard adjustment by acutely hazardous materials facilities following the Northridge earthquake. *Earthquake Spectra* 14:285–299.

Lindell, M. K., and R. W. Perry. 2000. Household adjustment to earthquake hazard. *Environment and Behavior* 32:590–630.

Lindell, M. K., and R. W. Perry. 2001. Community innovation in hazardous materials management: Progress in implementing SARA Title III in the United States. *Journal of Hazardous Materials* 88:169–194.

Lindell, M. K., and R. W. Perry. 2004. *Communicating environmental risk in multiethnic communities.* Thousand Oaks, CA: Sage.

Lindell, M. K., and R. W. Perry. Forthcoming, a. Planning and preparedness. In *Emergency management: Principles and practice for local government,* 2nd ed., ed. K. J. Tierney and W. F. Waugh, Jr. Washington, DC: International City/County Management Association.

Lindell, M. K., and R. W. Perry. Forthcoming, b. Onsite and offsite emergency preparedness. In *Encyclopedia of chemical processing,* ed. S. Lee. New York: Marcel Dekker.

Lindell, M. K., and C. S. Prater. 2000. Household adoption of seismic hazard adjustments: A comparison of residents in two states. *International Journal of Mass Emergencies and Disasters* 18:317–338.

Lindell, M. K., and C. S. Prater. 2002. Risk area residents' perceptions and adoption of seismic hazard adjustments. *Journal of Applied Social Psychology* 32:2377–2392.

Lindell, M. K., and C. S. Prater. 2003. Assessing community impacts of natural disasters. *Natural Hazards Review* 4:176–185.

Lindell, M. K., and C. S. Prater. 2005. *Critical behavioral assumptions in evacuation analysis: Examples from hurricane research and planning.* College Station, TX: Texas A&M Univ. Hazard Reduction and Recovery Center.

Lindell, M. K., C. S. Prater, J. C. Lu, S. Arlikatti, Y. Zhang, and J. E. Kang. 2004. *Hurricane Lili evacuation.* College Station TX: Texas A&M Univ. Hazard Reduction and Recovery Center.

Lindell, M. K., C. S. Prater, and W. G. Peacock. 2005. *Organizational communication and decision making in hurricane emergencies.* National Weather Service Hurricane Forecast Socioeconomic Workshop.

Lindell, M. K., Lu, J. C., and Prater, C. S. 2005. Household decision making and evacuation in response to Hurricane Lili. *Natural Hazards Review,* 6. 171–179.

Lindell, M. K., C. S. Prater, and R. W. Perry. 2005. *Planning for disaster recovery: A review of theory and practice.* College Station, TX: Texas A&M Univ. Hazard Reduction and Recovery Center.

Lindell, M. K., C. S. Prater, R. W. Perry, and J. Y. Wu. 2002. *EMBLEM: An empirically-based large scale evacuation time estimate model.* College Station, TX: Texas A&M Univ. Hazard Reduction and Recovery Center.

Lindell, M. K., C. S. Prater, W. G. Sanderson, Jr., H. M. Lee, Y. Zhang, A. Mohite, and S. N. Hwang. 2001. *Texas Gulf Coast residents' expectations and intentions regarding hurricane evacuation.* College Station, TX: Texas A&M Univ. Hazard Reduction and Recovery Center.

Lindell, M. K., C. S. Prater, and J. Y. Wu. 2002. *Hurricane evacuation time estimates for the Texas Gulf coast.* College Station, TX: Texas A&M Univ. Hazard Reduction and Recovery Center.

Lindell, M. K., W. G. Sanderson, and S. N. Hwang. 2002. Local government agencies' use of hazard analysis information. *International Journal of Mass Emergencies and Disasters* 20:29–39.

Lindell, M. K., and D. J. Whitney. 1995. Effects of organizational environment, internal structure and team climate on the effectiveness of local emergency planning committees. *Risk Analysis* 15:439–447.

Lindell, M. K., and D. J. Whitney. 2000. Correlates of seismic hazard adjustment adoption. *Risk Analysis* 20:13–25.

Lindell, M. K., D. J. Whitney, C. J. Futch, and C. S. Clause. 1996a. The local emergency planning committee: A better way to coordinate disaster planning. In *Disaster management in the U.S. and Canada: The politics, policymaking, administration and analysis of emergency management,* ed. R. T. Silves and W. L. Waugh, Jr., 234–249. Springfield, IL: Charles C. Thomas.

Lindell, M. K., D. J. Whitney, C. J. Futch, and C. S. Clause. 1996b. Multi-method assessment of organizational effectiveness in a local emergency planning committee. *International Journal of Mass Emergencies and Disasters* 14:195–220.

Lindell, M. K., J. A. Wise, A. E. Desrosiers, B. N. Griffin, and W. D. Meitzler. 1982. *Design basis for the NRC Operations Center.* Washington, DC: U.S. Nuclear Regulatory Commission.

Logue, J. N., H. Hansen, and E. Struening. 1979. Emotional and psychological stress following Hurricane Agnes. *Public Health Reports* 94:495–502.

Logue, J. N., H. Hansen, and E. Streuning. 1981. Some indications of the long-term health effects of a natural disaster. *Public Health Reports* 96:67–79.

Logue, J., M. Melick, and E. Struening. 1981. A study of health and mental status following a major natural disaster. In *Research in community and mental health,* ed. R. Simmons, 217–274. Greenwich, CT: JAI Press.

Lowrance, W. W. 1976. *Of acceptable risk.* Los Altos, CA: William Kaufman.

Macedo, E. S., A. T. Ogura, and J. Santero. 2002. *Landslide warning system in Serra do Mar slopes, São Paulo, Brazil.* São Paulo: Universidade de São Paulo Instituto de Pesquisas Tecnológicas.

Maeda, Y., and M. Miyahara. 2003. Determinants of trust in industry, government and citizen's groups in Japan. *Risk Analysis* 23:303–310.

Mallet, L. 2002. Should you be certified? *Contingency Planning and Management* 7 (March): 38–40.

Marcondes, C. R. 2003. *Defesa civil.* 2nd ed. São Paulo: Imprensa Oficial do Estado.

Marston, S. A. 1986. *Terminal disasters: Computer applications in emergency management.* Boulder: Univ. of Colorado Institute of Behavioral Science.

Martin, B., M. Capra, G. van der Heide, M. Stoneham, and M. Lucas. 2001. Are disaster management concepts relevant in developing countries? *Australian Journal of Emergency Management* 16:25–33.

Martin, H. W. 1993. Recent changes to seismic codes and standards: Are they coordinated or random events? In *Proceedings: 1993 National Earthquake Conference,* ed. Central United States Earthquake Consortium, 367–376. Memphis, TN: Central United States Earthquake Consortium.

Martínez, O. C., P. T. Arzayús, J. C. G. Bocanegra, L. A. M. Restrepo, and H. C. R. Ardila. 1997. *Presentácion genera ley de desarollo territorial: La política urbana del salto social.* Bogotá, Colombia: Fotolito Parra & Cia.

Mathieu, J., and D. Zajac. 1990. A review and meta-analysis of the antecedents, correlates, and consequences of organizational commitment. *Psychological Bulletin* 108:171–194.

May, P. J. 1993. Mandate design and implementation: Enhancing implementation efforts and shaping regulatory styles. *Journal of Policy Analysis and Management* 12:634–663.

May, P. J., and R. E. Deyle. 1998. Governing land use in hazardous areas with a patchwork system. In *Cooperating with nature: Confronting natural hazards with land-use planning for sustainable communities,* ed. R. J. Burby, 57–82. Washington, DC: Joseph Henry Press.

May, P. J., and W. Williams. 1985. *Disaster policy implementation: Managing programs under shared governance.* New York: Plenum Press.

Mazmanian, D. A., and P. A. Sabatier. 1989. *Implementation and public policy.* Lanham, MD: Univ. Press of America.

McCallum, D. B., and L. Anderson. 1991. Communicating about pesticides in drinking water. In *Communicating risks to the public: International perspectives,* ed. R. E. Kasperson and P. J. M. Stallen, 237–262. London: Kluwer Academic.

McEntire, D. 2003. Disaster preparedness. *ICMA IQ Reports* (Vol. 35, Item 11). Washington, DC: International City/County Management Association.

McGuire, W. J. 1969. The nature of attitudes and attitude change. In *Handbook of social psychology,* ed. G. Lindsey and E. Aronson, 329–348. Reading, MA: Addison-Wesley.

McGuire, W. J. 1985. The nature of attitudes and attitude change. In *Handbook of social psychology,* ed. G. Lindsey and E. Aronson, 233–256. New York: Random House.

McIntyre, R. M., and E. Salas. 1995. Measuring and managing for team performance: Lessons from complex environments. In *Team effectiveness and decision making in organizations,* ed. R. A. Guzzo, E. Salas, et al., 9–45. San Francisco: Jossey-Bass.

McKenna, T. 2000. Protective action recommendations based upon plant conditions. *Journal of Hazardous Materials* 75:145–164.

McNally, R. J., R. A. Bryant, and A. Ehlers. 2003. Does early psychological intervention promote recovery from posttraumatic stress? *Psychological Science in the Public Interest* 4:45–79.

Mechanic, D. 1962. Sources of power of lower participants in complex organizations. *Administrative Science Quarterly* 7:349–364.

Melick, M. 1985. The health of postdisaster populations. In *Perspectives on disaster recovery,* ed. J. Laube and S. Murphy, 179–209. New York: Appleton-Century-Crofts.

Meltsner, A. J. 1979. The communication of scientific information to the wider public: The case of seismology in California. *Minerva* 17:331–354.

Menninger, W. 1952. Psychological reactions in an emergency. *American Journal of Psychiatry* 109:128–130.

Meyer, E. 1977. *Chemistry of hazardous materials.* Englewood Cliffs, NJ: Prentice-Hall.

Meyer, J., S. Paunonen, I. Gellatly, R. Goffin, and D. Jackson. 1989. Organizational commitment and job performance: It's the nature of the commitment that counts. *Journal of Applied Psychology* 74:152–156.

Meyer, J. P., and N. Allen. 1984. Testing the "side bet theory" of organizational commitment: Some methodological considerations. *Journal of Applied Psychology* 69:372–378.

Michaels, J. V. 1996. *Technical risk management.* Upper Saddle River, NJ: Prentice Hall.

Midlarsky, E. 1968. Aiding responses: An analysis and review. *Merrill-Palmer Quarterly* 14:229–260.

Mileti, D. S. 1975. *Natural hazards warning systems in the United States.* Boulder: Univ. of Colorado Institute of Behavioral Science.

Mileti, D. S. 1983. Societal comparisons of organizational response to earthquake predictions. *International Journal of Mass Emergencies and Disasters* 1:399–413.

Mileti, D. S. 1993. Communicating public risk information. In *Prediction and perception of natural hazards,* ed. J. Nemec, J. Nigg, and F. Siccardi, 143–152. London: Kluwer Academic.

Mileti, D. S. 1999. *Disasters by design: A reassessment of natural hazards in the United States.* Washington, DC: Joseph Henry Press.

Mileti, D. S., and E. Beck. 1975. Communication in crisis. *Communication Research* 2:24–49.

Mileti, D. S., and J. D. Darlington. 1995. Societal response to revised earthquake probabilities in the San Francisco Bay area. *International Journal of Mass Emergencies and Disasters* 13:119–145.

Mileti, D. S., and J. D. Darlington. 1997. The role of searching in shaping reactions to earthquake risk information. *Social Problems* 44:89–103.

Mileti, D. S., J. D. Darlington, C. Fitzpatrick, and P. W. O'Brien. 1993. *Communicating earthquake risk: Societal response to revised probabilities in the Bay Area.* Fort Collins: Colorado State Univ. Hazards Assessment Laboratory and Department of Sociology.

Mileti, D. S., T. Drabek, and J. E. Haas. 1975. *Human systems in extreme environments.* Boulder: Univ. of Colorado Institute of Behavioral Science.

Mileti, D. S., and C. Fitzpatrick 1992. The causal sequence of risk communication in the Parkfield earthquake prediction experiment. *Risk Analysis* 12:393–400.

Mileti, D. S., C. Fitzpatrick, and B. C. Farhar. 1992. Fostering public preparations for natural hazards. *Environment* 34:16–39.

Mileti, D. S., and P. O'Brien. 1992. Warnings during disaster: Normalizing communicated risk. *Social Problems* 39:40–57.

Mileti, D. S., and L. Peek. 2000. The social psychology of public response to warnings of a nuclear power plant accident. *Journal of Hazardous Materials* 75:181–194.

Mileti, D. S., and J. H. Sorensen. 1987. Why people take precautions against natural disasters. In *Taking care: Why people take precautions,* ed. N. Weinstein, 296–320. New York: Cambridge Univ. Press.

Mileti, D. S., and J. H. Sorenson. 1988. Planning and implementing warning systems. In *Mental health response to mass emergencies,* ed. M. Lystad, 321–345. New York: Brunner/Mazel.

Mileti, D. S., J. H. Sorensen, and P. W. O'Brien. 1992. Toward an explanation of mass care shelter use in evacuations. *International Journal of Mass Emergencies and Disasters* 10:25–42.

Milliman, J. W. 1983. An agenda for economic research on flood hazard mitigation. In *A plan for research on floods and their mitigation in the United States,* ed. S. Chagnon. Champaign: Illinois State Water Survey.

Mitchell, J. T. 1983. When disaster strikes...The critical incident stress debriefing process. *Journal of Emergency Medical Services* 8:36–39.

Moeller, M., T. Urbanik, and A. Desrosiers. 1981. *CLEAR (calculates logical evacuation and response): A generic transportation network evacuation model for the calculation of evacuation time estimates.* Washington, DC: U.S. Nuclear Regulatory Commission.

Moore, H.E. 1958. *Tornadoes over Texas.* Austin: Univ. of Texas Press.

Moore, M. 1995. *Creating public value: Strategic management in government.* Cambridge, MA: Harvard Univ. Press.

Morrall, J. 1986. A review of the record. *Regulation* 10:25–34.

Morrow, B. H. 1997. Stretching the bonds: The families of Andrew. In *Hurricane Andrew: Ethnicity, gender and the sociology of disaster,* ed. W. G. Peacock, B. H. Morrow, and H. Gladwin, 141–170. London: Routledge.

Morrow, B. H., and W. G. Peacock. 1997. Disasters and social change: Hurricane Andrew and the reshaping of Miami. In *Hurricane Andrew: Ethnicity, gender and the sociology of disaster,* ed. W. G. Peacock, B. H. Morrow, and H. Gladwin, 226–242. London: Routledge.

Mosher, F. 1968. *Democracy and the public service.* Englewood Cliffs, NJ: Prentice-Hall.

Mulford, C. L., G. E. Klonglan, and J. P. Kopachevsky. 1973. *Securing community resources for social action.* Rural Sociology Rep. No. 112. Ames, IA: Iowa State Univ. Department of Sociology and Anthropology.

Mulford, C. L., G. E. Klonglan, and D. L. Tweed. 1973. *Profiles on effectiveness: A systems analysis.* Rural Sociology Rep. No. 110. Ames, IA: Iowa State Univ. Department of Sociology and Anthropology.

Multihazard Mitigation Council. 2005. *Natural hazard mitigation saves: An independent study to assess the future savings from mitigation activities.* Washington, DC: Multihazard Mitigation Council.

Murphy, S. 1984. Advanced practice implications of disaster stress research. *Journal of Psychosocial Nursing and Mental Health Services* 22:135–139.

National Emergency Management Association/Council of State Governments. 2001. *NEMA/CSG 2001 report on state emergency management funding and structures.* Lexington, KY: National Emergency Management Association/Council of State Governments.

National Fire Protection Association. 2004. *NFPA 1600: Standard on disaster/emergency management and business continuity programs.* Quincy, MA: National Fire Protection Association.

National Governors' Association. 1978. *Comprehensive emergency management.* Washington, DC: National Governors' Association.

National Governors' Association. 2001. *A governor's guide to emergency management, vol. 1: Natural disasters.* Washington, DC: National Governors' Association.

National Governors' Association. 2002. *A governor's guide to emergency management, vol. 2: Homeland security.* Washington, DC: National Governors' Association.

National Institute of Building Sciences. 1998. *HAZUS.* Washington, DC: National Institute of Building Sciences.

National Research Council. 2003. *The emergency manager of the future.* Washington, DC: National Academies Press.

National Response Team. 1987. *Hazardous materials emergency planning guide.* Washington, DC: National Response Team.

National Response Team. 1990. *Developing a hazardous materials exercise program: A handbook for state and local officials.* Washington, DC: National Response Team.

National Response Team. n.d. Incident Command System/Unified Command (ICS/UC) technical assistance document. http://www.nrt.org/Production/NRT/NRTwen.nsf/PagesByLevelCat/Level2ICS/UC?Opendocument.

National Safety Council. 1995. *User's manual for CAMEO: Computer-aided management of emergency operations.* Chicago: National Safety Council.

National Science and Technology Council. 1996. *Natural disaster reduction: A plan for the nation.* Washington, DC: National Science and Technology Council.

National Wildfire Coordinating Group. 1994. *Incident Command System national training curriculum.* http://www.nwcg.gov/pms/forms/ics_cours/ics_courses.htm.

National Wildland/Urban Interface Protection Program. n.d. *Wildland/urban interface fire hazard assessment methodology.* http://www.firewise.org.

Natural Hazards Research and Applications Information Center. 2001. *Holistic disaster recovery: Ideas for building local sustainability after a natural disaster.* Fairfax, VA: Public Entity Research Institute.

Nelson, L. S., and R. W. Perry. 1991. Organizing public education for technological emergencies. *Disaster Management* 4:21–26.

Neuwirth, K., S. Dunwoody, and R. J. Griffin. 2000. Protection motivation and risk communication. *Risk Analysis* 20:721–734.

Nicholson, W. C. 1999a. Beating the system to death: A case study in incident command and mutual aid. *Fire Engineering* 152 (October): 129–30.

Nicholson, W. C. 1999b. Standard operating procedures: The anchor of on-scene safety. *Our Watch* (Winter): 2.

Nicholson, W. C. 2003a. Litigation mitigation: Proactive risk management in the wake of the West Warwick Club fire. *Journal of Emergency Management* 2:14–18.

Nicholson, W. C. 2003b. Legal issues in emergency response to terrorism incidents involving hazardous materials: The hazardous waste operations and emergency response (HAZWOPER) standard, standard operating procedures, mutual aid and the incident command system. *Widener Symposium Law Journal* 9 (295): 298–300.

Nielsen, J. 2000. *Designing web usability.* Indianapolis: New Riders Press.

Nigg, J. M. 1982. Awareness and behavior: Public response to prediction awareness. In *Perspectives on increasing hazard awareness,* ed. T. F. Saarinen, 36–51. Boulder: Univ. of Colorado Institute of Behavioral Science.

Nigg, J. M. 1995. Disaster recovery as a social process. In *Wellington after the quake: The challenge of rebuilding,* 81–92. Wellington, New Zealand: Earthquake Commission.

Nigro, F. A., and L. G. Nigro. 1980. *Modern public administration.* 5th ed. New York: Harper and Row.

Nikoluk, E. 2000. A defesa civil em São Paulo. *ISDR Informs* 2:53.

Nisbett, R., and L. Ross. 1980. *Human inference: Strategies and shortcomings of social judgment.* Englewood Cliffs, NJ: Prentice Hall

Noji, E. K. 1997a. *The public health consequences of disasters.* New York: Oxford Univ. Press.

Noji, E. K. 1997b. The nature of disaster: General characteristics and public health effects. In *The public health consequences of disasters,* ed. E. K. Noji, 3–20. New York: Oxford Univ. Press.

Nordenson, G. J. P. 1993. Seismic codes. In *Mitigation of damage to the built environment,* ed. Central United States Earthquake Consortium, 89–114. Memphis, TN: Central United States Earthquake Consortium.

Oak Ridge National Laboratory. 1998. *OREMS: Oak Ridge Evacuation Management System.* Oak Ridge, TN: Oak Ridge National Laboratory.

O'Keefe, D. 1990. *Persuasion.* Thousand Oaks, CA: Sage.

Ollendick, G., and M. Hoffman. 1982. Assessment of psychological reaction in disaster victims. *Journal of Community Psychology* 10:157–167.

Olson, R., H. Lagorio, and S. Scott. 1990. *Knowledge transfer in earthquake engineering.* Berkeley: Univ. of California Press.

Olson, R. A., and R. Olson. 1985. *Urban heavy rescue.* Tempe: Arizona State Univ. School of Public Affairs.

Olson, R. S., and A. C. Drury. 1997. Untherapeutic communities: A cross-national analysis of post-disaster political unrest. *International Journal of Mass Emergencies and Disasters* 15:221–238.

Olson, R. S., R. A. Olson, and V. T. Gawronski. 1998. Night and day: Mitigation policymaking in Oakland, California, before and after the Loma Prieta earthquake. *International Journal of Mass Emergencies and Disasters* 16:145–179.

Organization for Economic Cooperation and Development. 2003. *Emerging risks in the 21st century: An agenda for action.* Paris: Organization for Economic Cooperation and Development.

Osborne, D., and P. Plastrik. 1998. *Banishing bureaucracy: The five strategies for re-inventing government.* New York: Plume.

Otway, H. 1973. Risk estimation and evaluation. In *Proceedings of the IIASA Planning Conference on Energy Systems,* 11–19. Laxenburg, Austria: International Institute for Applied Systems Analysis.

Oyola-Yemaiel, A., and J. Wilson. 2005. Three essential strategies for emergency management professionalization in the U.S. *International Journal of Mass Emergencies and Disasters* 23:77–84.

Ozer, E. J., and D. S. Weiss. 2004. Who develops post-traumatic stress disorder? *Current Directions in Psychological Science* 13:169–172.

Palm, R., M. Hodgson, R. D. Blanchard, and D. Lyons. 1990. *Earthquake insurance in California.* Boulder, CO: Westview Press.

Parker, D. J. 1999. Disaster response in London: A case of learning constrained by history and experience. In *Crucibles of hazard: Megacities and disasters in transition,* ed. J. K. Mitchell, 186–247. Tokyo: United Nations Univ. Press.

Partners in Protection. 1999. *FireSmart: Protecting your community from wildfire.* Edmonton, Alberta: Partners in Protection.

Peacock, W. G., and C. Girard. 1997. Ethnic and racial inequalities in disaster damage and insurance settlements. In *Hurricane Andrew: Ethnicity, gender and the sociology of disaster,* ed. W. G. Peacock, B. H. Morrow, and H. Gladwin, 171–190. London: Routledge.

Peacock, W. G., L. M. Killian, and F. L. Bates. 1987. The effects of disaster damage and housing on household recovery following the 1976 Guatemala earthquake. *International Journal of Mass Emergencies and Disasters* 5:63–88.

Peacock, W. G., B. H. Morrow, and H. Gladwin. 1997. *Hurricane Andrew: Ethnicity, gender and the sociology of disaster.* London: Routledge.

Peacock, W. G., and A. K. Ragsdale. 1997. Social systems, ecological networks and disasters: Toward a socio-political ecology of disasters. In *Hurricane Andrew: Ethnicity, gender and the sociology of disaster,* ed. W. G. Peacock, B. H. Morrow, and H. Gladwin, 20–35. London: Routledge.

Pearce, J. 1983. Job attitude and motivation differences between volunteers and employees from comparable organizations. *Journal of Applied Psychology* 68:646–652.

Pearce, L. 2003. Disaster management, community planning, and public participation: How to achieve sustainable hazard mitigation. *Natural Hazards* 28:211–228.

Peek-Asa, C., J. F. Kraus, L. B. Bourque, D. Vimalachandra, J. Yu, and J. Abrams. 1998. Fatal and hospitalized injuries resulting from the 1994 Northridge earthquake. *International Journal of Epidemiology* 27:459–465.

Pennebaker, J. W., and K. D. Harber. 1993. A social stage model for collective coping: The Loma Prieta earthquake and the Persian Gulf war. *Journal of Social Issues* 49:125–146.

Pennings, J. M. 1981. Strategically interdependent organizations. In *Handbook of organizational design,* vol. 1, ed. P. C. Nystrom and W. H. Starbuck, 433–455. New York: Oxford Univ. Press.

Perry, R. W. 1979a. Evacuation decision making in natural disaster. *Mass Emergencies* 4:25–38.

Perry, R. W. 1979b. Incentives for evacuation in natural disaster. *Journal of the American Planning Association* 45:440–447.

Perry, R. W. 1982. *The social psychology of civil defense.* Lexington, MA: Heath.

Perry, R. W. 1985. *Comprehensive emergency management: Evacuating threatened populations.* Greenwich, CT: JAI.

Perry, R. W. 1987. Racial and ethnic minority citizens in disasters. In *The sociology of disasters,* ed. R. Dynes and C. Pelanda, 87–99. Gorizia, Italy: Franco Angelli.

Perry, R. W. 1989. Taxonomy and model building for emergency warning response. *International Journal of Mass Emergencies and Disasters* 9:305–328.

Perry, R. W. 1991. Managing disaster response operations. In *Emergency management: Principles and practice for local government,* ed. T. E. Drabek and G. Hoetmer, 201–223. Washington, DC: International City/County Management Association.

Perry, R. W. 1995. The structure and function of emergency operating centers. *International Journal of Disaster Prevention and Management* 4:37–41.

Perry, R. W. 1998. Definitions and the development of a theoretical superstructure for disaster research. In *What is a disaster?,* ed. E. L. Quarantelli, 197–215. London: Routledge.

Perry, R. W., and M. Greene. 1983. *Citizen response to volcanic eruptions.* New York: Irvington.

Perry, R. W., and M. K. Lindell. 1978. The psychological consequences of natural disaster: A review of research on American communities. *Mass Emergencies* 3:105–115.

Perry, R. W., and M. K. Lindell. 1990. *Living with Mt. St. Helens: Human adjustment to volcano hazards.* Pullman, WA: Washington State Univ. Press.

Perry, R. W., and M. K. Lindell. 1997a. Aged citizens in the warning phase of disasters. *International Journal of Aging and Human Development* 44:257–267.

Perry, R. W., and M. K. Lindell. 1997b. Principles for managing community relocation as a hazard mitigation measure. *Journal of Contingencies and Crisis Management* 5:49–60.

Perry, R. W., and M. K. Lindell. 1997c. Earthquake planning for governmental continuity. *Environmental Management* 21:89–96.

Perry, R. W., and M. K. Lindell. 2003. Understanding citizen response to disasters with implications for terrorism. *Journal of Contingencies and Crisis Management* 11:49–60.

Perry, R. W., M. K. Lindell, and M. Greene. 1981. *Evacuation planning in emergency management.* Lexington, MA: Heath-Lexington.

Perry, R. W., M. K. Lindell, and M. R. Greene. 1982. Threat perception and public response to volcano hazard. *Journal of Social Psychology* 116:199–204.

Perry, R. W., and M. Montiel. 1997. Conceptualizando riesgo para disastres sociales. *Desastres Sociedad* 6:67–72.

Perry, R. W., and A. Mushkatel. 1986. *Minority citizens in disasters.* Athens, GA: Univ. of Georgia Press.

Perry, R. W., and L. Nelson. 1991. Ethnicity and hazard information dissemination. *Environmental Management* 15:581–587.

Perry, R. W., and E. L. Quarantelli. 2004. *What is a disaster?: New answers to old questions.* Philadelphia: Xlibris.

Peters, B. G. 1995. *The politics of bureaucracy.* White Plains, NY: Longman.

Petrucelli, U. 2003. Urban evacuation in seismic emergency conditions. *ITE Journal* 73 (8): 34–38.

Phillips, B. D. 1993a. Cultural diversity in disasters: Sheltering, housing, and long-term recovery. *International Journal of Mass Emergencies and Disasters* 11:99–110.

Phillips, B. D. 1993b. Disaster as a discipline: The status of emergency management education in the U.S. *International Journal of Mass Emergencies and Disasters* 23:111–139.

Pidd, M., F. de Silva, and R. Eglese. 1996. A simulation model for emergency evacuation. *European Journal of Operational Research* 90:413–419.

Pielke, R. A., Jr., and C. W. Landsea. 1998. Normalized hurricane damages in the United States: 1925–95. *Weather and Forecasting* 13:621–631.

Pine, J. C. 1991. Liability issues. In *Emergency management: Principles and practice for local government,* ed. T. E. Drabek and G. J. Hoetmer, 289–310. Washington, DC: International City Management Association.

Platt, R. H. 1998. Planning and land use adjustments in historical perspective. In *Cooperating with nature: Confronting natural hazards with land-use planning for sustainable communities,* ed. R. J. Burby, 29–56. Washington, DC: Joseph Henry Press.

Poplin, D. E. 1972. *Communities: A survey of theories and methods of research.* New York: Macmillan.

Porter, L., R. Steers, R. Mowday, and P. Boulian. 1974. Organizational commitment, job satisfaction, and turnover among psychiatric technicians. *Journal of Applied Psychology* 59:603–609.

Post, Buckley, Schuh, and Jernigan. 1999. *Hurricane Georges assessment.* Tallahasseee, FL: Author.

Prater, C., D. Wenger, and K. Grady. 2000. *Hurricane Bret post-storm assessment: A review of the utilization of hurricane evacuation studies and information dissemination.* College Station, TX: Texas A&M Univ. Hazard Reduction and Recovery Center.

Prater, C., and J. Y. Wu. 2002. The politics of emergency response and recovery: Preliminary observations on Taiwan's 921 earthquake. *Australian Journal of Emergency Management* 17:48–59.

Prater, C. S. 2001. *Project Impact: An evaluation.* College Station, TX: Texas A&M Univ. Hazard Reduction and Recovery Center.

Prater, C. S., and M. K. Lindell. 2000. The politics of hazard mitigation. *Natural Hazards Review* 1:73–82.

Prater, C. S., and M. K. Lindell. 2002. *Local jurisdictions' evacuation guidance and transportation support for hurricane evacuees.* College Station, TX: Texas A&M Univ. Hazard Reduction and Recovery Center.

Prater, C. S., W. G. Peacock, M. K. Lindell, Y. Zhang, and J. C. Lu. 2004. *A social vulnerability approach to estimating potential socioeconomic impacts of earthquakes.* College Station, TX: Texas A&M Univ. Hazard Reduction and Recovery Center.

President's Council on Sustainable Development. 1996. *Sustainable America: A new consensus for prosperity, opportunity, and a healthy environment.* Washington, DC: U.S. Government Printing Office.

Prestby, J. E., A. Wandersman, P. Florin, R. C. Rich, and D. M. Chavis. 1990. Benefits, costs, incentive management and participation in voluntary organizations: A means to understanding and promoting empowerment. *American Journal of Community Psychology* 18:117–150.

Prince, S. H. 1920. *Catastrophe and social change.* New York: Columbia Univ. Faculty of Political Science.

Quarantelli, E. L. 1954. The nature and conditions of panic. *American Journal of Sociology* 60:267–275.

Quarantelli, E. L. 1957. The behavior of panic participants. *Sociology and Social Research* 41:187–194.

Quarantelli, E. L. 1960. A note on the protective function of the family in disasters. *Marriage and Family Living* 22:263–264.

Quarantelli, E. L. 1977. Social aspects of disasters and their relevance to pre-disaster planning. *Disasters* 1:98–107.

Quarantelli, E. L. 1980. *Evacuation behavior and problems.* Columbus: Ohio State Univ. Disaster Research Center.

Quarantelli, E. L. 1981a. Disaster planning: Small and large—past, present and future. In *Proceedings: American Red Cross EFO Division Disaster Conference.* Alexandria, VA: American Red Cross Eastern Field Office.

Quarantelli, E. L. 1981b. Panic behavior in fire situations. In *Proceedings of the first conference and workshop on fire casualties,* ed. B. Halpin, 99–112. Baltimore: Johns Hopkins Univ. Applied Physics Laboratory.

Quarantelli, E. L. 1982a. *Sheltering and housing after major community disasters.* Columbus: Ohio State Univ. Disaster Research Center.

Quarantelli, E. L. 1982b. Ten research-derived principles of disaster planning. *Disaster Management* 2:23–25.

Quarantelli, E. L. 1982c. Social and organizational problems in a major community emergency. *Emergency Planning Digest* 9 (January): 7–10, 21.

Quarantelli, E. L. 1984. *Organizational behavior in disasters and implications for disaster planning.* Emmitsburg, MD: Federal Emergency Management Agency National Emergency Training Center.

Quarantelli, E. L. 1987. What should we study? *International Journal of Mass Emergencies and Disasters* 5:7–32.

Quarantelli, E. L. 1983. Delivery of emergency medical services in disasters: Assumptions and realities. New York: Irvington.

Quarantelli, E. L. 1992. The case for a generic rather than agent-specific approach to disasters. *Disaster Management* 2:191–96.

Quarantelli, E. L. 1997. Problematical aspects of the information/communication revolution for disaster planning and research: Ten non-technical issues and questions. *Disaster Prevention and Management* 6:94–106.

Quarantelli, E. L., and R. Dynes. 1972. When disaster strikes. *Psychology Today* 5:67–70.

Quarantelli, E. L., and R. Dynes. 1977. Response to social crisis and disaster. *Annual Review of Sociology* 2:23–49.

Quarantelli, E. L., and R. Dynes. 1985. Realities and mythologies in disaster films. *Communications* 11:31–43.

Quarantelli, E. L., C. Lawrence, K. J. Tierney, and Q. T. Johnson. 1979. Initial findings from a study of socio-behavioral preparations and planning for acute chemical hazard disasters. *Journal of Hazardous Materials* 3:79–90.

Radwan, A. E., A. G. Hobeika, and D. Sivasailam. 1985. A computer simulation model for rural network evacuation under natural disasters. *ITE Journal* 55 (9): 25–30.

Raven, B. 1965. Social influence and power. In *Current studies in social psychology,* ed. I. Steiner and M. Fishbein, 371–382. New York: Holt, Rinehart, and Winston.

Reader, I. 2000. *Religious violence in Japan: The case of Aum Shinrikyo.* London: Curzon Press.

Rees, W. E. 1992. Ecological footprints and appropriated carrying capacity: What urban economics leaves out. *Environment and Urbanization* 4:121–130.

Renn, O., and D. Levine. 1991. Credibility and trust in risk communication. In *Communicating risks to the public: International perspectives,* ed. R. E. Kasperson and P. J. M. Stallen, 175–218. London: Kluwer Academic.

República de Colombia, Ministério del Interior. 1997. Decreto Presidencial No. 93.

Riad, J. K., F. H. Norris, and R. B. Ruback. 1999. Predicting evacuation in two major disasters: Risk perception, social influence, and access to resources. *Journal of Applied Social Psychology* 29:918–934.

Ridente, J. L., A. T. Ogura, E. S. de Macedo, L. A. Gomes, N. C. Diniz, M. C. Alberto, and P. H. P. dos Santos. 2002. *Accidentes associados a movimentos gravitacionais de massa ocorridos no município de Campos de Jordão, SP, em Janeiro do Ano de 2000: Ações técnicas após o desastre.* São Paulo: Universidade de São Paulo Instituto de Pesquisas Tecnológicas.

Ridge, T. 2004. *Memorandum: National Incident Management System.* March 1. Washington, DC: U.S. Department of Homeland Security.

Rigger, S. 1999. *Politics in Taiwan: Voting for democracy.* New York: Routledge.

Rochefort, D. A., and R. W. Cobb. 1994. *The politics of problem definition: Shaping the policy agenda.* Lawrence, KS: Univ. of Kansas Press.

Rogers, G. O., and J. H. Sorensen. 1988. Diffusion of emergency warnings. *Environmental Professional* 10:185–198.

Rogers, G. O., A. P. Watson, J. H. Sorensen, R. D. Sharp, and S. A. Carnes. 1990. *Evaluating protective actions for chemical agent emergencies.* Oak Ridge, TN: Oak Ridge National Laboratory.

Rossi, P. H., J. D. Wright, E. Webber-Burdin, M. Pietras, and W. F. Diggins. 1982. *Natural hazards and public choice: The state and local politics of hazard mitigation.* New York: Academic Press.

Rubin, C. B. 1991. Recovery from disaster. In *Emergency management: Principles and practice for local government,* ed. T. E. Drabek and G. J. Hoetmer, 224–259. Washington, DC: International City Management Association.

Rubin, C. B., M. D. Saperstein, and D. G. Barbee. 1985. *Community recovery from a major natural disaster.* Monograph no. 41. Boulder: Univ. of Colorado Institute of Behavioral Science.

Ruch, C., and G. Schumann. 1997. *Corpus Christi study area hurricane contingency planning guide.* College Station, TX: Texas A&M Univ. Hazard Reduction and Recovery Center.

Ruch, C., and G. Schumann. 1998. *Houston/Galveston study area hurricane contingency planning guide.* College Station, TX: Texas A&M Univ. Hazard Reduction and Recovery Center.

Saarinen, T., and J. Sell. 1985. *Warning and response to the Mt. St. Helens eruption.* Albany: State Univ. of New York Press.

Salzer, M. S., and L. Bickman. 1999. The short- and long-term psychological impact of disasters: Implications for mental health interventions and policy. In *Response to disaster: Psychosocial, community, and ecological approaches,* ed. R. Gist and B. Lubin, 63–82. Philadelphia: Brunner Mazel.

Scachetti, E. N. n.d. *Civil defense preparedness plan.* São Paulo: Coordenadoria Estadual de Defesa Civil, Gabinete do Governador, Casa Militar, Governo do Estado de São Paulo.

Scawthorne, C. 1986. Use of damage simulation in earthquake planning and emergency response management. In *Terminal disasters: Computer applications in emergency management,* ed. S. A. Marston, 109–120. Boulder: Univ. of Colorado Institute of Behavioral Science.

Schmitt, N. W., and R. J. Klimoski. 1991. *Research methods in human resources management.* Cincinnati, OH: South-Western.

Schneider, B., and N. W. Schmitt. 1986. *Staffing organizations.* 2nd ed. Glenview, IL: Scott-Foresman.

Schwab, J., K. C. Topping, C. C. Eadie, R. E. Deyle, and R. A. Smith. 1998. *Planning for post-disaster recovery and reconstruction.* PAS Report 483/484. Chicago: American Planning Association.

Scientific Assessment and Strategy Team. 1994. *Science for floodplain management into the 21st century.* Washington, DC: Administration Floodplain Task Force.

Sen, A. 1981. *Poverty and famines: An essay on entitlement and depression.* Oxford: Oxford Univ. Press.

Shaw, R., and K. Goda. 2004. From disaster to sustainable civil society: The Kobe experience. *Disasters* 28:16–40.

Shaw, R., M. Gupta, and A. Sarma. 2003. Community recovery and its sustainability: Lessons from Gujarat earthquake of India. *Australian Journal of Emergency Management* 18: 2, 28–34.

Sheffi, Y., H. Mahmassani, and W. B. Powell. 1980. *NETVAC1: A transportation network evacuation model.* Working paper CTS 80–14. Cambridge, MA: MIT.

Sheffi, Y., H. Mahmassani, and W. B. Powell. 1981. Evacuation studies for nuclear power plant sites: A new challenge for transportation engineers. *ITE Journal* 57 (6): 25–28.

Sheffi, Y., H. Mahmassani, and W. B. Powell. 1982. A transportation network evacuation model. *Transportation Research* A (3): 209–218.

Shefner, J. 1999. Pre- and post-disaster political instability and contentious supporters: A case study of political ferment. *International Journal of Mass Emergencies and Disasters* 17:37–160.

Sherali, H. D., T. B. Carter, and A. G. Hobeika. 1991. A location-allocation model and algorithm for evacuation planning under hurricane/flood conditions. *Transportation Research* B (3): 439–452.

Shoaf, K. I., H. S. Sareen, L. H. Nguyen, and L. B. Bourque. 1998. Injuries as a result of California earthquakes in the past decade. *Disasters* 22:218–235.

Showalter, P. S., and M. F. Myers. 1994. Natural disasters in the United States as release agents of oil, chemicals, or radiological materials between 1980–1990: Analysis and recommendations. *Risk Analysis* 14:169–182.

Siegel, J. M., L. B. Bourque, and K. I. Shoaf. 1999. Victimization after a natural disaster: Social disorganization or community cohesion? *International Journal of Mass Emergencies and Disasters* 17:265–294.

Sierra Club. 2000. *Permitting disaster in America.* San Francisco: Sierra Club Books.

Simpson, D. A. 2001. Community emergency response training (CERT): A recent history and review. *Natural Hazards Review* 2:54–63.

Sinclair, A. 2003. *An anatomy of terror.* London: Macmillan.

Singer, T. 1982. An introduction to disaster. *Aviation, Space, and Environmental Medicine* 53:245–250.

Sinuani-Stern, Z., and E. Stern. 1993. Simulating the evacuation of a small city: The effect of traffic factors. *Socio-Economic Planning Sciences* 27:97–108.

Slovic, P. 1987. Perception of risk. *Science* 236:280–285.

Slovic, P., B. Fischhoff, and S. Lichtenstein. 1980. Facts and fears: Understanding perceived risk. In *Societal risk assessment: How safe is safe enough?*, ed. R. Schwing and W. Albers, 161–178. New York: Plenum.

Slovic, P., H. Kunreuther, and G. F. White. 1974. Decision processes, rationality, and adjustment to natural hazards. In *Natural hazards: Local, national and global*, ed. G. F. White, 187–204. New York: Oxford Univ. Press.

Smith, D. I., J. Handmer, and W. Martin. 1980. *The effects of floods on health.* Canberra: Australian National Univ. Press.

Smith, E. 2001. *Environmental hazards.* 3rd ed. London: Routledge.

Smith, G. 2004. *Holistic disaster recovery: Creating a more sustainable future.* FEMA Emergency Management Higher Education Project upper division college course. Emmitsburg, MD: FEMA Emergency Management Institute. http://training.fema.gov/emiweb/edu/completeCourses.asp.

Smith, S. K., J. Tayman, and D. A. Swanson. 2001. *State and local population projections: Methodology and analysis.* New York: Kluwer.

Sorensen, J. H. 1986. *Evacuations due to chemical accidents.* Oak Ridge, TN: Oak Ridge National Laboratory.

Sorenson, J. H. 1991. When shall we leave?: Factors affecting the timing of evacuation departures. *International Journal of Mass Emergencies and Disasters* 9:153–164.

Sorensen, J. H. 2000. Hazard warning systems: Review of twenty years of progress. *Natural Hazards Review* 1:119–125.

Sorensen, J. H., and G. O. Rogers. 1989. Warning and response in two hazardous materials transportation accidents in the U.S. *Journal of Hazardous Materials* 22:57–74.

Sorensen, J. H., B. L. Shumpert, and B. M. Vogt. 2004. Planning for protective action decision making: Evacuate or shelter in-place? *Journal of Hazardous Materials* A (109): 1–11.

Southworth, F. 1991. *Regional evacuation modeling: A state-of-the-art review.* Oak Ridge, TN: Oak Ridge National Laboratory.

Southworth, F., and S. M. Chin. 1987. Network evacuation modeling for flooding as a result of dam failure. *Environment and Planning* A (19): 1543–1558.

Spennemann, D. H. R., and D. W. Look. 1998. *Disaster management programs for historic sites.* San Francisco: Association for Preservation Technology.

Spiewak, D. L. 2006. Common body of knowledge: An interim report. *IAEM Bulletin Special Edition,* June 1–7.

Stallen, P. J. M. 1991. Developing communications about risks of major industrial accidents in the Netherlands. In *Communicating risks to the public: International perspectives,* ed. R. E. Kasperson and P. J. M. Stallen, 55–66. London: Kluwer Academic.

Stallings, R. A. 1978. The structural patterns of four types of organizations in disaster. In *Disasters: Theory and research,* ed. E. L. Quarantelli, 87–103. Beverly Hills, CA: Sage.

Stallings, R. A. 1991. Ending evacuations. *International Journal of Mass Emergencies and Disasters* 9:183–200.

Stallings, R. A. 1995. *Promoting risk: Constructing the earthquake threat.* New York: Aldine de Gruyter.

Stallings, R. A., and E. L. Quarantelli. 1985. Emergent groups and emergency management. *Public Administration Review* 45:93–100.

Staples, R. 1976. *Introduction to Black sociology.* New York: McGraw-Hill.

Staples, R., and A. Mirande. 1980. Racial and cultural variations among American families: A decennial review of the literature on minority families. *Journal of Marriage and the Family* 42:887–903.

State of Florida Department of Community Affairs/Division of Emergency Management and University of Florida School of Building Construction. 1997. *State of Florida model hurricane evacuation shelter selection guidelines: Student manual.* Tallahassee, FL: State of Florida Department of Community Affairs/Division of Emergency Management and University of Florida School of Building Construction.

Stogdill, R. 1963. *Manual for the leader behavior description questionnaire-Form XII: An experimental revision.* Columbus: Ohio State Univ. Bureau of Business Research.

Stone, D. 2000. Non-governmental policy transfer: The strategies of independent policy institutes. *Governance: An International Journal of Policy and Administration* 13:45–62.

Sullivan, M. 2003. Integrated emergency management: A new way of looking at a delicate process. *Australian Journal of Emergency Management* 18:4–27.

Sutphen, S., and V. Bott. 1990. Issue salience and preparedness as perceived by city managers. *In Cities and disaster: North American studies in emergency management,* ed. R. T. Silves and W. L. Waugh, Jr. Springfield, IL: Charles C. Thomas.

Swanson, H. D. 2000. The delicate art of practicing municipal law under conditions of hell and high water. *Notre Dame Law Review* 76:487.

Sylves, R. 1996. The politics and budgeting of federal emergency management. *In Disaster management in the U.S. and Canada,* ed. R. T. Sylves and W. L. Waugh, Jr., 26–45. Springfield, IL: Charles C. Thomas.

Sylves, R. T. 1991. Adopting integrated emergency management in the United States. *International Journal of Mass Emergencies and Disasters* 9:423–438.

Sylves, R. T. 1998. *Disasters and coastal states: A policy analysis of presidential declarations of disasters 1953–1997*. Newark, DE: Univ. of Delaware Press.

Taylor, J. C. 1978. The safe transportation of radioactive material shipping containers including accident and response experience. In *Proceedings of the Fifth International Symposium on Packaging and Transportation of Radioactive Materials*, 602–611. Washington, DC: U.S. Department of Transportation.

Taylor, V. 1977. Good news about disaster. *Psychology Today* 93:94–96.

Tewdwr-Jones, M. 2002. *The planning polity: Planning, government and the policy process*. New York: Routledge.

Texas Governor's Division of Emergency Management. 2002a. *Hurricane contingency planning guide: Lake Sabine study area*. Austin: Texas Governor's Division of Emergency Management.

Texas Governor's Division of Emergency Management. 2002b. *Hurricane contingency planning guide: Valley study area*. Austin: Texas Governor's Division of Emergency Management. http://www.txdps.state.tx.us/dem/pages/planning.htm.

Texas Governor's Division of Emergency Management. 2004. *Hurricane evacuee estimates and destinations*. Austin: Texas Governor's Division of Emergency Management.

Texas Governor's Division of Emergency Management. 2004. *Coastal Bend study area hurricane storm atlas*. Austin, TX: Texas Governor's Division of Emergency Management.

Thomas, D., and D. S. Mileti. 2004. *Designing educational opportunities for the hazards manager of the 21st century*. Emmitsburg, MD: Federal Emergency Management Agency Emergency Management Institute.

Thomas, K. W. 1992. Conflict and negotiation processes in organizations. In *Handbook of industrial and organizational psychology*, 2nd ed., vol. 3, ed. M. D. Dunnette and L. M. Hough, 651–717. Palo Alto, CA: Consulting Psychologists Press.

Thompson, A. A., Jr., and A. J. Strickland III. 1996. *Strategic management: Concepts and cases*. Chicago: Irwin.

Tichner, J. L. 1988. Clinical intervention after natural and technological disasters. In *Mental health response to mass emergencies*, ed. M. Lystad, 160–181. New York: Brunner-Mazel.

Tierney, K., M. K. Lindell, and R. W. Perry. 2001. *Facing the unexpected: Disaster preparedness and response in the United States*. Washington, DC: Joseph Henry Press.

Tierney, K. J. 1980. *A primer for preparedness for acute chemical emergencies*. Columbus: Ohio State Univ. Disaster Research Center.

Tierney, K. J. 1995. Social aspects of the Northridge earthquake. In *The Northridge, California, earthquake of 17 January 1994,* ed. M. C. Woods and R. W. Seiple, 255–262. Sacramento: California Department of Conservation, Division of Mines and Geology.

Tierney, K. J. 1997a. Business impacts of the Northridge earthquake. *Journal of Contingencies and Crisis Management* 5:87–97.

Tierney, K. J. 1997b. Impacts of recent disasters on business: The 1993 Midwest floods and the 1994 Northridge earthquake. In *Economic consequences of earthquakes: Preparing for the unexpected,* ed. B. G. Jones, 189–222. Berkeley, CA: National Center for Earthquake Engineering Research.

Tierney, K. J., and J. M. Dahlhamer. 1998. *Earthquake vulnerability and emergency preparedness among businesses.* Newark, DE: Univ. of Delaware Disaster Research Center. http://www.udel.edu/DRC/publications.html.

Tierney, K. J., and J. A. Nigg. 1995. *Business vulnerability to disaster-related lifeline disruption.* Newark, DE: Univ. of Delaware Disaster Research Center.

Tobin, G. A., and B. E. Montz. 1997. *Natural hazards: Explanation and integration.* New York: Guilford Press.

Tomeh, A. 1973. Formal voluntary organizations: Participation correlates and interrelationships. *Sociological Inquiry* 43:80–122.

Trank, C., and S. Rynes. 2003. Who moved our cheese?: Reclaiming professionalism in business education. *Academy of Management Learning and Education* 2:189–205.

Travis, R., and W. Riebsame. 1979. Communicating uncertainty: The nature of weather forecasts. *Journal of Geography* 78:168–172.

Trumbo, C. W., and K. A. McComas. 2003. The function of credibility in information processing for risk perception. *Risk Analysis* 23:343–353.

Tufecki, S., and T. Kisko. 1991. Regional evacuation modeling system (REMS): A decision support system for emergency area evacuations. *Computers and Industrial Engineering* 21:89–93.

Turner, R., J. Nigg, and D. Heller-Paz. 1986. *Waiting for disaster.* Los Angeles: Univ. of California Press.

Tweedie, S. W., J. R. Rowland, S. J. Walsh, R. P. Rhoten, and P. I. Hagle. 1986. A methodology for estimating emergency evacuation times. *Social Science Journal* 23:189–204.

Tyhurst, J. S. 1957. Psychological and social aspects of civilian disaster. *Canadian Medical Association Journal* 76:385–393.

United Nations Disaster Relief Organization. 1984. *Disaster prevention and mitigation: A compendium of current knowledge.* Vol. 11. Geneva: Office of the United Nations Disaster Relief Coordinator.

United Nations International Strategy for Disaster Reduction. 2002. *Living with risk: A global review of disaster reduction initiatives.* New York: United Nations. http://www.unisdr.org.

United Nations World Commission on Environment and Development. 1987. *Report of the United Nations World Commission on Environment and Development: Our common future.* New York: United Nations.

U.S. Department of Homeland Security. 2004a. *Statement of requirements for public safety wireless communications and interoperability: The SAFECOM program.* Washington, DC: U.S. Department of Homeland Security.

U.S. Department of Homeland Security. 2004b. *National incident management system.* Washington, DC: U.S. Department of Homeland Security.

U.S. Department of Homeland Security. 2004c. *National response plan.* Washington, DC: U.S. Department of Homeland Security.

U.S. Department of Homeland Security. 2004d. *Urban areas security initiative grant program.* Washington, DC: U.S. Department of Homeland Security.

U.S. Department of Justice. 2002. *Crisis information management software (CIMS) feature comparison report.* Washington, DC: U.S. Department of Justice.

U.S. Department of Transportation, Transport Canada, and Secretariat of Transport and Communications. 2000. *Emergency response guidebook.* Washington, DC: U.S. Department of Transportation, Transport Canada, and Secretariat of Transport and Communications.

U.S. Environmental Protection Agency. 1987. *Technical guidance for hazards analysis: Emergency planning for extremely hazardous substances.* Washington, DC: U.S. Environmental Protection Agency.

U.S. General Accounting Office. 2003. *Bioterrorism: Preparedness varied across state and local jurisdictions.* Washington, DC: U.S. Government Printing Office.

U.S. Nuclear Regulatory Commission. 1981. *Criteria for evaluation of emergency response facilities.* NUREG-0814. Washington, DC: U.S. Nuclear Regulatory Commission.

U.S. Nuclear Regulatory Commission/Environmental Protection Agency. 1978. *Planning basis for the development of state and local government radiological emergency response plans in support of light water nuclear power plants.* NUREG-0396, EPA 520/1–78–016. Washington, DC: U.S. Nuclear Regulatory Commission/Environmental Protection Agency.

U.S. Nuclear Regulatory Commission/Federal Emergency Management Agency. 1980. *Criteria for preparation and evaluation of radiological emergency response plans and preparedness in support of nuclear power plants.* NUREG-0654/FEMA-REP-1. Washington, DC: U.S. Nuclear Regulatory Commission/Federal Emergency Management Agency.

U.S. Office of Emergency Preparedness. 1972. *Report to the Congress: Disaster preparedness.* Washington, DC: U.S. Government Printing Office.

Urbanik, T. 2000. Evacuation time estimates for nuclear power plants. *Journal of Hazardous Materials* 75:165–180.

Urbanik, T., A. Desrosiers, M. K. Lindell, and C. R. Schuller. 1980. *An analysis of techniques for estimating evacuation times for emergency planning zones.* NUREG/CR-1745. Washington, DC: U.S. Nuclear Regulatory Commission.

Urbanik, T., M. P. Moeller, and K. Barnes. 1988a. *Benchmark study of the I-DYNEV evacuation time estimate computer* code. NUREG/CR-4873. Washington, DC: U.S. Nuclear Regulatory Commission.

Urbanik, T., M. P. Moeller, and K. Barnes. 1988b. *The sensitivity of evacuation time estimates to changes in input parameters for the I-DYNEV computer code.* NUREG/CR-4874. Washington, DC: U.S. Nuclear Regulatory Commission.

Vallance, T., and A. D'Augelli. 1982. The helpers community. *American Journal of Community Psychology* 10:197–205.

Van Meter, D. S., and C. E. Van Horn. 1975. The policy implementation process: A conceptual framework. *Administration and Society* 6:445–488.

Vatsa, K. S., and J. Joseph. 2003. Disaster management plan for the state of Maharashtra, India: Evolutionary process. *Natural Hazards Review* 4:206–212.

Velázquez, D. R. 1986. La Organización Popular ante el reto de la reconstrucción. *Revista Mexicana de Ciências Políticas e Sociales* 123:59–79.

Vernberg, E. M., A. M. LaGreca, W. K. Silverman, and M. J. Prinstein. 1996. Prediction of post-traumatic stress symptoms in children after Hurricane Andrew. *Journal of Abnormal Psychology* 105:237–248.

Vogt, B. M. 1991. Issues in nursing home evacuations. *International Journal of Mass Emergencies and Disasters* 9:247–265.

Wachernagel, M., and W. E. Rees. 1995 *Our ecological footprint: Reducing impact on the earth.* Gabriola Island, British Columbia: New Society.

Wachtendorf, T. 2004. *Improvising 9/11: Organizational improvisation following the World Trade Center disaster.* Newark, DE: Univ. of Delaware Disaster Research Center.

Walker, A. G. 1998. *Development of specialized accreditation for emergency management degree programs.* Emmitsburg, MD: FEMA Emergency Management Institute.

Wallace, A. F. C. 1957. Mazeway disintegration. *Human Organization* 16:23–27.

Warrick, R. A., J. Anderson, T. Downing, J. Lyons, J. Ressler, M. Warrick, and T. Warrick. 1981. *Four communities under ash after Mount St. Helens.* Boulder: Univ. of Colorado Institute of Behavioral Science.

Waugh, W. L., Jr. 1988. Current policy and implementation issues in disaster preparedness. In *Managing disaster: Strategies and policy perspectives,* ed. L. K. Comfort, 111–125. Durham, NC: Duke Univ. Press.

Waugh, W. L., Jr. 2001. Managing terrorism as an environmental hazard. In *Handbook of crisis and emergency management,* ed. A. Farazmand, 659–676. New York: Marcel Dekker.

Webb, G. R., K. J. Tierney, and J. M. Dahlhamer. 2000. Business and disasters: Empirical patterns and unanswered questions. *Natural Hazards Review* 1:83–90.

Webb, G. R., K. J. Tierney, and J. M. Dahlhamer. 2002. Predicting long-term business recovery from disasters: A comparison of the Loma Prieta earthquake and Hurricane Andrew. *Environmental Hazards* 4:45–58.

Weinstein, N. 1980. Unrealistic optimism about future life events. *Journal of Personality and Social Psychology* 39:806–820.

Wenger, D. E. 1972. DRC studies of community functioning. In *Proceedings of the Japan-United States Disaster Research Seminar,* 29–73. Columbus: Ohio State Univ. Disaster Research Center.

Wenger, D. E. 1978. Community response to disaster: Functional and structural alterations. In *Disasters: Theory and research,* ed. E. L. Quarantelli, 18–47. Thousand Oaks, CA: Sage.

Wenger, D. E. 1980. Empirical observations concerning the relationship between the mass media and disaster knowledge. In *Disasters and the mass media,* ed. Committee on Disasters and the Mass Media, 241–253. Washington, DC: National Academy Press.

Wenger, D. E., C. E. Faupel, and T. F. James. 1980. *Disaster beliefs and emergency planning.* Newark. DE: Univ. of Delaware Disaster Research Center.

Wenger, D. E., and T. F. James. 1994. The convergence of volunteers in a consensus crisis. In *Disasters, collective behavior and social organization,* ed. R. Dynes and K. Tierney, 229–243. Newark, DE: Univ. of Delaware Press.

Wert, B. J. 1979. Stress due to nuclear accident. *Occupational Health Nursing* 27:16–24.

White, G. F., and J. E. Haas. 1975. *An assessment of research on natural hazards.* Cambridge, MA: MIT Press.

Whitney, D. J., A. Dickerson, and M. K. Lindell. 2001. Non-structural seismic preparedness of Southern California hospitals. *Earthquake Spectra* 17:153–171.

Whitney, D. J., and M. K. Lindell. 2000. Member commitment and participation in local emergency planning committees. *Policy Studies Journal* 28:467–484.

Whitney, D. J., M. K. Lindell, and D. H. Nguyen. 2004. Earthquake beliefs and adoption of seismic hazard adjustments. *Risk Analysis* 24:87–102.

Wiley, J. A., J. P. Robinson, T. Piazza, K. Garrett, K. Cirksena, Y. T. Cheng, and G. Martin. 1991. *Activity patterns of California residents.* California Environmental Protection Agency Air Resources Board. http://www.arb.ca.gov/homepage.htm.

Wilkinson, D. 1999. Reframing family ethnicity in America. In *Family ethnicity,* ed. H. McAdoo, 16–62. Thousand Oaks, CA: Sage.

Wilmer, H. 1958. Toward a definition of the therapeutic community. *American Journal of Psychiatry* 114:824–834.

Wilson, D. J. 1987. Stay indoors or evacuate to avoid exposure to toxic gas? *Emergency Preparedness Digest (Canada)* 14:19–24.

Wilson, D. J. 1989. Variation of indoor shelter effectiveness caused by air leakage variability of houses in Canada and the USA. In *Proceedings of the conference on in-place protection during chemical emergencies,* ed. T. Glickman and A. Ujihara. Washington, DC: Resources for the Future.

Wilson, R. C. 1991. *The Loma Prieta quake: What one city learned.* Washington, DC: International City Management Association.

Windham, G., E. Posey, P. Ross, and B. Spencer. 1977. *Reactions to storm threat during Hurricane Eloise.* State College, MS: Mississippi State Univ.

Winslow, F. E. 2001. Planning for weapons of mass destruction/nuclear, biological, and chemical agents: A local/federal partnership. In *Handbook of crisis and emergency management,* ed. A. Farazmand, 667–692. New York: Marcel Dekker.

Withey, S. 1962. Reaction to uncertain threat. In *Man and society in disaster,* ed. G. Baker and D. Chapman, 93–123. New York: Basic Books.

Witt, J. L. 1995. *Keynote address: National mitigation conference.* Arlington, VA: Federal Emergency Management Agency.

Witt, J. L. 1997. *Strategic plan: Planning for a safer future.* Washington, DC: Federal Emergency Management Agency.

Wolensky, R. P., and K. C. Wolensky. 1990. American local government and the disaster management problem. *Local Government Studies* 20:15–32.

Wolshon, B. 2001. "One-way-out": Contraflow freeway operation for hurricane operation. *Natural Hazards Review* 2:105–112.

Wolshon, B. 2002. Planning for the evacuation of New Orleans. *ITE Journal* 72 (2): 44–49.

World Health Organization. 2004. *Public health response to biological and chemical weapons: WHO guidance.* 2nd ed. Geneva: World Health Organization.

World Health Organization/Pan American Health Organization. 2004. *Health aspects of biological and chemical weapons.* Geneva: World Health Organization/Pan American Health Organization. http://www.who.int/csr/delibepidemics/biochemguide/en/index.html.

Wright, J. D., P. H. Rossi, S. R. Wright, and E. Weber-Burdin. 1979. *After the clean-up: Long-range effects of natural disasters.* Beverly Hills, CA: Sage.

Wu, J. Y., and M. K. Lindell. 2004. Housing reconstruction after two major earthquakes: The 1994 Northridge earthquake in the United States and the 1999 Chi-Chi earthquake in Taiwan. *Disasters* 28:63–81.

Yates, J. 1990. *Judgment and decision making.* Englewood Cliffs, NJ: Prentice Hall.

Yelvington, K. A. 1997. Coping in a temporary way: The tent cities. In *Hurricane Andrew:* Ethnicity, gender and the sociology of disaster, ed. W. G. Peacock, B. H. Morrow, and H. Gladwin, 92–115. London: Routledge.

Yoshpe, H. B. 1981. *Our missing shield: The U.S. civil defense program in historical perspective.* Washington, DC: Civil Defense Preparedness Agency.

Zeigler, D., S. Brunn, and J. Johnson. 1981. Evacuation from a nuclear technological disaster. *Geographical Review* 71:1–16.

Zhang, Y., M. K. Lindell, and C. S. Prater. 2004. *Modeling and managing the vulnerability of community businesses to environmental disasters.* College Station, TX: Texas A&M Univ. Hazard Reduction and Recovery Center.

Zhang, Y., C. S. Prater, and M. K. Lindell. 2004. Risk area accuracy and evacuation from Hurricane Bret. *Natural Hazards Review* 5:115–120.

GLOSSARY

100-year flood An arbitrary standard of safety that reflects a compromise between the goals of providing long-term safety and developing economically valuable land.

Adaptive plan The answer to the question "What is the best method of protection?" Those at risk generally have at least two options—taking protective action or continuing normal activities.

Adverse selection The tendency for hazard insurance to be purchased mostly by those who are at the greatest risk of filing a claim for losses.

Agricultural vulnerability The vulnerabilities of all species of plants and animals.

Alert A class of a nuclear power plant emergency defined by the NRC and FEMA that involves substantial degradation of plant safety. Releases are expected to be well below EPA exposure limits.

Apparent temperature The combination of temperature and humidity into a heat index.

Behavior criteria A standard for judging the success of a training program that refers to trainees' ability to apply the new knowledge and skills to their jobs. This includes performance during drills, exercises, and incidents that take place after training is completed.

Budget narrative A document that accompanies the budget and includes a request for additional money. The narrative is submitted in written format and can include graphics to explain budget items.

Business interruption The loss of revenue due to disruption of a business's normal production of goods and services in exchange for money.

Capability assessment An evaluation of the degree to which your jurisdiction's resources are sufficient to meet the disaster demands identified in the hazard vulnerability analysis.

Capability assessment for readiness (CAR) program A state program that describes a self-assessment process for state emergency management agencies.

Capability shortfall The difference between the level of resources a jurisdiction currently has and the level it will need to meet the disaster demands identified in the hazard vulnerability analysis.

Capacity A measurement of an organization's ability to implement policy that includes budget allocations, staffing levels, and staff members' knowledge and skills.

Capital improvements program (CIP) A program used to plan community infrastructure and critical facilities.

Carcinogens Chemicals that cause cancer.

Certification An assurance that an individual has mastered the knowledge and the methods used to solve specific problems.

Channelization The process of deepening and straightening stream channels.

Civil society A society that includes all groups that are independent of the government, including religious groups, civic clubs, political parties, and other groups with specific interests.

Committed The fact that contamination by radioactive material on the skin or absorbed into the body will continue to administer a dose until it decays or is removed.

Community A specific geographic area that is frequently considered to be a town, city, or county with a government. A community also has stronger psychological ties and social interaction among its members than with outsiders.

Community emergency response teams (CERTs) Homeowners organized as groups to perform emergency management tasks in their neighborhoods. CERTs may also be known as neighborhood emergency response teams or other similar names, but they all organize and train neighborhood volunteers to perform basic emergency response tasks, such as search and rescue and first aid.

Competition The effort of two parties striving toward a goal that only one can achieve. In fair competition, the parties use legitimate methods.

Compressed gases Gases that are cooled to a liquid state so they occupy a small enough volume to be transported at a reasonable economic cost.

Concept of operations A summary statement of what emergency functions are to be performed and how they are accomplished.

Conflict The opposition that occurs when one party attempts to directly frustrate the goal achievement of another.

Contigency fund A sum of money in the budget that addresses the costs of resources that will be needed in case of an emergency.

Controller The person who provides information from a scenario. Drills are relatively simple, so the same person can serve as the evaluator.

Cooperation Activities that result in mutual benefit.

Core Molten rock at the center of the earth.

Corrosives Substances that destroy living tissue at the point of contact because they are either acidic or alkaline.

Crust Solid rock and other materials at the earth's surface that is defined by large plates floating on the mantle and moving gradually in different directions over time.

Damage assessment An evaluation that begins by identifying the boundaries of the impact area and then proceeds to estimating the total amount of damage to buildings and infrastructure in the impact area. This information is used to support a request for a presidential disaster declaration.

Dams Elevated barriers sited *across* a streambed that increase surface storage of floodwater in reservoirs upstream from them.

Development A function of an appraisal that focuses on improving an employee's *ability* to do a job. In this context, performance appraisal can be used to guide decisions about training, reassignment, or termination.

Disaster An event that produces greater losses than a community can handle, including casualties, property damage, and significant environmental damage.

Disaster subculture Behavioral patterns among groups of residents who adopt routines to prepare for disasters. These groups have usually experienced disasters and have resolved to better prepare for them in the future.

Discharge The volume of water passing a specific point per unit of time.

Drills A training exercise innvolving one or few people who must perform a specific task in response to a hypothetical scenario.

Earthquake A sudden release of energy that has been built up as two tectonic plates attempt to move past each other.

Economic groups Business stakeholders that organize the flow of goods and services and who are affected anytime there is an interruption to business caused by a disaster.

Elevating on continuous foundation walls A method used to raise a house slightly higher than the base flood, increasing the height of the basement walls and providing secure storage.

Elevating on open foundations A method used in which a structure's foundation only supports the structure at critical points, allowing high velocity water flow and breaking waves to pass under the structure with minimal resistance.

Emergency Alert System. A new alert system that replaced the Emergency Broadcast System and provides a greater range of capabilities in the digital age.

Emergency classification system A method of organizing a large number of potential incidents into a small set of categories. These categories link the threat assessment to the level of activation of the responding organization.

Emergency manager A person who manages a comprehensive program for hazards and disasters. An emergency manager is responsible for aspects of the program involving mitigation, preparation, response, and recovery.

Emergency preparedness practices Preimpact actions that provide the human and material resources needed to support active responses at the time of hazard impact.

Emergency responders People who directly respond to a disaster. They attack the threat to reduce the potential or actual losses of a disaster.

Emergency response A minor event that can cause a few casualties and a limited amount of property damage *or* an imminent event that requires prompt and effective action.

Emergency shelter An unplanned location that is intended only to provide protection from ordinary weather conditions of temperature, wind, and rain.

Emergent multiorganizational networks (EMONs) A group of organizations whose interactions develop in response to the demands of a disaster rather than being planned beforehand.

Emergent organizations A disaster response organization that performs novel tasks within novel organizations.

Eminent domain Power held by the government that can force private owners to sell their property to the government at a fair market value if the property is to be used for a public purpose.

Entity A public or private sector organization that is responsible for emergency/disaster management or continuity of operations.

Epicenter A point on the earth's surface directly above the hypocenter.

Escalating crisis A situation in which there is a significantly increased probability of an incident occurring that will threaten the public's health, safety, or property.

Established organizations A disaster response organization that performs normal tasks within normal organizations.

Evacuation trip generation The number and location of vehicles evacuating from a risk area.

Evaluator The person who observes the player's performance and notes any deviations from the EOP or its procedures.

Expanding organizations A disaster response organization that perform normal tasks within novel organizations.

Expert power Power that is based on someone's expertise on a particular topic.

Explosives Compounds or mixtures that undergo a rapid chemical transformation that is faster than the speed of sound.

Extending organizations A disaster response organization that performs novel tasks within normal organizations.

Eye of the hurricane The area of calm conditions that has a 10 to 20 mile radius. The eye is surrounded by bands of high wind and rain that spiral and form a ring around the eye.

Eyewall The spiral the forms a ring around the eye of a hurricane.

Firestorms Fires that are distinguished from other wildfires because they burn so intensely that they create their own local weather and are virtually impossible to extinguish.

Flammable liquids Liquids that evolve flammable vapors at 80°F or less, thus posing a threat similar to flammable gases.

Flammable solids Solids that self-ignite through friction, absorption of moisture, or spontaneous chemical changes.

Flood An event in which abnormally large amount of water accumulates in an area in which it is usually not found.

Floodwalls Water barriers that are built of strong materials such as concrete. They are more expensive than levees, but they are also stronger.

Focusing event A natural or technological disaster that draws public attention to the need for local disaster planning and hazard mitigation.

Full-scale exercise A training exercise that simulates a community-wide disaster by testing multiple functions at the same time. It also tests the coordination among these functions. The complexity of full-scale exercises requires thorough planning of the scenario. It also requires coordination among the many controllers and evaluators. There is also a need for training the controllers.

Functional annex The part of an EOP that describes how the emergency response organization will perform a function needed to respond to disaster demands. The annexes of an all-hazards EOP should collectively list all of the emergency response functions needed to respond to all hazards.

Functional exercise A training exercise that differs from a drill by involving more people. This makes functional exercises more comprehensive than drills. In addition, the scenario is usually more complex because it involves more tasks and equipment.

Gas A substance that expands to fill the available volume in a space.

General emergency A class of a nuclear power plant emergency defined by the NRC and FEMA that involves substantial core degradation and the possibility of radioactive material escaping from the containment building. Releases might exceed EPA limits offsite.

Governmental groups Stakeholders who are part of the government's structure. The foundation of the government structure is the town or the city followed by the county. The third level is the state.

Cities and counties have varying levels of power from one state to another because states differ in the powers they grant. Most emergency management policies are set at the federal and state levels.

Hardscape Impermeable surfaces, such as building roofs, streets, and parking lots.

Hazard A source of danger. Hazards have the potential to affect people's health and safety, their property, and the natural environment.

Hazard adjustments Actions that can reduce vulnerability to disasters. These include actions such as purchasing hazard insurance, living in safer locations, and renting or buying homes that are resistant to disaster.

Hazard exposure Living, working, or otherwise being in places that can be affected by hazard impacts.

Hazard mitigation Preimpact actions that provide *passive* protection at the time of disaster impact so there is less need for emergency response actions.

Hazard mitigation practices Actions that protect passively at the time of impact.

Hazards US-Multi Hazard (HAZUS-MH) A computer program that predicts losses from earthquakes, floods, and hurricane winds. The program estimates casualties, damage, and economic losses.

Hazmat Hazardous materials that are "capable of posing unreasonable risk to health, safety, and property."

Homeland Security Act (HSA) An act signed in November 2002 that restructured emergency management by integrating many agencies having emergency- or security-related functions into the Department of Homeland Security.

Human vulnerability People's susceptibility to death, injury, or illness from extreme levels of environmental hazards.

Hurricane The most severe type of tropical storm.

Hypocenter A point deep within the earth from which an earthquake's energy is released.

Industrial hazard controls Community protection works that are used to confine hazardous materials flows.

Information need A need that results from the question, "What information do I need to answer my question?"

Information power Power that involves true, new, and relevant facts or arguments. Information power can be exercised by either introducing or withholding information.

Information search plan A plan that results from addressing the question, "Where and how can I obtain this information?"

Intensity The measure of energy release at a given impact location, which can be assessed either by behavioral effects or physical measurements.

Interface fires Fires that burn into areas containing a mixture of natural vegetation and built structures.

Internal research A function of an appraisal that lets an emergency manager know what skills his or her employees have. Internal research also lets emergency managers know if they were looking for the right qualities when hiring for the position.

La Red A network of Latin American social scientists that publishes scholarly work on disasters in the region. Their work highlights the issues of social vulnerability and sustainable development.

Land-use practices Alternative ways in which people use the land. Residential, commercial, and industrial development of urbanized areas are especially important in determining disaster impacts.

Landslide The downward displacement of rock or soil because of gravitational forces.

Landslide controls Methods for reducing shear stress, increasing shear resistance, or a combination of these two.

Learning criteria A standard for judging the success of a training program that is defined by performance on written tests or performance tests of skills addressed in the training program.

Legal protection A function of an appraisal that is achieved when an organization conducts performance appraisals according to generally acceptable procedures.

Legitimate power Power that arises from one person's relationship to another and can come from a formal position. Any official elected by a fair voting process has legitimate power.

Levees Elevated barriers placed *along* a streambed that limit stream flow to the floodway.

Liquid A substance that spreads to cover the available area on a surface.

Local Emergency Management Committee (LEMC) A disaster-planning network that increases coordination among local agencies.

Magnitude The measure of energy release at the source. Earthquake magnitude is measured on the Richter scale where a one-unit increase represents a 10-fold increase in seismic wave amplitude and a 30-fold increase in energy release from the source.

Mantle An 1800 mile thick layer between the core and the crust.

Medical aid stations Off-hospital site medical care facilities.

Megadisaster An extreme event that costs hundreds of lives and millions of dollars in a small city but costs thousands of lives and billions of dollars in a major city.

Miscellaneous dangerous goods A diverse set of materials such as air bags, certain vegetable oils, polychlorinated biphenyls, and white asbestos.

Multi Hazard Identification and Risk Assessment A FEMA manual that describes exposure to many natural and technological hazards.

Multiyear development plan A plan that documents the specific steps for reducing the capability shortfall. The development plan is typically based on five years and should identify specific annual milestones and specific, measurable achievements to keep emergency managers on target.

Natural disaster An event that occurs in nature that results in casualties, property damage, and environmental damage. Natural disasters include earthquakes, floods, hurricanes, volcanic eruptions, and wildland fires.

Natural hazards Extreme events that originate in nature. Natural hazards are commonly categorized as meteorological, hydrological, or geophysical.

Nonconforming uses Structures that do not meet the zoning requirements for their geographic areas.

Normalcy bias People's tendency to delay recognition that an improbable event is occurring and affecting them.

North American Emergency Response Guidebook A manual that lists the chemicals commonly found in transportation. It details which one of its 172 emergency response guides provides the information needed to respond to a spill. It also helps you to determine how far from the spill location to shelter in-place or evacuate residents.

Overtopping The flow of water over the top of a levee. Once this happens, the water begins to erode a path that allows increasing amounts of water to flow through the opening.

Oxidizers and organic peroxides Chemicals that include halogens (chlorine and fluorine), peroxides (hydrogen peroxide and benzoyl peroxide), and hypochlorites. These chemicals destroy metals and organic substances and enhance the ignition of combustibles.

Pan American Health Organization The regional office of the World Health Organization. It emphasizes retrofitting hospitals and strengthening public health programs.

Panic An acute fear reaction marked by a loss of self-control which is followed by nonsocial and nonrational flight behavior.

Permanent housing Housing that reestablishes household routines in preferred locations.

Physical vulnerability Human, agricultural, or structural susceptibility to damage or injury from disasters.

Piping A penetration through a dam or levee that occurs when an animal burrow, rotted tree root, or other disturbance creates a long circular tunnel through or nearly through the structure.

Policy entrepreneur An issue champion who has the expertise and legitimacy to promote emergency planning.

Population monitoring and assessment A term used when identifying the size of the risk area population at any point in time.

Preliminary damage assessment Damage assessment that produces counts of destroyed, severely damaged, moderately damaged, and slightly damaged structures.

Profession An occupation that requires an advanced education and training.

Protection motivation A positive response to the question of whether there will be personal consequences if disaster occurs.

Radioactive materials Substances that undergo spontaneous decay, emitting radiation in the process.

Radionuclides Radioactive substances that vary in atomic weight.

Rapid damage assessment The first stage of damage assessment that provides you with immediate information about the magnitude of the impact. It defines the boundaries of the physical impact area and assesses the intensity of damage within that impact area.

Reaction criteria A standard for judging the success of a training program that consists of trainees' opinions of the training program. This includes evaluations of the trainers, the facilities and equipment, the material, their enjoyment of the class, and their desire to take another class from the instructor.

Recovery A hazard management strategy that has the goal of restoring the normal functioning of a community. Recovery begins as a disaster is ending and continues until the community is back to normal.

Referent power Power that is based on a person's desire to be like the power holder.

Response A hazard management strategy that has the goal of protecting the population, limiting damage from the impact of an event, and minimizing damage from secondary impacts. Response begins when a disaster event occurs.

Results criteria A standard for judging the success of a training program that refer to the consequences of trainees' performance on the job. Results criteria evaluates whether the training made a difference in the overall performance of the organization.

Reward and coercive power Power frequently referred to as the "carrot and the stick" approach. Coercive power can produce deception to avoid punishment. Moreover, punishment typically produces continuing hostility.

Reward A function of an appraisal that focuses on improving a person's *motivation* to do the job. Appraisals should have clear criteria that provide guidance to the employee about what is important to the organization.

Risk The *possibility* that people or property could be hurt. Risk is defined in terms of the likelihood that an event will occur at a given location within a given time period and will inflict casualties and damage. This risk must be effectively communicated to the people who are likely to be affected.

Risk assessment An evaluation of what will be the personal consequences if the disaster occurs.

RTK provisions A legal requirement that requires handlers of dangerous chemicals to inform neighboring communities when they store hazardous substances in amounts that are greater than EPA thresholds.

Secondary impacts Disasters caused by a disaster, including events such as hazardous materials caused by earthquakes.

Seepage erosion A form of erosion that occurs when the height of the water in the river puts pressure on water that has seeped into the riverbed, under the levee, and into the soil on the landward side of the levee. The resulting flow of water can eventually cause boils of muddy water that erode a path for the water to flow underneath and then behind the levee.

Severe storms A storm whose wind speed exceeds 58 mph, that produces a tornado, or that releases hail with a diameter of three-quarters of an inch or greater.

Signal value An indication that a previously unnoticed threat warrants attention.

Site area emergency A class of a nuclear power plant emergency defined by the NRC and FEMA that involves major failures of plant safety functions. Releases might exceed EPA limits onsite but not offsite.

Site assessment Damage assessment that is meant to produce detailed estimates of the cost to repair or replace each affected structure. This information is used to support requests for federal assistance to the owners of the damaged property. It includes estimates of losses to residential, commercial, industrial and public property.

Social groups Stakeholders that are primarily defined by households, who control a substantial amount of the assets (buildings and their contents) that are at risk from disasters. Social groups also include neighborhood, service, and environmental organizations.

Social vulnerability Lack of psychological, social, economic, and political resources to cope with disaster impacts.

Source term The mix of chemicals or radionuclides involved in a given release.

Stage The height of water above a defined level that is used by emergency managers to predict the level of flood casualties and damage.

Stakeholder Someone who has, or thinks they have, something to lose or gain in a situation. An emergency management stakeholder is affected by the decisions made (or not made) by emergency managers and policy makers.

Storm surge An increased height of a body of water that exceeds the normal tide.

Structural vulnerability The susceptibility of a structure, such as a building, to be damaged or destroyed by environmental events.

Sustainable development A concept stating that the needs of the present must be met without compromising the ability of future generations to meet their own needs.

Table-top exercise A training exercise involving a group of senior personnel who are usually branch or departmental administrators and serve as the directors of their functions. Scenarios for these exercises vary in their complexity.

Task A specific activity carried out for a distinct purpose.

Taskwork The ability to perform each separate step of the response.

Teamwork The ability to schedule tasks and allocate resources among team members to achieve a performance that is efficient, effective, and timely.

Technical Guidance for Hazards Analysis A guide that lists extremely hazardous substances and describes a simple method for calculating VZs.

Technological disasters Events that result from the accidental failures of technologies, such as the release of hazardous materials from facilities where they are normally contained.

Technological hazards Hazards that originate in human-controlled processes but are released into the air and water. The most important technological hazards are explosives, flammable materials, toxic chemicals, radiological materials, and biological hazards.

Temporary housing Housing that allows victims to reestablish household routines in nonpreferred locations.

Temporary shelter Housing that includes food preparation and sleeping facilities that are sought from friends and relatives or are found in hotels, motels or mass care facilities.

Terrorist disaster A deliberate attack that is intended to achieve political objectives by inflicting damage and casualties. Also referred to as terrorism.

Tornadoes Windstorms that form when cold air from the north collides with a warmer air mass.

Trusses Engineered systems that are internally braced to provide maximum strength at minimum weight.

Tsunamis Sea waves that are usually generated by undersea earthquakes. Tsunamis can also be caused by volcanic eruptions or landslides.

Unmet needs committee An emergent organization that is designed to serve those whose needs are not being addressed.

Unusual event A class of a nuclear power plant emergency defined by the NRC and FEMA that involves potential degradation of plant safety. No releases are expected unless other events occur.

Vapor The molecules that are in a gaseous state of a substance that is a liquid at normal temperature and pressure.

Victims' needs assessment An evaluation of the psychological, demographic, and economic impacts of disasters on victims.

Volcanoes Geological structures that transport a column of molten rock from the earth's mantle to the surface.

Vulnerable zone (VZ) The area surrounding a given source in which a chemical release is likely to produce death, injury, or illness. Stands for vulnerable zone.

Warning A risk communication about an imminent event that is intended to produce an appropriate disaster response.

Wave action A destructive condition that causes levee failure by attacking the face of the levee and scouring away the material from which it is constructed.

Weather Radio A system that provides tone-activated notification of emergencies throughout most of the United States.

Wildland fires Fires that burn areas with nothing but natural vegetation for fuel.

Window of opportunity The time during which local emergency managers are most likely to be able to influence policy. A window of opportunity usually opens immediately after a focusing event has drawn attention to hazard and closes after attention moves on to other public issues.

PHOTO CREDITS

Figure 1-1: Photo by Dane Golden/FEMA, FEMA Web site, Image Number 9424

Figure 1-2: Photo by Andrea Booher/FEMA, FEMA Web site, Image Number 5509

Figure 3-3: Photo by Jason Pack/FEMA, FEMA Web site, Image Number 9172

Figure 4-1: Photo by Win Henderson/FEMA, FEMA Web site, Image Number 14286

Figure 4-2: Photo by Win Henderson/FEMA, FEMA Web site, Image Number 14535

Figure 5-1: Photo by Jocelyn Augustino/FEMA, FEMA Web site, Image Number 14976

Figure 5-5: Photo by John Shea/FEMA, FEMA Web site, Image Number 12435

Figure 6-2: Photo by Win Henderson/FEMA, FEMA Web site, Image Number 16933

Figure 6-3: Photo by Ed Edahl/FEMA, FEMA Web site, Image Number 12992

Figure 8-4: Photo by Petty Officer 3rd Class Jay C. Pugh, Department of Defense; DoD Web site Image number 050905-N-0535P-014

Figure 10-1: Photo by Liz Roll/FEMA, FEMA Web site, Image Number 14857

Figure 11-2: Photo by Leif Skoogfors/FEMA, FEMA Web site, Image Number 11562

Figure 12-1: Photo by Mark Wolfe/FEMA, FEMA Web site, Image Number 170066

Figure 12-2: Photo by Jason Pack/FEMA, FEMA Web site, Image Number 9150

Figure 13-3: Courtesy of World Vision International.

Figure 13-4: Courtesy of the Buddhist Compassion Relief Tz' Chi Foundation

Figure 14-1: Photo by Jocelyn Augustino/FEMA, FEMA Web site, Image Number 14873

Figure 14-2: Photo by George Armstrong/FEMA, FEMA Web site, Image Number 24000

Figure 15-1: Photo by Win Henderson/FEMA, FEMA Web site, Image Number 15531

Figure 15-2: Photo by Andrea Booher/FEMA, FEMA Web site, Image Number 5484

INDEX

D

F

J

K

L

Q

R

S

T

U